T0192564

Bayesian Methods in Epidemiology

Chapman & Hall/CRC Biostatistics Series

Editor-in-Chief

Shein-Chung Chow, Ph.D.
Professor
Department of Biostatistics and Bioinformatics
Duke University School of Medicine
Durham, North Carolina

Series Editors

Byron Jones
Biometrical Fellow
Statistical Methodology
Integrated Information Sciences
Novartis Pharma AG
Basel, Switzerland

Jen-pei Liu
Professor
Division of Biometry
Department of Agronomy
National Taiwan University
Taipei, Taiwan

Karl E. Peace
Georgia Cancer Coalition
Distinguished Cancer Scholar
Senior Research Scientist and
Professor of Biostatistics
Jiann-Ping Hsu College of Public Health
Georgia Southern University
Statesboro, Georgia

Bruce W. Turnbull
Professor
School of Operations Research
and Industrial Engineering
Cornell University
Ithaca, New York

Chapman & Hall/CRC Biostatistics Series

Adaptive Design Methods in
Clinical Trials, Second Edition
Shein-Chung Chow and Mark Chang

Adaptive Design Theory and
Implementation Using SAS and R
Mark Chang

Advanced Bayesian Methods for Medical
Test Accuracy
Lyle D. Broemeling

Advances in Clinical Trial Biostatistics
Nancy L. Geller

Applied Meta-Analysis with R
Ding-Geng (Din) Chen and Karl E. Peace

Basic Statistics and Pharmaceutical
Statistical Applications, Second Edition
James E. De Muth

Bayesian Adaptive Methods for
Clinical Trials
Scott M. Berry, Bradley P. Carlin,
J. Jack Lee, and Peter Muller

Bayesian Analysis Made Simple: An Excel
GUI for WinBUGS
Phil Woodward

Bayesian Methods for Measures of
Agreement
Lyle D. Broemeling

Bayesian Methods in Epidemiology
Lyle D. Broemeling

Bayesian Methods in Health Economics
Gianluca Baio

Bayesian Missing Data Problems: EM,
Data Augmentation and Noniterative
Computation
Ming T. Tan, Guo-Liang Tian,
and Kai Wang Ng

Bayesian Modeling in Bioinformatics
Dipak K. Dey, Samiran Ghosh,
and Bani K. Mallick

Biosimilars: Design and Analysis of
Follow-on Biologics
Shein-Chung Chow

Biostatistics: A Computing Approach
Stewart J. Anderson

Causal Analysis in Biomedicine and
Epidemiology: Based on Minimal
Sufficient Causation
Mikel Aickin

Clinical Trial Data Analysis using R
Ding-Geng (Din) Chen and Karl E. Peace

Clinical Trial Methodology
Karl E. Peace and Ding-Geng (Din) Chen

Computational Methods in Biomedical
Research
Ravindra Khattree and Dayanand N. Naik

Computational Pharmacokinetics
Anders Källén

Confidence Intervals for Proportions and
Related Measures of Effect Size
Robert G. Newcombe

Controversial Statistical Issues in
Clinical Trials
Shein-Chung Chow

Data and Safety Monitoring Committees
in Clinical Trials
Jay Herson

Design and Analysis of Animal Studies in
Pharmaceutical Development
Shein-Chung Chow and Jen-pei Liu

Design and Analysis of Bioavailability and
Bioequivalence Studies, Third Edition
Shein-Chung Chow and Jen-pei Liu

Design and Analysis of Bridging Studies
Jen-pei Liu, Shein-Chung Chow,
and Chin-Fu Hsiao

Design and Analysis of Clinical Trials with
Time-to-Event Endpoints
Karl E. Peace

Design and Analysis of Non-Inferiority
Trials
Mark D. Rothmann, Brian L. Wiens,
and Ivan S. F. Chan

Difference Equations with Public Health Applications
Lemuel A. Moyé and Asha Seth Kapadia

DNA Methylation Microarrays: Experimental Design and Statistical Analysis
Sun-Chong Wang and Arturas Petronis

DNA Microarrays and Related Genomics Techniques: Design, Analysis, and Interpretation of Experiments
David B. Allison, Grier P. Page, T. Mark Beasley, and Jode W. Edwards

Dose Finding by the Continual Reassessment Method
Ying Kuen Cheung

Elementary Bayesian Biostatistics
Lemuel A. Moyé

Frailty Models in Survival Analysis
Andreas Wienke

Generalized Linear Models: A Bayesian Perspective
Dipak K. Dey, Sujit K. Ghosh, and Bani K. Mallick

Handbook of Regression and Modeling: Applications for the Clinical and Pharmaceutical Industries
Daryl S. Paulson

Interval-Censored Time-to-Event Data: Methods and Applications
Ding-Geng (Din) Chen, Jianguo Sun, and Karl E. Peace

Joint Models for Longitudinal and Time-to-Event Data: With Applications in R
Dimitris Rizopoulos

Measures of Interobserver Agreement and Reliability, Second Edition
Mohamed M. Shoukri

Medical Biostatistics, Third Edition
A. Indrayan

Meta-Analysis in Medicine and Health Policy
Dalene Stangl and Donald A. Berry

Monte Carlo Simulation for the Pharmaceutical Industry: Concepts, Algorithms, and Case Studies
Mark Chang

Multiple Testing Problems in Pharmaceutical Statistics
Alex Dmitrienko, Ajit C. Tamhane, and Frank Bretz

Optimal Design for Nonlinear Response Models
Valerii V. Fedorov and Sergei L. Leonov

Randomized Clinical Trials of Nonpharmacological Treatments
Isabelle Boutron, Philippe Ravaud, and David Moher

Randomized Phase II Cancer Clinical Trials
Sin-Ho Jung

Sample Size Calculations in Clinical Research, Second Edition
Shein-Chung Chow, Jun Shao and Hansheng Wang

Statistical Design and Analysis of Stability Studies
Shein-Chung Chow

Statistical Evaluation of Diagnostic Performance: Topics in ROC Analysis
Kelly H. Zou, Aiyi Liu, Andriy Bandos, Lucila Ohno-Machado, and Howard Rockette

Statistical Methods for Clinical Trials
Mark X. Norleans

Statistics in Drug Research: Methodologies and Recent Developments
Shein-Chung Chow and Jun Shao

Statistics in the Pharmaceutical Industry, Third Edition
Ralph Buncher and Jia-Yeong Tsay

Survival Analysis in Medicine and Genetics
Jialiang Li and Shuangge Ma

Translational Medicine: Strategies and Statistical Methods
Dennis Cosmatos and Shein-Chung Chow

Chapman & Hall/CRC Biostatistics Series

Bayesian Methods in Epidemiology

Lyle D. Broemeling

Broemeling and Associates

Medical Lake, Washington, USA

CRC Press
Taylor & Francis Group
Boca Raton London New York

CRC Press is an imprint of the
Taylor & Francis Group, an **informa** business

A CHAPMAN & HALL BOOK

CRC Press
Taylor & Francis Group
6000 Broken Sound Parkway NW, Suite 300
Boca Raton, FL 33487-2742

First issued in paperback 2020

© 2014 by Taylor & Francis Group, LLC
CRC Press is an imprint of Taylor & Francis Group, an Informa business

No claim to original U.S. Government works

Version Date: 20130710

ISBN 13: 978-0-367-57634-9 (pbk)
ISBN 13: 978-1-4665-6497-8 (hbk)

This book contains information obtained from authentic and highly regarded sources. Reasonable efforts have been made to publish reliable data and information, but the author and publisher cannot assume responsibility for the validity of all materials or the consequences of their use. The authors and publishers have attempted to trace the copyright holders of all material reproduced in this publication and apologize to copyright holders if permission to publish in this form has not been obtained. If any copyright material has not been acknowledged please write and let us know so we may rectify in any future reprint.

Except as permitted under U.S. Copyright Law, no part of this book may be reprinted, reproduced, transmitted, or utilized in any form by any electronic, mechanical, or other means, now known or hereafter invented, including photocopying, microfilming, and recording, or in any information storage or retrieval system, without written permission from the publishers.

For permission to photocopy or use material electronically from this work, please access www.copyright.com (http://www.copyright.com/) or contact the Copyright Clearance Center, Inc. (CCC), 222 Rosewood Drive, Danvers, MA 01923, 978-750-8400. CCC is a not-for-profit organization that provides licenses and registration for a variety of users. For organizations that have been granted a photocopy license by the CCC, a separate system of payment has been arranged.

Trademark Notice: Product or corporate names may be trademarks or registered trademarks, and are used only for identification and explanation without intent to infringe.

Visit the Taylor & Francis Web site at
http://www.taylorandfrancis.com

and the CRC Press Web site at
http://www.crcpress.com

Contents

1. Introduction to Bayesian Methods in Epidemiology 1
    1.1   Introduction 1
    1.2   Review of Statistical Methods in Epidemiology 1
    1.3   Preview of the Book 4
            1.3.1   Chapter 2: A Bayesian Perspective of Association
                    between Risk Exposure and Disease 4
            1.3.2   Chapter 3: Bayesian Methods of Adjustment of Data 9
            1.3.3   Chapter 4: Regression Methods for Adjustment 16
            1.3.4   Chapter 5: A Bayesian Approach to Life Tables 21
            1.3.5   Chapter 6: A Bayesian Approach to Survival Analysis 27
            1.3.6   Chapter 7: Screening for Disease 32
            1.3.7   Chapter 8: Statistical Models for Epidemiology 36
    1.4   Preview of the Appendices 45
            1.4.1   Appendix A: Introduction to Bayesian Statistics 45
            1.4.2   Appendix B: Introduction to WinBUGS 47
    1.5   Comments and Conclusions 48
    Exercises 48
    References 49

2. A Bayesian Perspective of Association between Risk
   Exposure and Disease 53
    2.1   Introduction 53
    2.2   Incidence and Prevalence for Mortality and Morbidity 54
    2.3   Association between Risk and Disease in Cohort Studies 57
    2.4   Retrospective Studies: Association between Risk and
           Disease in Case–Control Studies 61
    2.5   Cross-Sectional Studies 65
    2.6   Attributable Risk 69
    2.7   Comments and Conclusions 73
    Exercises 75
    References 80

3. Bayesian Methods of Adjustment of Data 81
    3.1   Introduction 81
    3.2   Direct Adjustment of Data 82
    3.3   Indirect Standardization Adjustment 90
            3.3.1   Introduction 90
            3.3.2   Indirect Standardization 91
            3.3.3   Bayesian Inferences for Indirect Adjustment 92
            3.3.4   Example of Indirect Standardization 93

3.4   Stratification and Association between Disease and
      Risk Exposure 96
      3.4.1   Introduction 96
      3.4.2   Interaction and Stratification 97
      3.4.3   An Example of Stratification 101
3.5   Mantel–Haenszel Estimator of Association 103
3.6   Matching to Adjust Data in Case–Control Studies 107
3.7   Comments and Conclusions 109
Exercises 110
References 120

4. Regression Methods for Adjustment 121
4.1   Introduction 121
4.2   Logistic Regression 123
      4.2.1   Introduction 123
      4.2.2   An Example of Heart Disease 124
      4.2.3   An Example with Several Independent Variables 131
      4.2.4   Goodness of Fit 133
4.3   Linear Regression Models 134
      4.3.1   Introduction 134
      4.3.2   Simple Linear Regression 135
      4.3.3   Another Example of Simple Linear Regression 138
      4.3.4   More on Multiple Linear Regression 141
      4.3.5   An Example for Public Health 146
4.4   Weighted Regression 149
4.5   Ordinal and Other Regression Models 156
4.6   Comments and Conclusions 156
Exercises 158
References 168

5. A Bayesian Approach to Life Tables 169
5.1   Introduction 169
5.2   Basic Life Table 170
      5.2.1   Life Table Generalized 174
      5.2.2   Another Generalization of the Life Table 177
5.3   Disease-Specific Life Tables 178
5.4   Life Tables for Medical Studies 181
      5.4.1   Introduction 181
      5.4.2   California Tumor Registry 1942–1963 183
5.5   Comparing Survival 187
      5.5.1   Introduction 187
      5.5.2   Direct Bayesian Approach for Comparison of Survival ... 188
      5.5.3   Indirect Bayesian Comparison of Survival 190
              5.5.3.1   Introduction 190
              5.5.3.2   Mantel–Haenszel Odds Ratio 191

5.6   Kaplan–Meier Test 194
5.7   Comments and Conclusions 201
Exercises 202
References 210

6   A Bayesian Approach to Survival Analysis 213
6.1   Introduction 213
6.2   Notation and Basic Table for Survival 214
6.3   Kaplan–Meier Survival Curves 217
        6.3.1   Introduction 217
        6.3.2   Bayesian Kaplan–Meier Method 221
        6.3.3   Kaplan–Meier Plots for Recurrence of Leukemia
                    Patients 225
        6.3.4   Log-Rank Test for Difference in Recurrence Times 226
6.4   Survival Analysis 232
        6.4.1   Introduction 232
        6.4.2   Parametric Models for Survival Analysis 233
        6.4.3   Cox Proportional Hazards Model 248
        6.4.4   Cox Model with Covariates 253
        6.4.5   Testing for Proportional Hazards in the Cox Model 257
6.5   Comments and Conclusions 261
Exercises 263
References 276

7   Screening for Disease 279
7.1   Introduction 279
7.2   Principles of Screening 280
7.3   Evaluation of Screening Programs 281
        7.3.1   Introduction 281
        7.3.2   Classification Probabilities 283
        7.3.3   Predictive Values 286
        7.3.4   Diagnostic Likelihood Ratios 287
        7.3.5   ROC Curve 288
        7.3.6   UK Trial for Early Detection 292
7.4   HIP Study (Health Insurance Plan of Greater New York) 296
        7.4.1   Introduction 296
        7.4.2   Descriptive Statistics 299
        7.4.3   Estimating the Lead Time 304
        7.4.4   Estimating and Comparing Survival 309
                    7.4.4.1   Life Tables 309
                    7.4.4.2   Survival Models 322
7.5.   Comments and Conclusions 330
Exercises 333
References 340

8 Statistical Models for Epidemiology 343
    8.1   Introduction 343
    8.2   Review of Models for Epidemiology 343
    8.3   Categorical Regression Models 346
    8.4   Nonlinear Regression Models 355
    8.5   Repeated Measures Model 365
    8.6   Spatial Models for Epidemiology 375
    8.7   Comments and Conclusions 396
    Exercises 398
    References 404

Appendix A: Introduction to Bayesian Statistics 407

Appendix B: Introduction to WinBUGS 437

Index 447

1

Introduction to Bayesian Methods in Epidemiology

1.1 Introduction

Our journey to Bayesian methods for epidemiology begins with this chapter. This chapter consists of two parts: a review of the statistical methods used in epidemiology, and a preview of the other seven chapters and two appendices. Statistical methods in epidemiology appear in two ways: in those books that more or less center around statistics, and those epidemiology books that have some data analysis methods. A brief review of statistics books with methods for epidemiology and of epidemiology books with statistical methods will be conducted. The latter part of the chapter presents a detailed preview of the following chapters, and the last section contains some comments on the future of statistics in epidemiology, with an emphasis on Bayesian methods.

1.2 Review of Statistical Methods in Epidemiology

Two types of books will be reviewed as to their content relating to statistical methods for epidemiology: (1) statistics books whose aim is to provide statistical methods for epidemiology and (2) books whose main aim is epidemiology but also include statistical methods.

In the first category, Kahn and Sempos[1] provided the author with much material with their emphasis on statistical methods (non-Bayesian). For the "usual" problems in epidemiology, I used some of their material, but gave it a Bayesian flavor. The book begins with a review of the elementary notions of mean and variance formulas for grouped data and attribute data. Various forms of sampling are presented, including simple random sampling, stratified random sampling, and systematic and cluster sampling. The following chapter of the book introduces methods to estimate the association between exposure to various risk factors, with a focus on the case–control and cohort designs. The two types of design use different ways to estimate the association, namely, the odds ratio for the case–control design and relative risk for cohort studies. An important aspect of epidemiology is the so-called adjustment of data with two approaches, the direct and indirect methods.

To compare two groups with different age distributions, direct and indirect adjustment techniques allow the user to make a fair comparison. For example, the mortality of two states, say California and Florida, are to be compared over several age intervals, but the states have quiet different age distributions. To make a fair comparison, the mortality is compared relative to a common age distribution, the so-called standard. The standard could be the age distribution of the total U.S. population or some other standard, and Kahn and Sempos[1] explain various approaches to choosing the standard. Many examples illustrate direct and indirect ways to adjust the data.

For establishing an association between disease and risk exposure, regression analysis is another approach to adjust data, and to this end, simple and multiple regression for normally distributed dependent variables (which measure disease or disease morbidity) are fully portrayed. When the dependent variable is binary, simple and multiple logistic regression is employed to establish an association between exposure and disease, adjusted for various risk factors.

Life table techniques were developed by epidemiologists to relate the survival experience of a group of subjects and to compare the survival experience of two or more groups of subjects, and Kahn and Sempos[1] supply a detailed exposition of the subject. The simplest life table observes a group of subjects over several time intervals, and for each interval records the number who enter each interval alive and the number who die in each interval. Various generalizations of the life table to more complex scenarios are introduced and depicted with several examples. One such scenario is when dropouts or withdrawals are taken into account in estimating the survival experience. Another generalization is reported by using person-years as the main end point of survival, and estimating survival by descriptive techniques. Of course, when using person-years, statistical models can be utilized, and the authors briefly mention the Cox proportional hazards model for estimating survival. As will be seen, epidemiology books vary widely in their treatment of statistical methods.

For example, Selvin[2] does not discuss life tables but does depict topics not presented by Kahn and Sempos[1] and does report on some topics not reported by the former. For example, he introduces the student to maximum likelihood estimation, a somewhat advanced topic. He then discusses odds ratios for 2 by 2 and 2 by 1 tables for case–control studies. Continuing, he puts a lot of stress on regression models including linear logistic and Poisson regression. His presentation to regression analysis is somewhat similar to the Kahn and Sempos[1] approach, but Kahn and Sempos report more examples applicable to epidemiology, namely, regression techniques to adjust the association between exposure and disease. Some advanced techniques such as nonparametric regression and classification techniques are also portrayed.

One of the better statistical books with a focus on epidemiology is by Jewell,[3] who defines the disease process and measures of disease prevalence and incidence. This is followed by an in-depth study of probability and its relevance to epidemiology and appears to be unique among such books. As with Kahn and Sempos[1] and Selvin,[2] Jewell provides techniques for the estimation of the association between disease and risk exposure through relative risk and odds ratios for cohort and case–control designs, respectively. Regression models such as the logistic and log-linear are also meticulously narrated. In summary, one can say with confidence that the three books reviewed so far vary a lot, but with some common epidemiology themes.

Now to be reviewed are primarily epidemiology books that contain some statistical methods, and the first is the popular book by Rothman, Greenland, and Lash.[4] With regard to statistical techniques, they present case–control and cohort designs with the corresponding odds ratio and relative risk as measures of association between exposure and disease occurrence, which is followed by issues of cause and effect and estimation of interaction for stratified studies. Also presented are regression models for surveillance, using secondary data; ecologic studies; and environmental, genetic, and molecular epidemiology.

A more traditional epidemiology book is Mausner and Kramer,[5] which is a good book for introducing epidemiology to the statistician. Traditional topics are (1) epidemiologic concepts, (2) measurement of mortality and morbidity, (3) sources of data on community health, (4) selected indices of health, (5) descriptive epidemiology, (6) analytical studies, (7) screening in the detection of disease, (8) population dynamics and health, (9) occupational epidemiology, and (10) selected statistical topics.

Selected statistical topics introduce the beginning student to survival analysis, adjustment rates, cohort analysis of mortality, and sample size determination. As can be seen, most of the emphasis is on epidemiology with little in the way of statistics. I recommend this book for the statistician who has little knowledge of epidemiology.

Spatial epidemiology is a way to discover disease etiology by mapping the relative risk of disease and exposure over distinct geographic units. It is a very powerful tool for epidemiology, and the following references play an important role in the last chapter of this book. Two books on spatial epidemiology will be reviewed. The first by Elliot, Wakefield, Best, and Briggs[6], which is a good introduction to the subject and includes the following topics: (1) use of population data for spatial epidemiology, (2) bias and confounding, (3) overview of statistical methods, (4) Bayesian approach, (5) detecting clusters, (6) spatial variation and correlation, (7) geostatistical methods, (8) the history of disease mapping, (9), mapping mortality data, (10) exposure assessment, (11) geographical studies of risk assessment, and (12) water quality and health. Another quite relevant book is Bayesian by Lawson[7] with the following content: (1) Bayesian inference and modeling, (2) computational issues including a description of the Monte Carlo Markov chain (MCMC), Gibbs sampling, and the Metropolis–Hasting algorithm, (3) residuals and goodness of fit, (4) disease mapping and reconstructing relative risk, (5) disease cluster detection, (6) ecological analysis, (7) multivariate disease analysis, (8) spatial survival and longitudinal studies, and (9) spatial-temporal disease mapping.

The preceding discussion is a brief review of statistical references for epidemiology, and the present work is based on most of them, but with a Bayesian approach for the statistical analysis of epidemiologic data.

1.3 Preview of the Book

At this point, Chapters 2–8 will be reviewed followed by a review of the two appendices. Chapters 2–8 present the reader with a review of those Bayesian methods that apply to epidemiology. Recall that epidemiology is the study (or the science of the study) of the patterns, causes, and effects of health and disease conditions in defined populations. It is the cornerstone of public health that influences policy decisions and evidence-based medicine by identifying risk factors for disease and targets for preventive medicine. Epidemiologists help with study design, collection, and statistical analysis of data, and interpretation and dissemination of results (including peer review and occasional systematic review). Epidemiology has developed methodology used in clinical research, public health studies, and, to some extent, basic research in the biological sciences.

From the preceding discussion, one sees the role of statistical methodology in various aspects of epidemiology. For example, in public health studies and clinical research, life table methods were developed to compare various therapies for the study of disease treatment.

Major areas of epidemiological study include disease etiology, outbreak investigation, disease surveillance and screening, biomonitoring, and comparisons of treatment effects such as in clinical trials. Epidemiologists rely on other scientific disciplines like biology to better understand disease processes, statistics to make efficient use of the data and draw appropriate conclusions, social sciences to better understand proximate and distal causes, and engineering for exposure assessment.

Of special importance for disease etiology, epidemiologists and statisticians have developed spatial models to draw disease maps, which more clearly reveals the association between exposure and disease etiology.

1.3.1 Chapter 2: A Bayesian Perspective of Association between Risk Exposure and Disease

A formal beginning for statistical methods in epidemiology is Chapter 2, which relays information about the association between exposure to risk factors and the occurrence of disease. Recall that epidemiology investigates the association between exposure to various risk factors and the occurrence of disease, and two designs are used to study that association. The first design is the cohort study with two groups, where one group of subjects is exposed to the risk factor and the other group of subjects is not exposed. Such a study is called prospective because once the subjects are identified as exposed or unexposed, they are followed through time until the disease manifests itself. A good example of this is following a group of smokers until disease develops. The other common design to study association is the case–control study, where two groups of subjects are identified, those subjects with the disease and those without. For those with the disease, the history of the subject's exposure to risk factors is identified, and the same determination followed for those without disease. The basic measurement in such designs is the incidence of disease in the case of the cohort study or the incidence of the risk factor for the case–control study. Such studies are not easy to implement and need to be executed with care. Incidence is the number of new cases of the disease over a given period of time.

For the cohort study with two groups, one of exposed subjects and the other without, the association between disease and risk factor is estimated by the relative risk. Relative risk is a ratio, with the numerator being the proportion of those with the risk factor who develop disease, while the denominator is the proportion who develop disease among those not exposed to the risk factor. It is assumed that a random sample of patients is taken from those exposed and another random sample is selected from those not exposed, thus, the relative risk is a ratio of probabilities, the numerator the probability of an exposed subject developing disease, and the denominator the probability of disease for an unexposed subject. Cohort studies are prospective in the sense that patients are followed for a period of time and their disease status determined.

From a Bayesian viewpoint, the posterior distribution of the relative risk is the posterior distribution of the ratio

$$\theta_{RR} = \frac{\theta_{++}}{\theta_{-+}} \tag{1.1}$$

where the cell probabilities of the cohort study are defined in Table 1.1.

TABLE 1.1

A Prospective Cohort Study

Risk Factor	Disease Status	
	+	−
+	θ_{++}, n_{++}	θ_{+-}, n_{+-}
−	θ_{-+}, n_{-+}	θ_{--}, n_{--}

Note that θ_{++} is the probability of disease among those exposed to the risk factor and θ_{-+} is the probability of disease among those not exposed. In a cohort study, it is assumed that a random sample of size of size $n_{++} + n_{+-}$ is taken from the exposed population among which n_{++} develop disease. Also note that $\theta_{+-} = 1 - \theta_{++}$ and θ_{--} is $1 - \theta_{-+}$. If very little is known about the present type of study, an improper prior could be used for θ_{++}, namely,

$$\zeta(\theta_{++}) = \frac{1}{\theta_{++}} \tag{1.2}$$

where $0 < \theta_{++} < 1$, and the posterior distribution of θ_{++} would be beta (n_{++}, n_{+-}) with posterior mean $n_{++}/(n_{++} + n_{+-})$, which is the usual estimator of θ_{++}. Of course, a similar situation is true for θ_{-+}, which will have a beta (n_{-+}, n_{--}) posterior distribution; therefore, the relative risk will have a posterior distribution, which is a ratio of two beta distributions. The Bayesian analysis for a cohort study is demonstrated with a 12-year study of stroke among males based on the work by Abbott, Yin, Reed, and Yano.[8] Among 3435 smokers, 171 had a stroke, while among 4437, 117 had a stroke. Is there an association between smoking and stroke?

Thus, the incidence rate of stroke among smokers is $171/(171 + 3264) = .04978$ and among nonsmokers, the incidence rate is $117/(117 + 4320) = .026369$, and it appears that there is an association between smokers and stroke.

Assuming an improper prior distribution (Equation 1.2) for θ_{++} and similarly for θ_{-+}, the posterior mean of the relative risk is 1.904 with a posterior standard deviation .2253 and 95% credible interval of (1.505, 2.387). Since the credible interval does not include 1, it appears that there is a strong association between smoking and the occurrence of stroke. For the MCMC simulation, 55,000 observations are generated for the posterior distributions, with a burn of 5000 and a refresh, and the posterior density of the relative risk is portrayed in Figure 1.1.

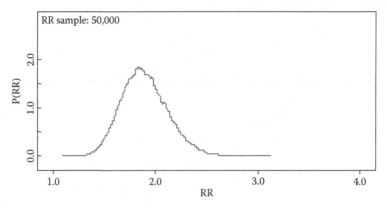

FIGURE 1.1
Posterior density of relative risk. Smokers and stroke association.

TABLE 1.2

A Case–Control Study

	Cases (Disease)	Controls (No Disease)
Risk Factor	+	−
+	$\theta_{++},\, n_{++}$	$\theta_{+-},\, n_{+-}$
−	$\theta_{-+},\, n_{-+}$	$\theta_{--},\, n_{--}$

The design of a case–control study is shown in Table 1.2, where a random sample of size $n_{++} + n_{-+}$ of cases is selected and a random sample of size $n_{+-} + n_{--}$ of controls is selected. Among the controls, n_{++} have the risk factor and n_{--} do not, and so on. For such a scenario, the odds ratio

$$\theta_{OR} = \frac{\left(\dfrac{\theta_{++}}{\theta_{-+}}\right)}{\left(\dfrac{\theta_{+-}}{\theta_{--}}\right)} \tag{1.3}$$

is employed to measure the association between disease and exposure to risk. Note the numerator of the odds ratio is the odds of having the disease and the denominator is the odds of not having the disease, and larger and smaller values are evidence of an association. A case–control study is usually retrospective in the sense that once a case or control has been identified, their history of exposure to risk is determined.

Note the numerator of the odds ratio is the odds of having the disease and the denominator is the odds of not having the disease. In Chapter 2, the case–control design is illustrated with the Johnson and Johnson[9] study of the association between tonsillectomy and Hodgkin's disease, where there are 174 cases (Hodgkin's disease) among which 90 had tonsillectomy, and 472 controls, among which 165 had tonsillectomy. Using BUGS CODE 3.4, with 55,000 observations for the simulation, the posterior analysis determined a posterior mean of 2.05 for the odds ratio with a 95% credible interval of (1.551, 2.5730), indicating a strong association. Thus, it appears that tonsillectomy is a risk factor of Hodgkin's disease, which is quite a surprise to me.

Cross-sectional studies are the next interesting topic of Chapter 2 and differ from the cohort and case–control studies in the way the sample is selected: for cross-sectional studies, a random sample of size

$$n = n_{++} + n_{+-} + n_{-+} + n_{--}$$

is selected and the cell frequencies n_{ij} have a multinomial distribution with parameter $(\theta_{++}, \theta_{+-}, \theta_{-+}, \theta_{--})$.

Cell frequencies of Table 1.3 are from the Shields Heart Study conducted in Spokane, Washington, at the Shields CT Imaging Center, where among 130 with heart disease, 119 have a positive CT scan for coronary artery calcium,

TABLE 1.3

The Cross-Sectional Design

Coronary Calcium	+ Disease (Heart Attack)	Nondisease (No Infarction)
+Positive	$\theta_{++}, n_{++} = 119$	$\theta_{+-}, n_{+-} = 2461$
−Negative	$\theta_{-+}, n_{-+} = 11$	$\theta_{--}, n_{--} = 1798$

and among 4259 without heart disease, 2461 tested positive for coronary artery calcium, and additional details are available in the work by Mielke, Shields, and Broemeling.[10]

When a uniform prior

$$\zeta(\theta_{++}, \theta_{+-}, \theta_{-+}, \theta_{--}) = 1 \tag{1.4}$$

where the cell probabilities vary between 0 and 1 and add to 1, the posterior distribution of the cell probabilities $(\theta_{++}, \theta_{+-}, \theta_{-+}, \theta_{--})$ is Dirichlet with parameter (120, 2462, 12, 1799). For cross-sectional studies, one has a look at the disease risk factor status at one time point, that is, one does not usually know the exact time at which the disease status and risk factor status are exactly known. One can compute the relative risk as for a cohort study or the odds ratio as for a case–control study.

For the Bayesian method, the posterior distribution of the relative risk and odds ratio is determined by the joint Dirichlet distribution of the cell probabilities.

With 55,000 observations for the simulation, a burn-in of 5000, and a refresh of 100, the posterior median of the relative risk is 7.805 with a 95% credible interval of (4.378, 15.49), and for the odds ratio, the posterior mean is 8.709, with a 95% credible interval of (4.529, 16.21), and one would conclude that there is a very strong association between coronary artery calcium and heart disease.

Of course this is apparent from Table 1.3, where the probability of heart disease among those who test positive for calcium is 119/2580 = .0461, and among those with negative coronary calcium, the probability is 11/1809 = .0060807, giving a relative risk of 7.68.

Chapter 2 continues with the introduction of the idea of attributable risk, which is a somewhat different way to measure association between disease and exposure to risk factors. The attributable risk is

$$\theta_{AR} = \frac{(I_r - I_0)}{I_r} \tag{1.5}$$

where I_0 and I_r are the incidence rates among the unexposed and exposed populations, respectively. Leviton[11] proposed this definition and explains various definitions of attributable risk, but Equation 1.5 is usually employed in epidemiology. Thus, the attributable risk is the difference in the incidence rates for the exposed and unexposed populations, divided by the incidence rate among the exposed, and can be expressed as

$$\theta_{AR} = \frac{(\theta_{RR} - 1)}{\theta_{RR}} \tag{1.6}$$

where θ_{RR} is the relative risk. According to Kahn and Sempos,[1] the attributable risk is the proportion of the incidence rate among those exposed to the risk factor due to the association with the risk factor. Recall the Abbott, Yin, Reed, and Yano[8] cohort study that investigates the association between smokers and stroke. Among the 3435 smokers, $n_{++} = 171$ had a stroke, and among the 4437 nonsmokers, $n_{-+} = 117$ had a stroke. Therefore, using an improper prior distribution for θ_{++} and θ_{-+}, the posterior median of the attributable risk (Equation 1.6) is .4702 and the 95% credible interval is (.3319, .5807). The Bayesian estimate coincides with the usual estimate. See Table 1.1 for the study design of a cohort study and the definition of the cell probabilities and frequencies. The Bayesian analysis is based on an improper prior distribution and is implemented with BUGS CODE 2.6.

The discussion of the attributable risk for a case–control design is illustrated with the Hiller and Kahn[12] study for the association between diabetes and cataracts, where the attributable risk is based on the odds ratio and defined as

$$\theta_{AROR} = \frac{(\theta_{OR} - 1)}{\theta_{OR}} \tag{1.7}$$

The last part of the chapter illustrates the estimation of attributable risk for the Shields Heart Study, see Table 1.3, where, based on the relative risk, the attributable risk has a posterior mean of .8672 and a 95% credible interval (.7716, .9354), but on the contrary, based on the odds ratio, the attributable risk is estimated by the posterior mean as .9092 with a 95% credible interval of (.8067, .9834). Thus, both estimates imply a strong association between heart disease and coronary artery calcium. Chapter 2 concludes with 16 exercises for the student, and the reader should gain a good understanding of the fundamental ideas of estimating the association between disease and exposure using the basic designs of the cohort, case–control, and cross-sectional designs.

1.3.2 Chapter 3: Bayesian Methods of Adjustment of Data

A standard epidemiologic technique is the adjustment of data. This is done so that a fair comparison can be made between the mortality of one state versus that of another state when the age distribution of the two states is quite different. That is to say, the mortality of the states is adjusted for age, by assuming both states have the same age distribution. Under this assumption, the mortality of the two states can computed and then compared with statistical techniques. Of course, in addition to mortality, one can adjust other variables, such as incidence rates of disease and prevalence rates of risk factors. There are many ways to adjust data, including the direct method, the indirect method, and regression techniques. Regression methods of adjustment are the subject of Chapter 4.

For example, consider the problem of comparing the mortality of California with Florida. Can a fair comparison be made? For a fair comparison, the U.S. population given in Table 1.4 will be used to compare the mortality between the two states given in Table 1.6. Note that age (years) is partitioned into the following age intervals: <5, 5–24, 25–44, 45–64, 65–84, and ≥85.

Table 1.5 portrays the total number of deaths for 2007 in each age category for the United States (area r), California (area 1), and Florida (area 2), and the information in Tables 1.4 and 1.5 is sufficient to use the direct method of comparing mortality between the two states, where the adjustment uses the U.S. population of 2010.

TABLE 1.4

U.S., California, and Florida Population by Age

	United States (Area r)		California (Area 1)		Florida (Area 2)	
Age (years)	Number	Percent	Number	Percent	Number	Percent
Total	308,745,538	100	37,253,956	100	18,801,310	100
<5	20,201,362	6.5430	2,531,333	6.7948	1,073,506	5.7097
5–24	84,652,193	33.9611	10,686,658	28.6859	4,668,242	24.8293
25–44	82,134,554	26.6026	10,500,587	28.1864	4,720,799	25.1088
45–64	81,489,445	26.3937	9,288,864	24.9338	5,079,161	27.0149
65–84	34,774,551	11.2631	3,645,546	9.7856	2,825,477	15.0280
≥85	5,493,433	1.7792	600,968	1.6131	434,125	2.3090

Source: U.S. Census Bureau 2010 Census Summary. Table QP-Q1. http://factfinder2.census. gov/faces/tableservices/jsf/pages/productview.xhtml?fpt=table.

TABLE 1.5

Number of Deaths for United States, California, and Florida by Age

	United States (Area r)	California (Area 1)	Florida (Area 2)
Age (years)	Mortality	Mortality	Mortality
<5	28,869	3446	1987
5–24	34,790	4137	1940
25–44	112,178	10,038	3993
45–64	493,566	29,511	9823
65–84	1,031,816	31,918	12,092
≥85	764,582	16,332	5225
Total	2,465,801	95,383	35,060

Source: Work Table 28R from The National Center of Health Statistics. http://www.cdc.gov/ nchs/data/dvs/MortFinal2007_Worktable308.pdf.

Consider two areas 1 and 2, and let r denote the reference area used for the standard age distribution to be used for the direct adjustment of mortality. The following notation is used for this scenario:

1: area 1

2: area 2

r: refers to the area of the reference standard

m_{i1}: the number of individuals in the i-th age category of area 1

m_{i2}: the number of individuals of area 2 in the i-th age category

m_{ir}: the number of subjects of the reference area r in the i-th age category

y_{i1}: the number of deaths in area 1 of the i-th age category

y_{i2}: the number of deaths in area 2 of the i-th age category

p_{i1}: the mortality rate for the i-th age category of area 1

The observed mortality rate for area 1 of the i-th age category is

$$p_{i1} = \frac{y_{i1}}{m_{i1}}, \quad \text{for} \quad i = 1,2,\ldots,k \tag{1.8}$$

with similar expressions for the death rates of area 2.

The directly standardized mortality rate for area 1 is

$$DSR_1 = \frac{\sum_{i=1}^{i=k} \theta_{i1} m_{ir}}{\sum_{i=1}^{i=k} m_{ir}} \tag{1.9}$$

where θ_{i1} is the unknown "true" mortality of the i-th category for area 1. If it is assumed that the number of individuals is fixed, Equation 1.9 is an unknown parameter, which can be estimated by Bayesian techniques. To employ Bayesian methods, it is assumed that the number of deaths occurring in area 1 for category i has a binomial distribution, that is,

$$y_{i1} \sim \text{binomial}\left(\theta_{i1}, m_{i1}\right) \tag{1.10}$$

for $i = 1, 2, \ldots, k$.

Note the standardized mortality for area 1, DSR_1, is a weighted average of the unknown "true" mortalities θ_{i1}, which is weighted by the corresponding number of subjects m_{ir} in the reference standard of area r. In a similar way, the standardized mortality rate for area 2 is

$$DSR_2 = \frac{\sum_{i=1}^{i=k} \theta_{i2} m_{ir}}{\sum_{i=1}^{i=k} m_{ir}} \tag{1.11}$$

and it is assumed that

$$y_{i2} \sim \text{binomial}(\theta_{i2}, m_{i2}) \tag{1.12}$$

for i = 1, 2, ..., k.

By assuming the number of deaths in a given age group have a binomial distribution, one is also assuming the probability of death of each individual in that age category is the same.

To illustrate Bayesian inference for comparing the standardized mortality rates for two areas, the mortality rates for California will be compared to that of Florida, using the U.S. census data for 2010 as the reference population.

What are the standardized mortality rates for California and Florida, using the 2010 census population figures? Using the notation defined earlier, the vector of deaths for the reference standard is

$$y_r = (y_{1r}, y_{2r}, \ldots, y_{6r})$$
$$= (28869, 34790, 112178, 493566, 1031816, 764, 582)$$

and the vector of deaths for area 1 (California) is

$$y_1 = (y_{11}, y_{21}, \ldots, y_{61})$$
$$= (3446, 4137, 10038, 29511, 31918, 16332)$$

As for Florida, the vector of deaths is

$$y_2 = (y_{12}, y_{22}, \ldots, y_{62})$$
$$= (1987, 1940, 3993, 9823, 12092, 5225)$$

In a similar fashion, the population for the reference standard by the six age categories is

$$m_r = (m_{1r}, m_{2r}, \ldots, m_{6r})$$
$$= (20201362, 84652193, 82134554, 81489445, 34774551, 5493433)$$

For California the vector is

$$m_1 = (m_{11}, m_{21}, \ldots, m_{61})$$
$$= (2531333, 10686658, 10500587, 9288864, 3645546, 600968)$$

and for Florida it is

$$m_2 = (m_{12}, m_{22}, \ldots, m_{62})$$
$$= (1073506, 4668242, 4720799, 5079161, 2825477, 434125)$$

Recall that for California, it is assumed $y_{i1} \sim \text{binomial}(\theta_{i1}, m_{i1})$ for $i = 1, 2, \ldots, 6$; therefore, in particular, for the five and under age group, the number of deaths is distributed as a binomial with parameters θ_{11} and m_{11}, that is

$$y_{11} \sim \text{binomial}(\theta_{11}, m_{11})$$
$$\sim \text{binomial}(\theta_{11}, 2531333) \tag{1.13}$$

where the number of observed deaths is 3446. Thus, one is assuming that 3466 is a sample from a binomial distribution with parameter vector (θ_{11}, 2, 531, 333). It follows that if one assumes an improper prior distribution for θ_{11}, the posterior distribution of θ_{11} is beta with parameter vector (3466, 2527867), and the posterior mean of θ_{11} is .001369239, the estimated probability of death for a person who is under age five. This is equivalent to a mortality rate of 137 per 100,000 in that age group. Our goal is to determine the posterior distribution of the directly standardized mortality rate for California

$$\text{DSR}_1 = \frac{\sum\limits_{i=1}^{i=k} \theta_{i1} m_{ir}}{\sum\limits_{i=1}^{i=k} m_{ir}}$$

which depends on the posterior distribution of the probabilities θ_{i1} for $i = 1, 2, \ldots, 6$.

On the contrary, if an improper prior distribution is used, the posterior distribution of θ_{i1} is beta with vector $(y_{i1}, m_{i1} - y_{i1})$ for $i = 1, 2, \ldots, 6$. The improper prior for θ_{i1} is expressed as

$$f(\theta_{i1}) \propto [\theta_{i1}(1 - \theta_{i1})]^{-1} \tag{1.14}$$

for $0 < \theta_{i1} < 1$.

A Bayesian analysis using BUGS CODE 3.1 is performed, assuming improper prior distributions, with 55,000 observations for the simulation, a burn-in of 5000, and a refresh of 100, and the results are reported in Table 1.6.

TABLE 1.6

Bayesian Analysis for Directly Standardized Mortality Rates of California and Florida

Parameter	Mean	SD	Error	2½	Median	97½
d	.001091	.0000126	<.00000001	.001066	.001091	.001116
r	1.654	.01030	<.00001	1.634	1.654	1.675
DSR$_1$	.002758	<.000001	<.00000001	.00274	.002758	.002775
DSR$_2$	.001667	<.000001	<.00000001	.001649	.001667	.001685
θ_{11}	.001362	<.000001	<.0000001	.001317	.001362	.001407

Therefore, the mortality rate for California is 276 per 100,000, 167 per 100,000 for Florida, and their ratio (California to Florida) has a posterior mean of 1.654 with a 95% credible interval of (.634, 1.675). Note that the posterior mean of the probability of death (directly standardized mortality) for California is .002758, which implies a mortality rate of 276 per 100,000. Since the MCMC errors are extremely small, one has confidence 55,000 observations is sufficient to estimate the true posterior mean of the five parameters given in the table. Of interest is the 95% credible interval of r, the ratio of the directly standardized mortality of California to that of Florida. Since the interval does not include unity, one is inclined to believe that the two mortality rates are indeed not the same. The difference in the two standardized mortality rates has a posterior mean of .001091 with a 95% credible interval of (.001066, .001116), implying that there is indeed a difference in the two directly standardized mortality rates.

Chapter 3 continues by using other reference populations (other than the U.S. population). For example, based on a technique suggested by Kahn and Sempos,[1] the inverse of the variance of the posterior distribution of $\theta_{i1} - \theta_{i2}$ is used as the weights m_{ir} for the standard for i = 1, 2, …, k, where k is the number of age categories. Also presented is the indirect method of adjusting for comparison of mortality between two areas.

Factors other than the risk factor of interest may have an effect on the association between risk and disease. Consider the example of the Israeli Ischemic Heart Disease Study, which is analyzed by Kahn and Sempos,[1] where the main emphasis is on the association between myocardial infarction and systolic blood pressure, but where age may have an effect on both (p. 105). Chapter 3 describes how to study the association between blood pressure and myocardial infarction taking into account age. When a factor affects both the disease (or morbidity) and the risk factor, the factor is referred to as a confounder, and one would expect a different association between the risk factor and disease for different levels of the possible confounder. If this is true, interaction between the risk factor and confounder is said to exist. This should be the case for the Israeli Heart Disease Study, and it was formally shown that interaction does exist between systolic blood pressure and age. Based on BUGS CODE 3.4, an analysis is executed to estimate the odds ratio (the study design was case–control) separately for each of the two strata (age $\geq$60 vs. age <60), and it was found that the posterior mean of the odds ratio is 1.071 for age $\geq$60 and 1.919 for age <60. The 95% credible interval for the first stratum is (.4493, 2.357) and (1.188, 2.925) for the second, thus indicating an association between myocardial infarction and systolic blood pressure, but the association is not the same for both the strata. For age <60, the association appears to be stronger.

When analyzing the association between disease and a risk factor in the presence of a confounding factor, one should determine if interaction is present. If interaction is present, the odds ratios of the various strata vary, while on the contrary, if no interaction is present, the various odds ratios do not differ. It is important to know the status of interaction in a stratified study.

One way to estimate the association between disease and risk factor in a stratified study is with the Mantel–Haenszel[15] (MH) estimator. If interaction is present, the MH estimator is a weighted average of the odds ratio of the various strata, while if interaction is not present, the estimator estimates of the common association. Chapter 3 continues with a detailed description of the MH estimator, which is illustrated with the Israeli Heart Study. Using BUGS CODE 3.5, the posterior mean of the MH estimator is 1.636, with a 95% credible interval of (.9049, 2.735) indicating the overall association is strong between a heart attack and systolic blood pressure. Recall that there is interaction between the risk factor age and blood pressure; thus, the MH estimate of 1.636 takes into account the confounder age, because it is a weighted average of the two separate odds ratios. An alternative to the MH estimator is also described.

Chapter 3 is concluded with an adjustment technique through the MH estimator for a series of 2 by 2 tables of a case–control study.

Consider a series of 2 by 2 tables, where each table corresponds to a case of a case–control study. For example, in the Shields Heart Study, each patient with an infarction could be matched with a patient without an infarction, with age as a confounder. The two matched patients would have the same age or be within, say less than 1 year of each other in age. This approach can result in adjusting the effect of the confounder on the association between risk and disease. A layout for a matched study is provided in Table 1.7.

Cell entries are the cell probabilities and cell frequencies. For example, for the cell where the case has the risk factor present and for the corresponding control the risk factor is absent, there are n_{+-} such pairs and the probability of each pair is θ_{+-}. To analyze such a situation, the study is stratified so that each case corresponds to a stratum. It can be shown that the MH estimator MH is given by

$$\theta_{MH} = \frac{\theta_{+-}}{\theta_{-+}} \qquad (1.15)$$

Notice that the two cells where the status of the risk factor and the disease is the same play no part in the calculation of MH (Equation 1.15). Chapter 3 ends with 21 exercises that comprehensively cover the various topics presented in the chapter.

TABLE 1.7

Matched Case–Control Study

Cases	Controls		
	RF present	RF absent	Total
RF present	θ_{++}, n_{++}	θ_{+-}, n_{+-}	$n_{+.}$
RF absent	θ_{-+}, n_{-+}	θ_{--}, n_{--}	$n_{-.}$
Total	$n_{.+}$	$n_{.-}$	$n_{..}$

1.3.3 Chapter 4: Regression Methods for Adjustment

Data adjustment by regression allows the epidemiologist a powerful method to study the association between various risk factors and disease status.

Recall in Chapter 3 that introductory adjustment techniques are described and are confined to one confounder and one risk factor. When several risk factors and possible confounders are available, regression models are ideal and are a standard tool for the analysis of data. Regression analysis is an important topic in statistics and many textbooks are accessible.

For a good introduction, see Chatterjee and Price,[16] and for a Bayesian approach, see Ntzoufras.[17]

Several types of regression models are presented: (1) logistic regression models, (2) simple and multiple linear regression models, (3) categorical or ordinal regression models, and nonlinear models. The types of models differ in regard to the type of dependent variable. A regression model has a dependent variable and at least one independent variable. When the dependent variable is binary (two values), the association between disease and other variables can be modeled by a logistic regression. If the dependent variable is continuous or quantitative, normal theory simple linear and multiple regression models are applicable, but if the dependent variable is categorical (several values), multinomial regression models are appropriate. Of interest to the epidemiologist, the dependent variable is an indicator of disease status or morbidity, and the independent variables are various risk factors or confounders.

Logistic models are appropriate for the type of adjustment techniques (direct and indirect standardization, matched pair designs, etc.) encountered in Chapter 3. For example, for the Israeli Heart Disease Study, the dependent variable is the disease status (myocardial infarction), and the independent variables are age and systolic blood pressure, where age is a possible confounder and systolic blood pressure can be considered a risk factor. In this case, the independent variables are binary (take on two values), but they could be regarded as continuous if one uses the actual age and blood pressure values.

Using the logistic regression model, the odds ratios is used for measuring association between disease and risk factors and is computed with the logistic regression models, which allows one to reanalyze the examples of Chapter 3 with the logistic regression model.

Simple linear and multiple regression models are appropriate when the dependent variable (an indicator of disease status) is quantitative (often called continuous) and considered to have a normal distribution. A simple linear regression model has one independent variable, while a multiple linear regression model has more than one independent variable.

For example, for a population of diabetics, one might want to study the association between blood glucose values and age. One could use the simple linear regression model where blood glucose values constitute the dependent variable and age the independent variable, but if another variable such as gender is included as an independent variable, one would have a multiple linear regression model with two independent variables, where age and gender are risk factors. The word "linear" in regression refers to the fact that the average value of the dependent variable is a linear function of the regression coefficients. Obviously, a model does not have to be linear, but can be nonlinear, a topic that will be explained in more detail. For additional information about normal theory regression models, see Chatterjee and Price,[16] a book that will be frequently referred to throughout the chapter.

For categorical regression models, the dependent variable can have a small number of values. For example, disease status for cancer could be labeled as local and confined to the primary tumor, metastasized to the lymph nodes, or metastasized beyond the lymph nodes, in which case the dependent variable assumes three values. Special models are used in this situation and are also referred to as multinomial regression models. The independent variables can be either continuous or categorical. As before, the dependent variable is an indicator of disease status, while the independent variables are risk factors or confounders. For example, for cancer, possible independent variables are age, sex, type of treatment, and so on, and one wants to study their effect on the stage of disease.

When the dependent variable is survival time, special regression methods using the Cox proportional hazards model will be described in Chapter 6 on life tables.

The first regression model presented in Chapter 4 is logistic, described as follows: For n subjects, suppose for the i-th, the response y_i is binary, that is, $y_i = 0$ or 1, and y_i is distributed as a Bernoulli with parameter θ_i, that is to say, $\theta_i = P(y_i = 1)$ and $1 - \theta_i = P(y_i = 0)$.

In addition, suppose there is one independent variable x that assumes the value x_i for subject i. Then the logistic model is defined as

$$\text{logit}(\theta_i) = \alpha + \beta x_i \tag{1.16}$$

where

$$\text{logit}(\theta_i) = \ln\left[\frac{\theta_i}{(1-\theta_i)}\right] \tag{1.17}$$

The name logistic comes from the fact that Equation 1.16 is equivalent to the logistic transformation

$$\theta_i = \frac{[\exp(\alpha + \beta x_i)]}{[1 + \exp(\alpha + \beta x_i)]} \tag{1.18}$$

Of course, an additional p − 1 independent variables can be added, in which case,

$$\text{logit}(\theta_i) = \alpha + \beta_1 x_{1i} + \beta_2 x_{2i} + \ldots + \beta_p x_{ip} \tag{1.19}$$

Unknown regression coefficients in the model are $\beta_1, \beta_2, ..., \beta_p$, which are estimated from the data and for the Bayesian approach are estimated from their joint posterior distribution. The model is linear in the regression coefficients on the logit scale (Equation 1.19) but not for the probability parameter (Equation 1.18). Note also the odds that $y_i = 1$ is the antilog of Equation 1.19 or

$$\left[\frac{\theta_i}{(1-\theta_i)}\right] = \exp(\alpha + \beta x_i) \tag{1.20}$$

If x is a categorical variable with $x_i = 0 \text{ or } 1$, then the odds that $y_i = 1$ when $x_i = 1$ divided by the odds that $y_i = 1$ when $x_i = 0$ is the odds ratio $\exp(\beta)$. Formally we have that the odds ratio expressed as

$$\frac{\left[\frac{\theta_i}{(1-\theta_i)}\middle| x_i = 1\right]}{\left[\frac{\theta_i}{(1-\theta_i)}\middle| x_i = 0\right]} = \exp(\beta) \tag{1.21}$$

or that the log of the odds ratio is β. On the contrary, if x is continuous, then $\exp(\beta)$ is the odds ratio for a unit increase in x, that is, when x increases from a value of x to a value of x + 1.

The same interpretation applies for the multiple linear logistic model (Equation 1.19). Note that for a unit increase in x_{1i}, the odds that $y_i = 1$ given $x_{1i} = 1$ divided by the odds that $y_i = 1$ given $x_{1i} = 0$ is $\exp(\beta_1)$, assuming the other p − 1 variables are constant, for all possible values of the remaining independent variables.

It is important to remember the restriction that $\exp(\beta_1)$ is the odds ratio, assuming the other variables are constant for all possible values of those other variables. Bayesian inferences about the odds will be based on the posterior distribution of β_1, which induces the posterior distribution of the odds ratio $\exp(\beta_1)$.

Recall the example of the Israeli Heart Disease Study, presented in Section 1.3.2, where the association between heart attack and two independent variables age and systolic blood pressure is investigated. Age is categorized as age <60 and age ≥60 and systolic blood pressure (SBP) is categorized as SBP ≥140 or SBP <140. The occurrence of heart attack is modeled as y = 0 for no attack and y = 1 for heart attack. In a similar way, let $x_1 = 1$ if age is ≥60 and 0 otherwise, and $x_2 = 1$ if SBP ≥140 mm Hg and 0 otherwise. For the logistic regression, we have a model with two independent variables, where both are binary, thus β_1 is the odds of a heart attack for patients over 60 divided by the odds of a heart attack for patients under the age of 60, for all values of blood pressure. Every regression model must be tested for goodness of fit, whereby the predicted model values are compared to the actual values of the dependent variable. The explanation of logistic linear regression continues with a Bayesian analysis for the Israeli Heart Study where the odds ratio for age and blood pressure are estimated with the posterior mean and this section terminates with a detailed presentation of a logistic regression model that includes many independent variables.

Simple and multiple linear regression models are used in epidemiology to determine associations between a dependent and several independent variables and the subject is vast. If one refers to the latest issue of the *American Journal of Epidemiology*, one will mostly likely find a regression analysis that is employed to find some type of association between disease and various risk factors and cofounders. The models considered in this chapter differ from the logistic model in that the dependent variable is quantitative. Quantitative variables assume values like one would encounter in measuring blood glucose values, where, in principle, one can find another blood glucose value between any two values. Thus, measurements such as age, weight, and systolic blood pressure are examples of a quantitative variable.

The presentation of regression analysis is initiated with the definition of a simple linear regression model, which has one dependent variable and one independent variable, with the goal being to establish an association between the two. For example, the dependent variable might be systolic blood pressure, with the independent variable indicating two groups, where the subjects are with and without coronary artery disease. Simple linear regression has very strict assumptions, such as the dependent variable must be normally distributed, and the variance of the dependent variable must be constant over all values of the independent variable. These assumptions will be relaxed to some extent by allowing for unequal variances, where a weighted regression is appropriate. Of course, the approach is Bayesian, where the posterior distribution of the regression coefficients (intercept and slope) and the variance about the regression line is determined. Several examples relevant to epidemiologic studies are presented. One problem to be explained is that of interpreting the estimated value of the regression coefficient.

Simple linear regression models are generalized to multiple linear regression models, where the goal is to establish an association between one quantitative dependent variable and several (more than one) independent variables. For example, the dependent variables might be blood glucose values, and the dependent variables might be age, weight, gender, and subjects with and without diabetes.

The goal is to estimate the effect of age, weight, gender, and diabetes on the average blood glucose value. One challenge with such a regression analysis is the interpretation of the regression coefficients of the model, and this will be carefully explained with many examples relevant to epidemiology.

One uses the term "linear regression" with the emphasis on linear, which means that the average value of the dependent variable is linear in the unknown regression coefficients, which are to be estimated from their posterior distribution. In practice, the dependent and independent variables are often transformed to achieve the linear assumption and the assumption of constant variance.

The definition of simple linear regression is as follows:

$$y_i = \alpha + \beta x_i + e_i \qquad (1.22)$$

where the dependent variable y and independent variable x are paired as (x_i, y_i) for the i-th individual with i = 1, 2, ..., n. If one plots the n pairs of observations, one would expect a linear association to develop; however, the relationship would not appear exactly linear because of the error term e with n values e_i, which are assumed to be independent and normally distributed with mean zero and unknown variance σ^2. This implies that the average value of the dependent variable y is

$$Avg(y_i) = \alpha + \beta x_i \qquad (1.23)$$

Thus, the average value of y is linear in x, namely, $\alpha + \beta x_i$ for i = 1, 2, ..., n.

With simple linear regression, there are three unknown parameters, the intercept term α, the slope β, and the variance of the error term σ^2, which is also the variance of the dependent variable y.

A good example of simple linear regression examines the effect of age x on systolic blood pressure y, and such an example is taken from Woolson.[18] This example is a subset of a larger study investigating the effect of weight on systolic blood pressure adjusted for age (p. 298). The Bayesian analysis estimates the slope β and intercept α of the model (Equation 1.22) assuming uninformative prior distributions for the parameters. Examples of multiple linear regression with several independent variables include the investigation of the effect of weight and age on systolic blood pressure and an example taken from public health where the dependent variable is cigarette consumption and the independent variables are age, education, age, and fraction of female subjects.

When the usual assumption of constant variance is not satisfied, a weighted regression is in order and the procedure is described in detail and exemplified by several examples.

Ordinal regression models are appropriate when the dependent variable assumes several nominal or ordinal values. For example, Broemeling[19] performs an ordinal regression analysis that estimates the accuracy of sentinel lymph biopsy for assessing the extent of metastasis in melanoma patients (p. 117).

The dependent variable assumes five values: 1 indicates absolutely no evidence of metastasis, 2 no evidence, 3 very little evidence, 4 some evidence, and 5 definite evidence of metastasis. Independent variables are four radiologists who are assessing the degree of metastasis on a 5-point scale.

Of course, there are many examples of ordinal regression appropriate for epidemiology, and the topic will be explained in more detail in a Chapter 8 devoted to advanced modeling techniques. Also in that chapter, nonlinear regression is introduced and many examples are used to illustrate the Bayesian methodology. Many exercises about regression will assist the student in their mastery of the subject.

1.3.4 Chapter 5: A Bayesian Approach to Life Tables

A ubiquitous epidemiologic technique is the estimation of survival by the life table, and the method will be described with a Bayesian approach. Life tables play an important role in many areas, including estimating survival in clinical trials and screening tests, estimating survival for use in the insurance industry, and measuring survival for use of pension funds.

For example, many retired people take lifetime annuities in the form of a monthly or annual payment and it is of essential interest to the pension fund to have a good idea of the life expectancy of the recipient. Knowing the life expectancy of the pensioner is essential in setting the payout amount of the pension. Of course, the same problem is faced by the Social Security Administration and other insurance entities. When a person takes out a whole life policy, the insurance company needs to know the survival time to set the premiums of the policy. Needless to say, survival analysis is an industry by itself. I was employed by the University of Texas MD Anderson Cancer Center and was involved in designing Phase I and Phase II clinical trials. In a Phase II trial, the objective is to compare a therapy with a historical control in regard to the response to therapy, and life tables are employed to estimate the response rate.

Chapter 5 begins with a definition of the basic life table, where a certain number of individuals are followed for a fixed period of time and the survival is measured at the end of each interval, where there are many intervals. For example, a cohort is followed for 10 years and the mortality measured at the end of each year. The basic table bases mortality on the number who die during a given interval, assuming that there is no loss to follow-up and that each mortality is from the same cause and not some other factor. The basic life table is generalized to include subjects lost to follow-up and patients that die from other causes (not the disease of interest).

The Bayesian approach is to assume the number of deaths over a given interval follows a binomial distribution with an unknown probability of mortality; thus, the joint distribution of the number of deaths for all intervals has a multinomial distribution and the joint distribution of the probability of death for all intervals is Dirichlet (assuming a uniform prior density for the probabilities of mortality).

Various generalizations allow for more realistic survival studies. The first generalization is to assume a random number of individuals are lost to follow-up in each interval and that the probability of survival is the ratio of the number who died divided by the number alive at the beginning of the interval minus the number who withdrew (for various reasons) during that interval of time. The next generalization is to allow for those who withdraw and those that die from other causes, other than the cause of interest (e.g., lung cancer).

Several examples will illustrate Bayesian inferences for estimating the survival experience of a cohort of subjects. For example, a cohort of melanoma patients is followed over the course of therapy and for a fixed period after the termination of the trial the survival time is estimated by the life table technique. The estimation of mortality is based on the joint posterior of mortality for each interval, then the overall mortality is also easily estimated. As in Chapter 1, the foundation of the analysis is the WinBUGS code, which can be used by the reader to learn the fundamentals of an important Bayesian methodology.

A group of subjects is followed over time and their survival is measured at the end of each period. Consider the following notation for a life table:

n: the length of each period

t: the time at the beginning of the period

O_t: number under observation at exact time t

$_n m_t$: the mortality during the period t to t + n

$_n p_t$: the probability of surviving from time t to time t + n

$_n q_t$: the probability of dying from time to to time t + n.

$_n P_t$: the probability of surviving over the period t to t + n (for an interval larger than a single period)

Suppose a group of 1000 patients who had a heart attack is followed for 10 years and the length of each period is a year.

The information in Table 1.8 is based on a chart from the National Institutes of Health (NIH)[20] and can be accessed at the link cited in the reference. However, the information reported in Table 1.8 is somewhat hypothetical and is estimated from the NIH table. Using the notation defined above, the information is displayed as follows. The reader should be aware that the information is related to the NIH information but is hypothetical.

Note that the probability of surviving from time 0 (beginning of the year 1996) through the end of period 9 (the end of the year 2005) is given by

$$_9 P_0 = \prod_{i=1}^{i=10} {}_1 p_i = .955 \tag{1.24}$$

Equation 1.24 equals the product of the 10 values in last column of Table 1.9.

Thus, the overall 10-year survival is 95.5%. Why use the product of 10 numbers to calculate the 10-year survival, when it is obvious that the answer is given by 950/1000? The life table analysis assumes that there were no withdrawals. This assumption will be relaxed in future analyses of life tables. Note that it is also assumed that for a given period, the probability of death is the same for each individual entering that period and that the event of death is independent among th n individuals entering the cohort.

How is the Bayesian used to estimate the individual period survivals and the overall survival? One approach is to assume that the mortality$_n m_t$ (the

TABLE 1.8

Summarization of Survival for Subjects with
Coronary Heart Disease

Year	Under Observation at Start of Period	Mortality during Period
1996	1000	6
1997	994	5
1998	989	5
1999	984	5
2000	979	5
2001	974	4
2002	970	5
2003	965	5
2004	960	5
2005	955	5

Source: National Institutes of Health. National Heart, Lung, and Blood Institute. Chart 3-26, Death and Age-Adjusted Death Rates for Coronary Heart Disease, 1980–2006. http://www.nhlbi .nih.gov/resources/docs/2009_ChartBook.pdf.

TABLE 1.9

Life Table Calculations for Survival of Coronary Heart Disease Subjects

Time at Beginning of Period t	Under Observation at Time t O_t	Mortality during Period $_1m_t$	Probability of Dying in Interval $_1q_t$	Probability of Surviving through Period $_1p_t$
1 (1996)	1000	6	.006	.994
2 (1997)	994	5	.0050301810	.9949698
3 (1998)	989	5	.0050556117	.9949443
4 (1999)	984	5	.0050813008	.9949186
5 (2000)	979	5	.0051072522	.9948927
6 (2001)	974	4	.0041067761	.9958932
7 (2002)	970	5	.0051546391	.9948453
8 (2003)	965	5	.0051813471	.9948186
9 (2004)	960	5	.0052083333	.9947916
10 (2005)	955	5	.0052356020	.9947643

number that die in a given period) has a binomial distribution with parameters $(_nq_t, O_t)$, where O_t is the number under observation at the beginning of time t (for an interval of length n) and $_nq_t$ is the probability of death of an individual subject during the interval from time t to time t + n. Note that for coronary heart disease, n = 1 and t = 1, 2, ..., 10. The unknown parameters are the mortality probabilities $_n q_t$, and if one assumes an improper prior density, namely,

$$f(_1 q_1, _1 q_2, _1 q_3, \ldots, _1 q_{10}) \propto \frac{1}{_1 q_1 \, _1 q_2 \, _1 q_3 \ldots _1 q_{10}} \quad (1.25)$$

for $0 <_1 q_t < 1$, $t = 1, 2, \ldots, 10$, the posterior density of the mortality probability $_1 q_t$ is a beta distribution with parameter $(_1 m_t, O_t -_1 m_t)$, where O_t is the total number of subjects available at time t. That is to say, the mortality probabilities are jointly independent and

$$_1 q_t \sim \text{beta}(_1 m_t, O_t -_1 m_t) \quad (1.26)$$

$t = 1, 2, \ldots, 10$.

A Bayesian analysis is performed on the data of Table 1.8, using 55,000 observations with a burn-in of 5000 and a refresh of 100, and the results are given in Table 1.10. The reader should compare the second column (posterior mean) of Table 1.10 with the fourth column of Table 1.9 and notice the similarity. The results are quite similar, because an improper prior (Equation 1.25) was used for the probabilities of death.

The Bayesian analysis provides an estimate of uncertainty for each posterior distribution, namely, the posterior standard deviation, and the uncertainty is also reflected with the corresponding 95% credible interval. For example, the probability of a death for the first period (1996) is estimated as .005985 with a 95% credible interval of (.002214, .01164). Of primary interest is the overall 10-year survival P, estimated as .9501 with a 95% credible interval of (.9355, .9626).

As was stated earlier, the basic life table assumes that for a given period, the probability of the death is the same for all individuals and there are no

TABLE 1.10

Bayesian Life Table Analysis for 1000 Subjects with Coronary Artery Disease

Parameter	Mean	SD	Error	2½	Median	97½
$_1 q_1$	.005985	.002438	<.00001	.002214	.005654	.01164
$_1 q_2$	.005035	.002246	<.000001	.001639	.004708	.0103
$_1 q_3$	.005045	.00225	<.000001	.001619	.004726	.01026
$_1 q_4$	.005075	.002273	<.000001	.00164	.004743	.01042
$_1 q_5$	.005117	.002273	<.00001	.001677	.004785	.01042
$_1 q_6$	.004104	.00205	<.000001	.001119	.003768	.008996
$_1 q_7$	.005141	.002294	<.00001	.001657	.004797	.0106
$_1 q_8$	.00517	.0023	<.00001	.00169	.004827	.01056
$_1 q_9$	.00522	.002311	<.00001	.001692	.004903	.01062
$_1 q_{10}$	.00522	.002311	<.000001	.001707	.004877	.01063
P	.9501	.006929	<.00001	.9355	.9504	.9626

withdrawals. The second restriction is relaxed to allow for withdrawals from the study. When the cohort is followed through the study time, an individual can drop out for various reasons: moving away, leaving the study for a variety of reasons, and so on. Consider a hypothetical study that follows the postoperative experience of 450 lung cancer patients, where Table 1.11 portrays the number of deaths and number who withdraw in each time period.

Such a scenario is described in Chapter 4 and a Bayesian analysis of the results of Table 1.11 executed with 55,000 observations for the simulation.

For the preceding analyses, the cause of death was of no concern, but in many clinical trials, one is primarily interested in the number that die from a specific disease. For example, with a Phase II trial for melanoma, where the therapy is an immunotherapy, one wants to know the response rate of that therapy and also wants to know the time to recurrence of each patient. Also for the life table in Table 1.11, one would want to estimate the survival probabilities for death from lung cancer. Recall the study begins with a cohort of 450 patients, who underwent surgery to remove the primary tumor. Thus, returning to the lung cancer survival study, consider a column that records the number that die from complications of lung cancer, a column of those who die from other cancers, a column that records the number who die from noncancer diseases, and lastly a column that shows the number who withdraw. Chapter 5 demonstrates the Bayesian analysis of such a situation.

The latter part of Chapter 5 details Bayesian approaches for comparing the survival experience between two groups of subjects. For example, consider Table 1.12, which is similar to the group of lung cancer patients of Table 1.11. What is the Bayesian approach?

There are several ways to compare the survival experience of several life tables. Among them are the standard approaches including the log rank test, the MH test, and the Kaplan–Meier method. In what is to follow, Bayesian

TABLE 1.11

Postoperative Survival Experience of 450 Subjects with Lung Cancer (Medical Center 1)

Time at Beginning of Period t	Under Observation at Time t Q_t	Number of Deaths in Interval $_2m_t$	Number Withdrawn in Interval $_2w_t$	Adjusted $O_t(O_t -_2w_t/2)O_t^{'}$	Mortality during Interval $_2m_t/O_t\,_2q_t$	Survival $_2p_t$
1	450	207	8	446	.4641	.5358
3	235	41	10	230	.1782	.8217
5	184	9	9	179.5	.0501	.9498
7	166	7	11	160.5	.0436	.9563
9	148	18	12	142	.1267	.8732
11	118	10	9	113.5	.0881	.9118
	99					

TABLE 1.12

Postoperative Survival Experience of 410 Subjects with Lung Cancer (Medical Center 2)

Time at Beginning of Period t	Under Observation at Time t O_t	Number of Deaths in Interval $_2m_t$	Number Withdrawn in Interval $_2w_t$	Adjusted $O_t(O_t - _2w_t/2)O_t'$	Mortality during Interval $_2m_t/O_t'{_2q}$	Survival $_2p_t$
1	410	200	8	402	.4975	.5316
3	202	39	9	193	.2020	.7979
5	154	10	7	147	.0680	.9319
7	137	16	7	130	.1230	.8769
9	114	24	3	111	.2162	.7837
11	87	15	6	81	.1851	.8148
13	66					

adaptations of the standard approaches are taken, but the more direct Bayesian methods are in general much easier to interpret and implement.

This section begins with an explanation of the direct Bayesian approach, where the posterior distribution of the difference in the overall survival of two groups is determined. Bayesian interpretations of the standard tests will also be presented and all techniques illustrated with examples introduced earlier in Chapter 5.

The probability of survival is calculated for each time period (6 months) for both studies, then for each study, the overall probability of survival is calculated, then lastly the difference in the overall survival (for 7 years) between the two groups is calculated. Of course, the Bayesian analysis determines the posterior distribution of overall survival by calculating the posterior mean, median, standard deviation, and 95% credible interval. See Table 1.13, which reports a survival probability of .3103 for the first medical center (Table 1.11) and .2093 for the second (Table 1.12) resulting in a posterior mean of .101 for the difference with a 95% credible interval of (.03932, .1623). Note the time periods are 6 months with a 7-year follow-up.

Indirect methods are adaptations of classical epidemiologic methods. The first adaptation is the MH odds ratio, which is considered an unknown parameter. Recall that the MH odds ratio is applicable to studies with different strata, and in the case of a life table, the various time periods (or intervals) are considered strata. Thus, the comparison of two life tables is based on the odds ratio, where the odds of death from life Table 1 is compared to the odds of death of life Table 2. The two lung cancer groups of 450 and 410 patients of two medical centers are used to illustrate the Bayesian technique.

The chapter is concluded with a Bayesian analog of the Kaplan–Meier product limit method of estimating the survival curve. A group of patients is followed until the last patient dies or is lost to follow-up, and the method

TABLE 1.13

Comparing Survival among Lung Cancer Patients of Two Medical Centers

Parameter	Mean	SD	Error	2½	Median	97½
D	.101	.03155	<.0001	.03932	.101	.1623
s_1	.3103	.02323	<.00001	.2659	.31	.3567
s_2	.2093	.02143	<.00001	.1689	.2088	.2528

records the time of each event (death or withdrawal); thus the entire survival experience of each group is known. Bayesian determinations of the survival curve of each group allow one to plot the probabilities of survival for each group and to make a decision about the similarity of the two.

Chapter 5 ends with many exercises that provide the student with additional opportunities to reinforce what is being learned.

1.3.5 Chapter 6: A Bayesian Approach to Survival Analysis

Chapter 5 is now extended with another approach to estimating survival, where the main focus is on parametric models such as the Weibull and a nonparametric approach that includes the Cox proportional hazards model. These models use the survival time of each patient that experience an event (death, time to recurrence, etc.) and also the time of survival for those patients that are censored, namely, those that are lost to follow-up. The survival time up to the point when the patient withdraws (lost to follow-up) is known and utilized to estimate the survival experience of a group of interest.

The chapter begins with the Kaplan–Meier curve of the survival experience, which allows a formal definition of three basic ideas that are fundamental to understanding survival models: (1) the hazard function, (2) the density of the survival times, and (3) the survival function. The Kaplan–Meier curve was introduced in Chapter 5, but the presentation in Chapter 6 will give the student a deeper understanding of this important concept. In Chapter 5, the main focus is on the life table method of estimating survival, where the times of death are grouped into various periods over the range of a person's life. On the contrary, survival models including the Cox proportional hazards model utilize the actual time of the event and the time of censoring.

Chapter 6 continues with the Kaplan–Meir curve and the log-rank tests for testing for a difference in the survival experience of two or more groups. The Bayesian version of the log-rank test is based on the posterior distribution of the observed minus the expected number of events in a particular group. Next, a parametric model for survival based on the Weibull distribution is introduced, followed by the Cox proportional hazards model. This is followed by a detailed outline of how to estimate the survival and hazard functions. Several interesting examples are explained, including one involving the survival experience of two groups of leukemia patients, where one group receives a treatment to extend the remission time, and the other group is a placebo.

Evaluating the proportional hazards assumption is an essential part of executing the Bayesian analysis and is based on the estimated survival curves.

Latter parts of the chapter are focused on more specialized topics such as the stratified Cox procedure, which allows one to control by stratification when the proportional hazards assumption is not true. An interesting aspect of stratification is the test of the no interaction assumption. Every theme of this chapter is accompanied by examples that illustrate the important features of that theme. Exercises at the end of the chapter will develop a further understanding of the various topics.

The WinBUGS® package is utilized throughout for the Bayesian analyses, and the code is "borrowed" from several examples provided by the package. The following link will give the reader access to the several examples involving survival analysis: www.mrc-bsu.cam.ac.uk/bugs/winbugs/contents .shtml. Examples from the package are used for survival analyses based on the Weibull parametric model as well as the Cox proportional hazards model.

A good introduction to survival analysis is given by Kleinbaum[21] and more specialized Bayesian references are presented by Congdon.[22,23] I used the Congdon material to some extent because of the Bayesian nature of the approach. Newman[24] is especially appropriate for epidemiologists and develops a comprehensive technique to survival analysis that includes many realistic examples

To describe a survival study, several fundamental functions are defined: If T is the random variable that represents the survival time of a patient, the probability that a patient survives at least t years is given by

$$S(t) = P(T > t) \tag{1.27}$$

A plot of the $S(t_j)$ for j = 1, 2, ..., n is the survival curve for this group of n individuals. However, note that the times t_j are not necessarily ordered from smallest to largest, that is, t_n is not necessarily the last recorded time, either for a failure or for a censored observation. Thus, S(t) would be estimated by the proportion of n individuals who survive past time t. Another important function that is used in survival studies is the hazard function

$$h(t) = \lim_{\Delta t \to \infty} \frac{P(t \le T < t + \Delta t \mid T \ge t)}{\Delta t} \tag{1.28}$$

which is interpreted as the probability that a person will fail in the next instant, given they have survived up to time t. Another interpretation is that the hazard at time t is the instantaneous rate of failure. Thus, it appears that the hazard function and the survival function are inversely related, that is, as the survival decreases over time, the hazard function tends to increase with time.

The survival function is related to the distribution function of survival, namely,

$$\begin{aligned} F(t) &= 1 - S(t) \\ &= P[T \le t] \text{ for } t > 0 \end{aligned} \tag{1.29}$$

Thus, the probability that a patient will not survive after time t is given by F(t). The three functions involving survival are related, and in particular,

$$h(t) = \frac{[dS(t)/dt]}{S(t)} \qquad (1.30)$$

that is, the hazard function at time t is the derivative of the survival function with respect to t, divided by the survival function at time t.

Kaplan–Meier curves plot the survival of a cohort of patients taking into account the number of censored observations. Consider the following example of a group of leukemia patients that are treated for the disease from the study of Freireich, Gehan, Frei, and Schroeder,[25] where the information appears in Table 1.14.

The last column of Table 1.14 gives the estimated probabilities of survival using the product limit method of Kaplan–Meier, and the first column records the ordered times at which the recurrence time and the time of censored patients are recorded. For example, at 6 weeks, there were three patients that had a recurrence and one patient who was censored (lost to follow-up). The fourth column is the number of patients at the beginning of the time period, for example, 15 are available for observations at the beginning of week 13.

Consider Table 1.14 and the ordered time $t_{(4)} = 13$ weeks. The proportion of recurrences occurring after time 13 weeks is estimated as S(4) = (21/21) (18/21)(16/17)(14/15)(11/12) = .690. The Kaplan–Meier product limit method is based on the formula

$$S(t_{(j)}) = \prod_{i=1}^{i=j} P[T > t_{(i)} | T \ge t_{(i)}] \qquad (1.31)$$

for j = 1, 2, ..., n, and the estimated survival probability at time $t_{(j)}$ is the product of j estimated conditional probabilities.

TABLE 1.14

Times to Recurrence for Treatment Group

t_0	d_i is the # Recurrences	c_i is the # Censored	$R(t_{(j)})$	$S(t_{(j)})$	$P[T>t_{(j)}|T \ge t_{(j)}]$
$t_{(0)} = 0$	0	0	21	1	
$t_{(1)} = 6$	3	1	21	.8571	.8571
$t_{(2)} = 7$	1	1	17	.8067	.941
$t_{(3)} = 10$	1	2	15	.7529	.933
$t_{(4)} = 13$	1	0	12	.6901	.9166
$t_{(5)} = 16$	1	3	11	.6275	.909
$t_{(6)} = 22$	1	0	7	.5378	.857
$t_{(7)} = 23$	1	5	6	.4482	.833

What is the Bayesian approach to estimating the survival problems with the product-limit method? Referring to Table 1.14, one must specify a posterior distribution for the unknown parameters, which are the probabilities of recurrence for the various ordered recurrence and censoring times. To that end, it is assumed the number of recurrences at each of the ordered times has a binomial distribution, that is, for the i-th ordered time

$$d[i] \sim \text{binomial} \left(q[i], R[i] \right) \tag{1.32}$$

where $q[i]$ is the probability of recurrence and $R[i]$ is the number at risk at the beginning of the i-th ordered time. Assuming a beta prior for the $q[i]$, namely,

$$q[i] \sim \text{beta}(.01, .01) \tag{1.33}$$

which induces a posterior distribution for the probabilities of recurrence. By using a beta $(.01, .01)$ prior distribution, the posterior mean will be quite similar to the usual estimators of the recurrence probabilities $S(t_{(i)})$ and the conditional probabilities $P[T > t_{(i)} | T \geq t_{(i)}]$, for $i = 1, 2, \ldots, m$, where m is the number of order times (recurrence and censored). The number of censored observations is also given a binomial distribution, namely,

$$c[i] \sim \text{binomial} \left(qc[i], R[i] \right) \tag{1.34}$$

and

$$qc[i] \sim \text{beta}(.01, .01) \tag{1.35}$$

If $r[i]$ is the number at risk at time i, then

$$r[i] = r[i-1] - c[i-1] - d[i-1] \tag{1.36}$$

for $i = 2, 3, \ldots, m$, where m is the number of distinct ordered times.

Thus, distribution of the number of deaths and number of censored observations induces a distribution on the number at risk. Note also, the probability of an event and the probability of a censored observation are not the same at each time period, that is, it is not assumed that $q[i] = qc[i]$! This topic is continued with a Bayesian analysis based on BUGS CODE 6.1 and produces estimates of the product limit survival probabilities similar to the last column of Table 1.14. The Bayesian technique is illustrated with the treatment and placebo groups of leukemia patients, and the two times to recurrence curves are plotted. To compare the survival experience of two groups of subjects, a Bayesian version of the log-rank test is developed.

The latter part of Chapter 6 presents a modeling approach to survival analysis consisting of parametric and nonparametric scenarios. For the parametric scenario, the Bayesian methodology is based on the Weibull regression model, while for the nonparametric approach, the Bayesian technique is based on the Cox proportional regression model. Both approaches are illustrated with the leukemia study of Freireich, Gehan, Frei, and Schroeder.[25]

The Cox proportional hazards model is one of the most useful in biostatistics and appears in many of the major medical journals.

It is defined as

$$h(t, X) = h_0(t)e^{\sum_{i=1}^{i=p} \beta_i X_i} \tag{1.37}$$

where $h_0(t)$ is the baseline hazard function, the β_i are unknown regression parameters, and the X_i are known covariates or independent variables. Note that the baseline hazard function is a function of time only, but that the covariates are not functions of t. The time t is the time to the event of interest, which is usually the survival time of a group of patients or the time to recurrence, or some other event measured by time.

Recall that the regression function is defined in the terms of the hazard function

$$h(t) = \lim_{\Delta t \to \infty} \frac{P(t \leq T < t + \Delta t | T \geq t)}{\Delta t} \tag{1.38}$$

where T is denoted the survival time of a subject. In survival studies, the Cox regression model is expressed as a hazard, whereas the usual way to express a regression is more directly using T as a function of unknown regression coefficients. One reason the Cox model is so popular is its versatility: for example, if the actual survival time has, say, an exponential distribution, the Cox model will provide similar results, or if the survival time has a Weibull distribution, the Cox model will also give similar results.

With the Cox model, the time variable T is not assumed to have a specific distribution, thus the model is quite general in that it can be applied in a large variety of time to event studies. Also note that the p covariates $X_1, X_2, ..., X_p$ are not functions of t; however, there are cases where one would have time-dependent covariates, in which case, a more general Cox model is appropriate. This scenario will be described in a later section of this chapter. The important assumption of the Cox model is that if one is comparing the survival of two groups, the corresponding hazard functions must be proportional. Such a case will also be studied in a later section of the chapter.

The most important parameter in survival studies is the hazard ratio

$$HR = \frac{h(t, X^*)}{h(t, X)} \tag{1.39}$$

between two individuals, one with the covariate measurements X^*, and the other with the measurement X, on the p covariates. Note that it easy to show that the hazard ratio is equivalent to

$$HR = e^{\sum_{i=1}^{i=p} \beta_i (X_i^* - X_i)} \tag{1.40}$$

where both X^* and X are known. It is important to remember that to the Bayesian approach, the HR hazard ratio is an unknown parameter because it depends on p unknown parameters

$$\beta = (\beta_1, \beta_2, ..., \beta_p) \tag{1.41}$$

Thus, the Bayesian approach must specify a prior distribution for β, then, through Bayes' theorem, determine the posterior distribution of β and any function of β such as the hazard ratio.

As an example of the hazard ratio, consider the two groups of leukemia patients, where group 1 is the treatment group and group 2 the placebo, with one covariate X, where $X^* = 1$ denotes the treatment group and X = 0 the placebo. Then the hazard ratio reduces to

$$HR(\beta) = e^{\beta} \tag{1.42}$$

where one individual is a patient from the treatment group, and the other a patient from the placebo. Thus, Equation 1.42 expresses the effect of the treatment as a hazard ratio. Once one estimates β, one has an estimate of the hazard ratio, or for our interest, once one has the posterior distribution of β, one has the posterior distribution of HR. Additional topics involving the Cox model are presented, including using covariates in the model and testing for the proportional hazards assumption.

1.3.6 Chapter 7: Screening for Disease

Chapter 7 introduces an important topic that is part of the experience of many epidemiologists, and the topic is screening for disease among individuals that do not exhibit any symptoms of the disease. For example, the Morrison[26] book presents the analytical methods necessary to design and analyze screening programs, and the Shapiro, Venet, Strax, and Venet[27] analysis of the Health Insurance Plan of Greater New York (HIP) illustrates epidemiological methods for estimating the lead time and survival of the trial participants.

Chapter 7 continues by describing the fundamentals of a screening program and measures of test accuracy of the various modalities for screening. A modality is an instrument (e.g., an imaging device) that measures the health status of the participants of the screening program, while the several measures of test accuracy are statistical techniques that estimate the accuracy of the modality. For example, Miller, Chamberlain, Day, Hakama, and Prorok[28] describe the estimation of the specificity and sensitivity of the U.K. screening trial for breast cancer trial as reported by Chamberlain et al.[29] The positive and negative predictive values also estimate the test accuracy and are illustrated with a study to diagnose coronary artery disease with the exercise stress test. For screening programs with two groups, a study group and a control, the validity of the modality is often determined by estimating the lead time and the survival experience of the study group compared to that of the control.

The HIP study is one of the earlier screening programs and consisted of two groups, where participants were randomized into the study and control groups, each with about 30,000 subjects. It was well designed and analyzed with Bayesian techniques, where the analysis is composed of estimating the lead time (the time between the time the disease is detected with the screening device and the time the disease would have been clinically detected without screening) and estimating the survival times (from diagnosis) to death. Survival will be compared between the two groups with life table techniques, while lead time will be estimated by the so-called method of differences.

Bayesian methods for estimating lead time and survival will be fully explained and illustrated with the aid of WinBUGS. The student should be able to appreciate the importance of screening programs and the statistical methods that are used to evaluate them. The difference method to estimate lead time and the life table to compare survival between the study and control groups are unique to epidemiology, but will be interpreted with a Bayesian approach.

Screening for chronic disease gained interest in the 1960s as a result of clinical experience that showed that disease detected at an earlier stage had better prognosis than disease detected at a later stage. Thus, interest in screening is based on the hypothesis that it would shift the diagnosis to an earlier stage and treatment would have a better change to make an impact on the development of diseases such as breast cancer. According to Shapiro, Venet, Strax, and Venet,[27] reports began to appear in the early 1960s from many periodic examination programs on the detection of breast cancer by palpation. For example, Holleb, Venet, Day, and Hoyt[30] report that a larger fraction of patients were diagnosed with a localized disease than that of the general population of patients.

Of course there was considerable debate and doubt about the impact examinations could have toward the reduction of deaths (due to breast cancer) in the more general population. It was difficult to generalize the results of these early reports because they were not designed randomized studies but instead were based on patients who volunteered for the examinations, and the selection factors associated with these groups made a meaningful comparison difficult to perform.

In the early 1960s, advances in mammography played an import role in the emergence of screening for breast cancer. It is interesting to note that just after the discovery of x-ray by Roentgen, Saloman[31] reported the use of x-ray to examine the breast, and he indeed recognized the potential of the modality in visualizing mass densities, mass irregularities, and microcalcifications. Not much progress was made until after World War I when the application of radiography with emphasis on the potential in helping the clinician diagnose the disease in asymptomatic women is reported.

It was at the MD Anderson Hospital and Tumor Institute where acceptance of mammography as a valuable device for the diagnosis of breast cancer had an important impact on screening for the disease. Their studies reported mammography involving 2000 patients with the objective of preserving the maximum detail in the image with the lowest kilovolts for penetration, compensation for increased exposure of the radiation, and proper focus of the x-ray beam.

The developments at MD Anderson proved very impressive with the result that the National Cancer Institute initiated a long-term study of the value of mammography in reducing breast cancer mortality, which resulted in the first randomized study involving the HIP. HIP was a good choice to execute a screening study because it included an experienced research group, where many projects were financed by NIH and various private foundations. HIP was a prepaid comprehensive medical care plan with 31 affiliated medical groups located in New York City and Long Island. Approximately 700,000 members were group enrollments among city, state, and federal government employees. Other research projects showed that the membership constituted a wide range of socioeconomic, ethnic, and religious groups in New York City; however, there were some areas that included members with very high income. HIP also had other important assets including an electronic information system that included (1) a file with each member's ID number, name, sex, month and year of birth, size of the covered family, medical group membership, source of enrollment, date of enrollment, and data of termination from enrollment; and (2) a reporting system of all services provided by physicians in each medical group, from which diagnostic and therapeutic services received by each patients was easy to determine.

Based on incidence rates of breast cancer, during 1958–1961, it was decided to enroll women over the range from 40 to 64 years of age. Total enrollment in the 23 medical groups was about 490,000 among which 80,300 were aged 40–64. Within each of the medical groups, two systematic random samples were selected where every n-th women was placed in the study group, and the (n + 1)-st in the control group, resulting in about 31,000 in each. The scheduled date for the initial exam became the entry date to the study, where each control group woman was assigned the same date as the corresponding study group woman. The HIP was the first randomized study to determine the efficacy of mammography and clinical examination for screening of breast cancer. The main question was: Does screening for disease increase the survival for those that were screened compared to a comparable control group of women? Another important consideration is to estimate the lead time of the study.

What is the lead time? Consider the following progression of breast cancer, where S_0 is the disease-free phase, S_p is the preclinical phase, where mammography and clinical exam can detect the disease, T_d is the time where the disease is in fact detected by screening, and S_c is the clinical phase, where breast cancer becomes apparent (in the absence of screening) to the patient.

$$S_0 \rightarrow S_p \rightarrow T_d \rightarrow S_c$$

The lead time is $S_c - T_d$, the time by which the diagnosis is advanced by screening. Note the beginning of the preclinical phase S_p and the beginning of the clinical phase S_c are random, as is the lead time $S_c - T_d$.

To estimate the lead time for the HIP study, consider the diagnostic profile of the study group. I have assumed the time from entry to the study to the time of diagnosis is less than or equal to 60 months. Note that of the 229 study cohort patients, 59 were diagnosed with breast cancer by clinical (breast examination) while 44 were diagnosed with mammography. Of the 44, 27 were diagnosed by observing a lesion (mass) and 17 by observing microcalcifications on the image.

To continue, 29 were diagnosed with breast cancer by both mammography and breast examination, where 26 were observed with a tumor mass and 3 with both a mass and microcalcifications. Lastly, 97 were interval (diagnosed between screening examinations) diagnoses, and of those 45 were diagnosed within 12 months of the last examination, and 52 identified after 12 months from the last examination. To estimate the lead time, interval diagnosis is ignored (breast cancer diagnosed between examinations or after the patients' last examination). This last restriction is somewhat controversial because a patient's participation makes the patient more aware of the clinical symptoms of the disease; thus, an interval diagnosis can be an effect of screening. For the control group, the effect of screening is taken to be nil and the time to diagnosis from time to entry will also assumed to be no more than 60 months. Descriptive statistics for estimating the lead time are portrayed in Table 1.15.

Thus, one sees that the median time from entry to diagnosis is 436 days with a mean of 555 days, and the corresponding entries for the control group are 972 days and 945 days, respectively. Based on the sample median, the

TABLE 1.15

Time from Entry to Diagnosis by Cohort in Days

Cohort	Mean	SD	N	Median
Study	555	473	132	436
Control	945	558	303	972

Note: Time to diagnosis is ≤ 60 months.

lead time is estimated as 972–436 = 536 days or 1.468 years. Based on the sample mean, the lead time is estimated as 945–555 = 390 days or 1.06 years.

Curiously, this estimate of 1.06 years for the lead time is quite similar to that reported by Wu, Kafadar, Rosner, and Broemeling.[32] For the study group, the lead time is skewed to the right, but for the control, there is a very small left skewness. I would advise using the medians to estimate the lead time. Chapter 7 gives much more detail about the lead time than the preceding presentation.

Of course, it is important to compare the survival time of the study cohort to that of the control. Two approaches are taken to compare the survival of the two cohorts: a life table method, and the use of the Cox proportional hazards model. After estimating the survival times of both cohorts, they are compared through a Bayesian version of the log-rank test. Plots of the survival curves are based on the Bayesian MH presented in Chapter 6. The last part of the chapter presents a comparison of the two groups with the Cox proportional hazards model, and finally 19 exercises, closely related to the various concepts presented in the chapter, are provided for the student.

1.3.7 Chapter 8: Statistical Models for Epidemiology

Chapter 8 presents regression models that are useful in epidemiology. Examples of some elementary models were introduced in earlier chapters and included simple linear and multivariate models where the dependent variable has a normal distribution or where the dependent variable is binary. Also presented were models for survival, which included the Weibull and Cox models. The main focus of Chapter 8 is on models that generalize the models introduced in earlier chapters.

The first class of models is regression with an ordinal score as the dependent variable. Categorical and ordinal regression models are generalizations of the logistic model, where the dependent variable had two values. Thus regression models are presented with dependent variables that assume a small number of responses. The model is referred to as an ordinal regression when the responses of the dependent variable are ordinal (can be ordered from smallest to largest). An ordinal regression model is employed to estimate the receiver operating characteristic (ROC) area for medical tests with ordinal scores.

This particular formulation of regression uses an underlying latent scale assumption. The cumulative odds model is often expressed in terms of an underlying continuous response. The following specification of the ordinal model follows Congdon,[33] where the observed response score Y_i with possible values 1, 2, ..., K is taken to reflect an underlying continuous part of the cumulative probability (p. 102)

$$\gamma_{ij} = \Pr(Y_i \le j) = F(\theta_j - \mu_i) \tag{1.43}$$

where i = 1, 2, ..., N is the number of patients and j = 1, 2, ..., K − 1. It is noted that

$$\mu_i = \beta X_i \tag{1.44}$$

expresses the regression relationship between the ordinal responses and the covariates X_i for the i-th patient. F is a distribution function and the θ_j are the cut points corresponding to the j-th rank. For our purposes, F is usually given a logistic or probit link, where the former leads to a proportional odds model. Suppose p_{ij} is the probability that the i-th patients has response j, then

$$\gamma_{ij} = p_{i1} + p_{i2} + ... + p_{ij} \tag{1.45}$$

Of course, Equation 1.45 can be inverted to give

$$\begin{aligned} p_{i1} &= \gamma_{i1} \\ p_{ij} &= \gamma_{ij} - \gamma_{i,j-1} \end{aligned} \tag{1.46}$$

and

$$p_{i,K} = 1 - \gamma_{i,K-1}$$

Suppose F is the logistic distribution function and

$$\begin{aligned} C_{ij} &= \log it(\gamma_{ij}) \\ &= \theta_j - \beta X_i \end{aligned} \tag{1.47}$$

where β, the vector of unknown regression coefficients, is constant across response categories j. Then the θ_j are the logits of the probabilities of belonging to the categories 1, 2, ..., j as compared to belonging to the categories j + 1, ..., K for subjects with X = 0. The difference in cumulative logits for different values of X, say X_1 and X_2, is independent of j, which is called the proportional odds assumption, namely,

$$C_{1j} - C_{2j} = \beta(X_1 - X_2) \tag{1.48}$$

Using the preceding ordinal regression model, the posterior distribution of the individual probabilities p_{ij} can be determined, as can the probabilities q_j (j = 1, 2, ..., K) of the basic ordinal responses.

Once the posterior distribution of the basic responses is known, for the diseased and nondiseased groups, the posterior distribution of the area under the ROC curve can also be computed. Several scenarios will be displayed for a given example of ordinal regression: (1) the ROC area induced by all covariates or selected subsets of covariates and (2) the ROC area conditional on certain values of the covariates or subsets of covariates.

An example with ordinal scores, a study involving melanoma metastasis to the lymph nodes, is considered. A sentinel lymph node biopsy is performed on the patients to determine the degree of metastasis, where the diagnosis is made on the basis of the depth of the primary lesion, the Clark level of the primary lesion, and the age and gender of the patient. The procedure involves the cooperation of an oncologist, a surgical team that dissects the primary tumor, pathologists, and radiologists who perform the imaging aspect of the biopsy. A radiologist makes the primary determination of the degree of metastasis on a 5-point ordinal scale, where 1 designates absolutely no evidence of metastasis, 2 means no evidence of a biopsy, 3 indicates very little evidence of metastasis, 4 implies there is some evidence, and 5 denotes strong evidence of metastasis.

The Bayesian analysis is executed with BUGS CODE 8.1 and produces the posterior distribution of the relevant parameters, including the posterior distribution of the individual ROC curves, corresponding to the different radiologists. Chapter 8 continues with an introduction to nonlinear models.

Nonlinearity is a feature of many studies involving epidemiology. A nonlinear regression model is defined as

$$Y_n = f(x_n, \theta) + e_r \qquad (1.49)$$

where the n-th observation of the dependent variable Y is Y_n, x_n is the corresponding observation of the q by 1 independent variable x, and e_n is the corresponding error term corresponding to the n-th observation. It is assumed the N error terms e_n are independent random variables with a normal distribution with mean zero and unknown variance σ^2.

The N values of Y_n and the vector x_n are known, but the p by 1 vector

$$\theta = (\theta_1, \theta_2, ..., \theta_p) \qquad (1.50)$$

is assumed to be unknown.

Using a Bayesian approach, the objective is to examine the effect of the q exposures x on the dependent variable Y. A prior distribution is placed on the unknown parameters θ and σ^2, then through Bayes' theorem, the posterior distributions of θ and σ^2 are determined. As discussed earlier, the posterior analysis will be executed using WinBUGS.

To illustrate nonlinear regression, several examples will be presented as a five-step procedure:

1. Plots of the independent variables versus the dependent variable will be portrayed.
2. A model will be defined based on the plots of 1.
3. A prior distribution will be assumed for the unknown parameters θ and σ^2.

4. Based on WinBUGS, the posterior distribution of θ and σ^2 will be determined.

5. The goodness-of-fit of the model will be assessed by plotting the predicted values of Y (those predicted by the model) versus the corresponding actual values Y_n.

To investigate the polychlorinated biphenyls (PCB) concentration in fish from Cayuga Lake, NY, Bache, Serum, Youngs, and Lisk[34] conducted a study to see the effect of the age of the fish on PCB concentration. Since the fish are annually stocked as yearlings and distinctly marked as to year class, the ages of the fish are accurately known. The fish is mechanically chopped, ground, and mixed, and 5-g samples taken. Age is recorded in years and PCB concentration expressed as parts per million. PCB is a toxin, and it is important for public health to know its concentration in the environment. See the graph of log PCB versus the cube root of age in Figure 1.2.

Based on the plot, the regression model is assumed to be

$$\ln(Y_n) = \theta_1 + \theta_2 \times x_n + e_n \qquad (1.51)$$

where n = 1, 2, …, 28, and 28 is the total number of observations shown in Table 8.3. The dependent variable is the natural log of PCB = Y_n and the independent variable is cube root of age = x_n, and the association is linear with unknown regression coefficients $\theta = (\theta_1, \theta_2)$ and thus can be analyzed like a simple normal linear regression model.

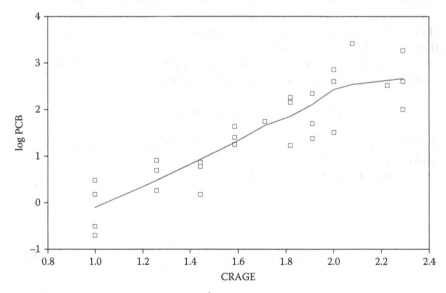

FIGURE 1.2
Log PCB versus cube root of age.

However, it should be noted that the association between PCB and age is nonlinear, namely,

$$Y_n = \exp\left(\beta_1 + \beta_2 \times \sqrt[3]{x_n}\right) \tag{1.52}$$

A Bayesian analysis gave the following estimates for the linear and nonlinear regressions: (1) θ_1 has posterior mean −2.398 and 95% credible interval (−3.21, 1.581), and for the slope θ_2, 2.307 is the posterior mean, while (1.831, 2.778) is the 95% credible interval, and (2) for the parameter β_1 of the nonlinear regression (Equation 1.52), the posterior mean is −1.67 with a 95% credible interval (−4.402, .05629), while the parameter β_2 has a 95% credible interval (1.151, 3.063) and posterior mean 1.981. Thus, for this example, the relationship between PCB and age can be examined with a linear model (Equation 1.51) and a nonlinear regression (Equation 1.52); however, it should be noted the models are not the same, but are related.

Repeated measures (or longitudinal models) play an important part of epidemiology. Our first encounter with a repeated measures study is an example involving Alzheimer's disease in a study done by Hand and Taylor,[35] where two groups of patients were compared. One group was given a placebo and the other group received lecithin. Each of the 26 patients in the placebo group and 22 in the treatment group were measured five times, where the measurement was the number of words the subject could recall from a list of words. Note that the same measurement is repeated on the same subject for a fixed number of occasions, and one would expect the measurements to be correlated. The unique aspect of a repeated measures study is the presence of correlation between measurements on the same subject. From a statistical point of view, this correlation is taken into account when estimating the other parameters of the model.

To analyze the Alzheimer's information, the following model is adopted.

Let the observation for the i-th subject on occasion j be

$$y_{ij} = \theta + \alpha_i + \beta_j + e_{ij} \tag{1.53}$$

where i = 1, 2, ..., n, j = 1, 2, ..., p, where n is the number of subjects and p the number of time points.

It is assumed that θ is a constant,

$$\alpha_i \sim nid(0, \tau_\alpha), \; i = 1, 2, \ldots n \tag{1.54}$$

$$\beta_j \sim nid(0, \tau_\beta), \; j = 1, 2, \ldots, p \tag{1.55}$$

and

$$e_{ij} \sim nid(0, \tau) \tag{1.56}$$

The variance of the α_i is $\sigma_\alpha^2 = 1/\tau_\alpha$, of the β_j is $\sigma_\beta^2 = 1/\tau_\beta$, and of the e_{ij} is $\sigma^2 = 1/\tau$, where the three tau variables are positive. The variance component $\sigma_\alpha^2 = 1/\tau_\alpha$ measures the variability of the observations between the various subjects, while the component $\sigma_\beta^2 = 1/\tau_\beta$ measures the variability between the several times (occasions), and $\sigma^2 = 1/\tau$ measures the overall variability of the $y(i, j)$ observations. Note that the θ parameter measures the overall mean of the observations.

Note that

$$\mathrm{cov}(y_{ij}, y_{ij}) = \sigma_\alpha^2 \tag{1.57}$$

and

$$\mathrm{cov}(y_{ij}, y_k) = \sigma_\alpha^2 + \sigma_\beta^2 + \sigma^2 \tag{1.58}$$

that is, observations of the same subject are correlated with covariance σ_α^2 and the common variance is $\sigma_\alpha^2 + \sigma_\beta^2 + \sigma^2$, which implies that the correlation between measurements of the same subject is

$$\rho = \frac{\sigma_\alpha^2}{(\sigma_\alpha^2 + \sigma_\beta^2 + \sigma^2)} \tag{1.59}$$

There are other patterns of correlation between observations of the same subject. For example, consider a Bayesian analysis for the placebo group of the Hand and Taylor[35] Alzheimer's study. The analysis is executed with 55,000 observations (Table 1.16).

Therefore, for the placebo group, the correlation between observations of the same subject is estimated as .7536 with a 95% credible interval of (.6195, .8629), and the average number of correctly recalled words is estimated as 9.3165 with the posterior mean. Note that σ_β^2 measures the variability of the observations between the time periods and has a posterior median of .00595, indicating very little variation compared to the variation between individuals, which is estimated at 23.2 with the posterior mean.

TABLE 1.16

Posterior Analysis of Alzheimer's Study (Placebo Group)

Parameter	Mean	SD	Error	2½	Median	97½
ρ	.7536	.06426	.000403	.6195	.7578	.8629
σ^2	6.978	.9967	.00585	5.296	6.889	9.202
σ_α^2	23.2	7.537	.04688	12.56	21.88	41.51
σ_β^2	.1256	.6324	.01072	.0000977	.005952	.9147
θ	9.315	.9729	.0238	7.41	9.317	11.22

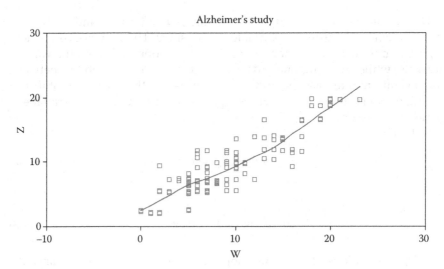

FIGURE 1.3
Predicted values Z versus actual values V number of correctly recalled words.

The Bayesian analysis includes a prediction Z of the number of correctly recalled words W, and the goodness of fit of the model is assessed by observing Figure 1.3. Does the model (Equation 1.53) give a good fit?

Chapter 8 is concluded with a description of the use of spatial models in epidemiology. Before introducing spatial models used in epidemiology, it is interesting to note the definition of epidemiology given by Mausner and Kramer:[5] "Epidemiology may be defined as the study of the distribution and the determinants of disease and injuries in the human population. That is, epidemiology is concerned with the frequencies and types of illnesses and injuries in groups of people and with the factors that influence their distribution" (p. 51). As will be seen in this section, spatial models allow us to investigate the distribution of disease in the spatial domain and its association with various exposures (risk factors).

The subject is introduced with an example of the number of cases of lip cancer diagnosed in 56 counties in Scotland, where the relative risk of the disease is estimated for each county, and the association between the number employed in outdoor jobs and the incidence of lip cancer is explored. Essentially, the model is a Poisson regression model where the spatial correlation between the lip cancer rates of one county and its neighbors is taken into account by a conditional autoregressive (CAR) model. One would expect the incidence of lip cancer of a county to be related to the incidence of its neighboring counties. One would also expect the same sort of association between an exposure in one county and its neighbors. That is, one would expect the incidence of lip cancer to be more related to the incidence with its neighbors, more so than with counties that are not neighbors. Thus, one sees the similarity between spatial models in epidemiology and the repeated measures models given in Section 1.3.6, in that adjacent observations tend to be correlated.

In small populations, maps of disease rates and exposure may give estimates of rates and risk factors that are unstable. Thus, our Bayesian analysis will be based on regression models where the spatial correlation is modeled by a CAR process. The effect will be to spatially smooth disease rates and risk estimates by allowing each site to borrow strength from its neighbors. Covariates that measure exposure can also be included in the model in such a way that a possible association between risk factors (exposures) and disease may be established. Thus, the lip cancer incidences will be smoothed over counties through a Bayesian analysis that employs MCMC techniques to estimate the model parameters.

Geographical epidemiology and medical geography are terms used for mapping the distribution of disease with respect to place and time. Maps take into account the spatial relationship that may be missed in descriptive tables. Good examples of this are the Palm[36] study of the spatial distribution of rickets, which established the association of the disease with the lack of sunlight, and in a similar fashion, the study of Lancaster[37] for the association between exposure to sunlight and melanoma. The maps revealed the spatial distribution of the diseases, which established the association between the disease and the relevant exposure.

A major concern is that the data values being mapped including estimates of relative risk can be very unstable when dealing with disease clusters, rare diseases, and small populations. The number of observed cases of disease are usually assumed to have a Poisson distribution, but extra-Poisson variability usually occurs (recall that the mean and variance of the Poisson are the same) and the extra variability is accounted for by including variables that follow a CAR distribution.

The topic of spatial models is vast and there have been many approaches to analyzing such data. For example, empirical Bayesian approaches with regression and CAR processes for estimating the association between disease and risk factors have been pursued by Clayton and Kaldor,[38] Cressie,[39] and Mollie and Richardson,[40] and this approach with some alteration will be implemented for the lip cancer example.

The rates of lip cancer in 56 counties in Scotland have been analyzed by Clayton and Kaldor. The form of the data includes the observed and expected cases (expected numbers based on the population and its age and sex distribution in the county), a covariate measuring the percentage of the population engaged in agriculture, fishing, or forestry, and the "position" of each county expressed as a list of adjacent counties.

We may smooth the raw standardized mortality rates (SMRs) by fitting a random-effects Poisson model allowing for spatial correlation, using the intrinsic CAR prior. For the lip cancer example, the model may be written as

$$O_i \sim \text{Poisson}(mu_i) \tag{1.60}$$

$$\log (mu_i) = \frac{\log E_i + \alpha_0 + \alpha_1 x_i}{10 + b_i} \qquad (1.61)$$

where α_0 is an intercept term representing the baseline (log) relative risk of disease across the study region, x_i is the covariate "percentage of the population engaged in agriculture, fishing, or forestry" in district i, with associated regression coefficient α_1, and b_i is an area-specific random effect capturing the residual or unexplained (log) relative risk of disease in area i.

We often think of bi as representing the effect of latent (unobserved) risk factors. Note that the O_i are the observed number of lip cancer cases and E_i the corresponding expected number of cases (expected numbers based on the population and its age and sex distribution in the county).

To allow for spatial dependence between the random effects b_i in nearby areas, we may assume a CAR prior for these terms. We give a brief description of the CAR process as follows:

Let the b_i, i = 1, 2, ..., n, be normal random variables where the index i denotes the i-th site (i-th county) such that the conditional distribution of b_i given b_j is denoted by

$$b_i \mid b_j \sim \text{norm}(\mu_i + \rho \sum_{j=1}^{j=n} w_{ij}(b_j - \mu_j), \sigma^2) \qquad (1.62)$$

where $w_{ii} = 0, w_{ij} = 1$ if i and j are adjacent neighbors, otherwise $w_{ij} = 0$ and ρ is a constant.

The objective is to estimate the relative risk RR[i] of each county and the parameters α_0 and α_1 of the Poisson regression (Equations 1.60 and 1.61) (Table 1.17).

The posterior mean of the intercept α_0, the baseline log relative risk, is −.3025, and that the effect of the covariate α_1 of the log of the average number of cases is .4595 with a credible interval of (.227, .6781). Both appear to have an effect on the average number of lip cancer cases. A map of the relative risk by county is portrayed in Figure 1.4.

Additional examples of spatial modeling are presented in Chapters 8 along with 13 problems that challenge the student for a better understanding of the subject.

TABLE 1.17

Bayesian Analysis for Parameters of the Poisson Regression Lip Cancer Study

Parameter	Mean	SD	Error	2½	Median	97½
α_0	−.3025	.1127	.001598	−.5202	−.3033	−.0783
α_1	.4595	.1158	.00182	.227	.4624	.6781

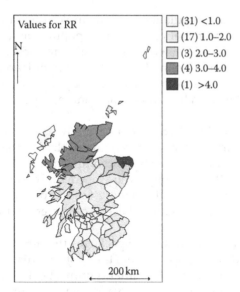

FIGURE 1.4
Relative risk by county for lip cancer study.

1.4 Preview of the Appendices

Before continuing to Chapter 2, the reader should read the two appendices, one that introduces Bayesian inference for epidemiology, and the other an introduction to the use of WinBUGS. If the student is not familiar with Bayesian inference, Appendix A is a must, and I assume most readers are not familiar with WinBUGS, thus Appendix B is a requirement.

1.4.1 Appendix A: Introduction to Bayesian Statistics

Bayesian methods will be employed to design and analyze studies in epidemiology and this chapter will introduce the theory that is necessary to describe Bayesian inference. Bayes' theorem, the foundation of the subject, is first introduced and followed by an explanation of the various components of Bayes' theorem: prior information; information from the sample given by the likelihood function; the posterior distribution, which is the basis of all inferential techniques; and lastly the Bayesian predictive distribution. A description of the main three elements of inference, namely, estimation, tests of hypotheses, and forecasting future observations, follows.

Of course, inferential procedures can only be applied if there is adequate computing available. If the posterior distribution is known, often analytical methods are quite sufficient to implement Bayesian inferences and will be demonstrated for the binomial, multinomial, and Poisson populations and several cases of normal populations. For example, when using a beta prior distribution for the parameter of a binomial population, the resulting beta posterior density has well-known characteristics, including its moments. In a similar fashion, when sampling from a normal population with unknown mean and precision and with a vague improper prior, the resulting posterior t-distribution for the mean has known moments and percentiles that can be used for inferences.

Posterior inferences by direct sampling methods are easily done if the relevant random number generators are available. On the contrary, if the posterior distribution is quite complicated and not recognized as a standard distribution, other techniques are needed. To solve this problem, MCMC techniques have been developing for the past 25 years and have been a major success in providing Bayesian inferences for quite complicated problems.

Minitab, S-Plus, and WinBUGS are packages that provide random number generators for direct sampling from the posterior distribution for many standard distributions, such as binomial, gamma, beta, and t-distributions. On occasion these will be used; however, my preference is WinBUGS, because it has been well accepted by other Bayesians. This is also true for indirect sampling, where WinBUGS is a good package and is the software of choice for the book; it is introduced in Appendix B. Many institutions provide special purpose software for specific Bayesian routines. For example, at MD Anderson Cancer Center, where Bayesian applications are routine, several special purpose programs are available for designing (including sample size justification) and analyzing clinical trials and will be described. The theoretical foundation for MCMC is introduced in the following sections.

Inferences for studies in epidemiology consist of testing hypotheses about unknown population parameters, estimation of those parameters, and forecasting future observations.

If the main focus is estimation of parameters, the posterior distribution is determined, and the mean, median, standard deviation, and credible intervals found, either analytically or by computation with WinBUGS. For example, when sampling from a normal population with unknown parameters and using a conjugate prior density, the posterior distribution of the mean is a t and will be derived algebraically. On the contrary, in observational studies, the experimental results are usually portrayed in a 2 by 2 table that gives the cell frequencies for the four combinations of exposure and disease status where the consequent posterior distributions are beta for the cell frequencies, and posterior inferences are provided both analytically and with WinBUGS. Of course, all analyses should be preceded by checking to determine if the model is appropriate, and this is where the predictive distribution comes into play. By comparing the observed results of the experiment (e.g., a case–control study) with those predicted, the model assumptions are tested. The most frequent use of the Bayesian predictive distribution is for forecasting future observation in time series studies, and time series in the form of cohort studies (repeated measures) is part of many epidemiologic studies.

A good introduction to the use of Bayesian inference in biostatistics is Woodworth.[41]

1.4.2 Appendix B: Introduction to WinBUGS

WinBUGS is the statistical package that is used for the book and it is important that the novice be introduced to the fundamentals of working in the language. This is a brief introduction to the package and for the first-time user it will be necessary to gain more knowledge and experience by practicing with the numerous examples provided in the download. WinBUGS is specifically designed for Bayesian analysis and is based on MCMC techniques for simulating samples from the posterior distribution of the parameters of the statistical model. It is quite versatile and once the user has gained some experience, there are many rewards.

Once the package has been downloaded, the essential features of the program are described, first by explaining the layout of the BUGS document. The program itself is made up of two parts, one part for the program statements, and the other for the input of the sample data and the initial values for the simulation. Next to be described are the details of executing the program code and what information is needed for the execution. Information needed for the simulation are the sample sizes of the MCMC simulation for the posterior distribution and the number of such observations that will apply to the posterior distribution.

After execution of the program statements, certain characteristics of the posterior distribution of the parameters are computed including the posterior mean, median, credible intervals, and plots of posterior densities of the parameters. In addition, WinBUGS provides information about the posterior distribution of the correlation between any two parameters and information about the simulation. For example, one may view the record of simulated values of each parameter and the estimated error of estimation of the process. These and other activities involving the simulation and interpretation of the output will be explained.

Examples based on accuracy studies illustrate the use of WinBUGS and include estimation of the true and false positive fractions for the exercise stress test and modeling for the ROC area. Of course, this is only a brief introduction, but should be sufficient for the beginner to begin the adventure of analyzing data. Because the book's examples provide the necessary code for all examples, the program can easily be executed by the user. After the book is completed by the dedicated student, they will have a good understanding of WinBUGS and the Bayesian approach to measuring test accuracy. Ntzoufras[17] is an excellent reference of Bayesian modeling with WinBUGS.

1.5 Comments and Conclusions

Chapter 1 introduces the fundamental ideas of Bayesian methods in epidemiology and previews the remaining 7 chapters of the book. The preview should give the reader a good idea of what to expect. Special emphasis is placed on the two appendices, which is required reading for those new to Bayesian statistics and WinBUGS.

Exercises

1. Read Appendix A.
2. Read Appendix B.
3. Describe a cohort design for estimating the association between exposure to a risk factor and the occurrence of disease.
4. Describe a case–control design for estimating the association between exposure to a risk factor and the occurrence of disease.
5. Refer to problem 3 and explain what measure is used for cohort designs to measure association between risk and disease.
6. Refer to problem 4 and explain what measure is used for case–control designs to measure association between risk and disease.
7. Explain how the sampling schemes differ between a cohort design and a case–control design.
8. Explain the posterior distribution of the parameter d in Table 1.6.
9. What does the MH estimator measure in a case–control study?
10. Explain the general use of regression models in epidemiology. What is the purpose of a regression model in epidemiology? Give an example.
11. A logistic regression model is defined in Equation 1.19. Describe how the logistic model is used to analyze the Israeli Heart Study.
12. The simple linear regression model is defined in Equation 1.22. Describe the use of this model to measure the effect of age on systolic blood pressure of the Woolson study.
13. In Table 1.9, explain how to calculate the probability of death in each time period.
14. Explain how the Bayesian analysis of Table 1.10 is related to Table 1.9 for the coronary artery study.

15. The Bayesian analysis of a life table is based on Equation 1.26. Explain this assumption about the distribution of the number of deaths in a given time period.

16. Equations 1.32 and 1.34 describe the distribution for the number of deaths and number of withdrawals for each time period of a life table. Explain the importance of the probability of death $q[i]$ and probability $qc[i]$ of a censored observation in the Bayesian analysis of a life table. What prior distribution is assigned to the $q[i]$ and $qc[i]$?

17. What is the basic assumption about the Cox proportional hazards model? See Equation 1.39.

18. In a screening study, define the lead time of the study. How is the lead time estimated? Explain Table 1.14 for the HIP study. What is the estimated lead time for mammography?

19. Describe the Bayesian analysis of Table 1.16. What is the posterior distribution of the correlation between different observations of the same subject in the placebo group of the Alzheimer's study?

20. Figure 1.3 is a plot of the observed versus predicted values for the Alzheimer's study. Does the plot indicate a good fit of the model to the data?

21. Equations 1.60 and 1.61 define the spatial model for the lip cancer data for counties in Scotland. Based on Table 1.17, describe the posterior distribution of α_0 and α_1.

References

1. Kahn, H. and Sempos, C.T. *Statistical Methods in Epidemiology*, Oxford University Press, 1989, New York.
2. Selvin, S. *Statistical Tools for Epidemiologic Research*, Oxford University Press, 2011, New York.
3. Jewell, N. *Statistics for Epidemiology*, Chapman & Hall/CRC, 2003, New York.
4. Rothman, K.J., Greenland, S., and Lash, T.L. *Modern Epidemiology*, Lippincott, Williams & Wilkins, 2008, New York.
5. Mausner, J.S. and Kramer, S.H. *Mausner and Bahn Epidemiolpogy: An Introductory Text*, Second edition, W.B. Saunders Company, 1985, London, UK.
6. Elliot, P., Wakefield, J., Best, N., and Briggs, D. *Spatial Epidemiology Methods and Applications*, Oxford University Press, 2001, New York.
7. Lawson, A. *Bayesian Disease Mapping, Hierarchical Modeling in Spatial Epidemiology*, Chapman and Hall/CRC, 2008, New York.
8. Abbott, R.D., Yin, Y., Reed, D.M., and Yano, K. Risk of stroke in male cigarette smokers, *N. Engl. J. Med.*, 315, 717–720, 1986.
9. Johnson, S.K. and Johnson, R.E. Tonsillectomy in Hodgkin's disease, *N. Engl. J. Med.*, 287, 1122–1125, 1972.

10. Mielke, C.H., Shields, J.P., and Broemeling, L.D. Coronary artery calcium scores for men and women of a large asymptomatic population, *CVD Prevention*, 2, 194–198, 1999.
11. Leviton, A. Definitions of attributable risk, *Am. J. Epidemiol.*, 98(3), 231, 1973.
12. Hiller, R.A. and Kahn, H.A. Senile cataract extraction and diabetes, *Br. J. Opthamol.*, 60, 283–286, 1976.
13. U.S. Census Bureau 2010 Census Summary. Table QP-Q1. http://factfinder2 .census.gov/faces/tableservices/jsf/pages/productview.xhtml?fpt=table.
14. Work Table 28R from The National Center of Health Statistics. http://www.cdc .gov/nchs/data/dvs/MortFinal2007_Worktable308.pdf.
15. Mantel, N. and Haenszel, W.J. Statistical aspects of the analysis of data from retrospective studies of disease, *Natl. Cancer Inst.*, 22, 719, 1959.
16. Chatterjee, S. and Price, B. *Regression Analysis by Example*, John Wiley & Sons Inc., 1991, New York.
17. Ntzoufras, I. *Bayesian Modeling Using WinBUGS*, John Wiley & Sons Inc., 2009, New York.
18. Woolson, R.F. *Statistical Methods for the Analysis of Biomedical Data*, John Wiley & Sons Inc., 1987, New York.
19. Broemeling, L.D. *Advanced Bayesian Methods for Medical Test Accuracy*, Francis & Taylor, 2012, Boca Raton, FL.
20. National Institutes of Health. National Heart, Lung, and Blood Institute. Chart 3–26, Death and Age-Adjusted Death Rates for Coronary Heart Disease, 1980–2006. http://www.nhlbi.nih.gov/resources/docs/2009_ChartBook.pdf.
21. Kleinbaum, D.G. *Survival Analysis a Self-Learning Text*, Springer Verlag, 1996, New York.
22. Congdon, P. *Bayesian Statistical Modelling*, John Wiley & Sons Inc., 2001, New York.
23. Congdon, P. *Applied Bayesian Modelling*, John Wiley & Sons Inc., 2003, New York.
24. Newman, S.C. *Biostatistical Methods in Epidemiology*, John Wiley & Sons Inc., 2001, New York.
25. Freireich, E.J., Gehan, E., Frei, E., and Schroeder, L.R. The effect of 6-mercapto-purine on the duration of remission in acute leukemia: A model for the evaluation of other potentially useful therapies, *Blood*, 21, 699–716, 1963.
26. Morrison, A.S. *Screening in Chronic Disease*, Second edition, Oxford University Press, 1992, New York.
27. Shapiro, S., Venet, W., Strax, P., and Venet, L. *Periodic Screening for Breast Cancer, The Health Insurance Project and its Sequelae, 1963–1986*, Johns Hopkins University Press, 1988, Baltimore, MD.
28. Miller, A.B., Chamberlain, J., Day, N.E., Hakama, M., and Prorok, P.C. *Cancer Screening*, Cambridge University Press, 1990, Cambridge, UK.
29. Chamberlain, J., Coleman, D., Ellman, R., et al. Sensitivity and specificity of screening in the UK trial of early detection of breast cancer, Article in *Cancer Screening*, pp. 3–17, Edited by Miller, A.B., Chamberlain, J., Day, N.E., Hakama, M., and Prorok, P.C., Cambridge University Press, 1990, Cambridge, UK.
30. Holleb, A., Venet, L., Day, E., and Hoyt, S. Breast cancer detected by routine examination, *New York State J. Med.*, 60, 823, 1960.
31. Saloman, A. Beitrage zur Pathologie und Klinik des Mammakarzinims [Contributions to the pathology and clinic of mammakarzinims]. *Arch. F. Kun. Chir.*, 101, 573, 1913.

32. Wu, D., Kafadar, K., Rosner, G.L., and Broemeling, L.D. The lead time distribution, when lifetime is a random variable in periodic cancer screening, *Int. J. Biostatistics*, 8(1), 1–14, 2012.
33. Congdon, P. *Applied Bayesian Modelling*, John Wiley & Sons Inc., 2003, New York.
34. Bache, C.A., Serum, J.W., Youngs, D.W., and Lisk, D.J. Polychlorinated biphenyl residues: Accumulation in Lake Cayuga trout with age, *Science* 117, 1192–1193, 1972.
35. Hand, D.J. and Taylor, C.C. *Multivariate Analysis of Variance and Repeated Measures*, Chapman & Hall, 1987, London, UK.
36. Palm, T.A. The geographical distribution and aetiology of rickets, *Practitioner* 45, 270–279, 1890.
37. Lancaster, H.O. Some geographical aspects of the mortality from melanoma in Europeans, *Med. J. Aust.*, 1, 1082–1087, 1956.
38. Clayton, D. and Kaldor, J. Empirical Bayes estimates of age-standardized relative risks for use in disease mapping, *Biometrics* 43, 671–681, 1987.
39. Cressie, N.A.C. Regional mapping of incidence rates using spatial Bayesian models, *Med. Care*, 31(supplement), YS60–YS65, 1993.
40. Mollie, A. and Richardson, S. Empirical Bayes estimates of cancer mortality rates using spatial models, *Stat. Med.*, 10, 95–112, 1991.
41. Woodworth, G.C. *Biostatistics, A Bayesian Introduction*, Wiley Interscience, John Wiley & Sons Inc., 2004, Hoboken, NJ.

2

A Bayesian Perspective of Association between Risk Exposure and Disease

2.1 Introduction

From a statistical point of view, epidemiology investigates the association between risk and disease by estimating the prevalence and/or incidence of disease and risk factor with various designs. The two most important designs are the cohort and case–control designs, where with the former, the relative risk (RR) parameter measures the association between disease and risk, while with the latter, the odds ratio (OR) parameter measures that association. In this chapter, the Bayesian approach is taken whereby the posterior distribution of the RR is determined for the cohort study, the OR for the case–control design, and both (RR and OR) for the cross-sectional study.

For the cohort design in a prospective study, the risk factor is identified and the subjects consist of two groups; one group consists of subjects exposed to the risk factor, while the other group consists of subjects unexposed to the risk factor. The main response is the incidence of disease of the two groups, and the two incidences are compared via the RR, which is estimated by its posterior distribution. A similar situation occurs with the retrospective study and the case–control design, where two groups are identified—subjects with disease and subjects without disease. Then looking into the past, the exposure history to risk is measured from the subjects' medical history. For the cross-sectional design, one sample is taken and the subject is classified according to their disease status (yes, no) and the presence of a risk factor (yes, no).

This chapter begins with the definition of the two most important parameters in epidemiology, namely, the incidence and prevalence rates for morbidity and mortality of disease. Then, the Bayesian approach begins in earnest with a description of the posterior distribution of the RR for a cohort study, and the ideas are illustrated by performing the posterior analysis for several examples (of cohort studies) executed with WinBUGS®. These ideas are extended to designs for retrospective studies such as for case–control studies, where the association between risk and disease is measured with the OR. Finally, the cross-sectional study is considered, where the disease status and the risk status are measured at the same time in all subjects. The latter type of study is shown in terms of the Shields Heart Study, where over 4300 subjects are measured for various risk factors that may affect the development of coronary artery disease.

To summarize, three designs will be considered in this chapter—the cohort study, the case–control study, and the cross-sectional study. From a statistical viewpoint, these three differ by the way samples are taken. For the cohort study, one sample is taken from those exposed to the risk factor and one sample from those who are not. Subjects are followed into the future for a given period. The main response of interest is disease incidence for those exposed versus those unexposed. For the case–control design, one sample is selected from the population of subjects with the disease and one selected from those without the disease. For the cross-sectional study, the disease status, and exposure experience of each subject is measured one, that is one sample is selected and two variables measured on each subject.

2.2 Incidence and Prevalence for Mortality and Morbidity

There are many ways to measure mortality and morbidity, but most of them are rates that measure prevalence and incidence. For example, for incidence, the rate is defined as a ratio, where the numerator is the number of new cases of a disease that occur over a given period, and the denominator is the number of people in the population at risk. To the statistician, the incidence is an unknown parameter that is estimated by the corresponding sample prevalence rate, and, in particular for the Bayesian, the incidence rate is estimated from the posterior distribution of the incidence rate parameter.

It is important to remember that the incidence includes only new cases of disease over a specified period, thus, the health status of each member of the population must be known. The denominator is the number of people in the population at risk and must be carefully defined. Recall that the number of people in the population at risk will vary over a given period because of death, birth, immigration, and migration. Thus, one must be careful in their definition of the denominator of the incidence rate. Often, the population is taken to be the number of people at risk at the midpoint of the observation period, but in any case, the numerator and denominator are estimates.

Suppose the incidence rate is defined as

$$\theta_{IR} = \frac{\text{Number of new cases of disease}}{\text{Population at risk}} \qquad (2.1)$$

over a given period. If the number of people in the population at risk is known, then assuming a random sample of individuals from the population, the numerator can be considered a binomial random variable with the two parameters θ_{IR} (for the probability parameter) and the number of subjects at risk for the second parameter. Using this approach, probability of disease is the same for each person in the population at risk.

Table 2.1, taken from the SEER database at the National Cancer Institute, gives the annual incidence of breast cancer for all women and for particular subgroups. For example, for all races, the incidence of breast cancer is estimated to be 124 annual cases per 100,000 women. Reading the details of the methodology given by the SEER report one can see that the incidence rates are calculated using the information taken from samples from 17 geographical areas of the United States over the period 2004–2008. It is not obvious how the numerator and denominator of the incidence parameter are calculated from the SEER database. The incidence rates for breast cancer given in Table 2.1 are computed using the statistical methodology of Howland et al.[1]

To illustrate the Bayesian approach, suppose a random sample of 100,000 women is taken from the population of U.S. women and that 124 new cases of breast cancer are diagnosed for 2006. Then the number of new cases can be considered a binomial random variable with parameters θ_{IR} and 100,000. From a Bayesian view point, the goal is to determine the posterior distribution of θ_{IR}. Assuming a uniform prior for the incidence rate, the Bayesian analysis is executed with the following code.

TABLE 2.1

Incidence Rates by Race

Race/Ethnicity	Female
All Races	124.0 per 100,000 women
White	127.3 per 100,000 women
Black	119.9 per 100,000 women
Asian/Pacific Islander	93.7 per 100,000 women
American Indian/Alaska Native	77.9 per 100,000 women
Hispanic	

Source: Methodology for calculating incidence at National Cancer Institute: http://seer.cancer.gov/statfacts/html/breast.html#incidence-.

BUGS CODE 2.1

```
model;
{
# binomial distribution of the number of new cases nc
nc~dbin(theta,100000)
# prior distribution of incidence theta is uniform
# note theta is the incidence rate parameter θ_IR
theta~dbeta(1,1)
}
# this is the data for the number of new cases nc = 124
list(nc = 124)
# initial value of incidence theta
list(theta =.5).
```

The Bayesian analysis is executed with 55,000 observations generated from the posterior distribution of θ_{IR} using a burn-in of 5000 and a refresh of 100, and the posterior distribution of the incidence has the following characteristics: mean = 0.001249, median = 0.001245, standard deviation (SD) = 0.0001122, and 95% credible interval (0.00104, 0.001477). The Markov chain Monte Carlo (MCMC) error is 0.0000005421 indicating that the estimate of 0.001249 for the posterior mean is within 0.0000005421 units of the "true" posterior mean. Thus, 55,000 observations appear to be quite sufficient for the MCMC simulation. Note that this is only an example, and is not based on the data used by the SEER approach given in Table 2.1.

In contrast to the incidence rate, the prevalence rate is defined as

$$\theta_P = \frac{\text{Number of existing cases of disease}}{\text{Total population}} \tag{2.2}$$

at a point in time. The prevalence can also be defined over a given period, where the numerator is the number of existing cases of a disease (over the specified period) and the denominator the average population size over that period.

The number of existing cases is the number of new cases plus the number of preexisting cases. For example, the SEER database reports the prevalence rate for breast cancer on January 1, 2008 in the United States as approximately 2,632.005 women alive who had a history of breast cancer. This includes any person alive on January 1, 2008 who had been diagnosed with breast cancer at any point prior to January 1, 2008, and includes women with active disease and those cured of the disease.

The methodology is based on the website http://surveillance.cancer.gov/ prevalence/ (Ref. [3]) and the reader is invited to upload this information and learn about the various ways to estimate prevalence including limited-duration prevalence and complete prevalence.

Continuing with the estimate of point prevalence for breast cancer on January 1, 2008, one would need to have an estimate of the denominator of

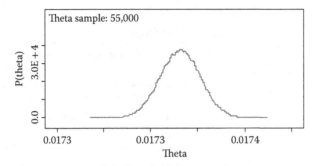

FIGURE 2.1
Posterior density of θ_P, the prevalence rate.

Equation 2.2, namely the size of the total female population on that data. From the U.S. Census Bureau,[4] see http://www.census.gov/popest/data/ state/totals/2008/tables/NST-EST2008-01.csv, the U.S. population on April 1, 2008 was 304,054,724. Thus, using 304,000,000 as the estimated population on January 1, 2008, the prevalence rate for breast cancer is estimated as $\dfrac{2,632,005}{152,000,000} = 0.017315882$, where the denominator is one half of the U.S. population size. Or stated another way, the prevalence rate for breast cancer on January 1, 2008 is 1737 per 100,000 women. From a Bayesian point of view, when assuming a uniform prior distribution for the prevalence rate, and assuming that the observed number of prevalent cases is binomially distributed with parameters θ_P and 152,000,000, the posterior distribution of θ_P is beta with posterior mean 0.01732, posterior median 0.01732, SD 0.00001065, and 95% credible interval (0.01729, 0.01734). The Bayesian analysis is performed with 55,000 observations generated from the posterior distribution, with a burn-in of 5000 and a refresh of 100. I executed the analysis using BUGS CODE 2.1, where the number of cases is 2,632,005 with a binomial distribution, and theta is the prevalence rate. It can be shown that the MCMC error is approximately 10^{-8} and that the posterior density is as shown in Figure 2.1.

Because the sample size (the U.S. female population on January 1, 2008) is very large, the posterior density is concentrated about the mean, and one is quite confident that the estimated prevalence of breast cancer of 0.01732 is accurate and agrees with the usual estimate.

2.3 Association between Risk and Disease in Cohort Studies

Consider Table 2.2. A cohort study is initiated by choosing samples from two populations, where all subjects are initially disease free. A sample is taken from a population of subjects that are exposed to the risk factor, while the

TABLE 2.2

A Prospective Cohort Study

Risk factor	Disease Status	
	+	−
+	θ_{++}, n_{++}	θ_{+-}, n_{+-}
−	θ_{-+}, n_{-+}	θ_{--}, n_{--}

second sample is taken from a population of subjects unexposed to the risk factor. Let θ_{++} be the probability that a subject with the risk factor has the disease, and let θ_{--} be the probability that a subject without the risk factor does not have the disease. Note that $\theta_{+-} = (1 - \theta_{++})$ and $\theta_{--} = (1 - \theta_{-+})$. The study is prospective in the sense that all subjects are followed into the future (for a given period) until their disease status is determined. Thus, the incidence rate (new cases) of disease for those with the risk factor is estimated by $n_{++}/(n_{++} + n_{+-})$. On the other hand, the incidence rate for those without the risk factor is $n_{-+}/(n_{-+} + n_{--})$.

It is important to note the nature of the two populations, because any statistical inference is relevant only to the populations described in the study. For example, the populations could be hospital based, community based, or national based. The disease status of individuals needs to be clearly described, because often the so-called disease status is not a disease per se but instead some morbidity condition such as high blood pressure.

From a statistical viewpoint, the number of diseased n_{++} with the risk factor has a binomial distribution with parameters θ_{++} and $n_{+.} = n_{++} + n_{+-}$, while the number of subjects with disease without being exposed to the risk factor have a binomial distribution with parameters θ_{-+} and $n_{-.} = n_{-+} + n_{--}$.

How should the Bayesian approach be performed? One must select prior distributions for θ_{++} and θ_{-+}, which are combined with the likelihood function

$$L(\theta_{++}, \theta_{-+}) \propto \theta_{++}^{n_{++}} (1 - \theta_{++})^{n_{+-}} \theta_{-+}^{n_{-+}} (1 - \theta_{-+1})^{n_{--}} \qquad (2.3)$$

where $0 \leq \theta_{++} \leq 1$ and $0 \leq \theta_{-+} \leq 1$.

If little prior information is available from related previous studies, one could use a uniform prior

$$g(\theta_{++}, \theta_{-+}) \propto 1 \qquad (2.4)$$

where $0 \leq \theta_{++} \leq 1, 0 \leq \theta_{-+} \leq 1$, and the posterior distribution of θ_{++} is beta with parameter $(n_{++} + 1, n_{+-} + 1)$. Similarly, the posterior distribution of θ_{-+} is beta with parameter $(n_{-+} + 1, n_{--} + 1)$.

Also, if little prior information is available, one could use the improper prior density

$$g(\theta_{++}, \theta_{-+}) \propto \frac{1}{\theta_{++}\theta_{-+}} \tag{2.5}$$

where $0 \le \theta_{++} \le 1$, $0 \le \theta_{-+} \le 1$, and the posterior distribution of θ_{++} is beta with parameter (n_{++}, n_{+-}). Thus the posterior distribution of θ_{-+} is beta with parameter (n_{-+}, n_{--}), and with this choice of a prior distribution, the posterior mean of θ_{-+} is $n_{-+}/(n_{-+} + n_{--})$, which is the "usual" estimator of the incidence rate for those subjects unexposed to the risk factor.

To compare the incidence rate of the diseased population of subjects exposed to the risk factor to that of the nondiseased population of subjects unexposed to the risk factor, the RR parameter is defined as

$$\theta_{RR} = \frac{\theta_{++}}{\theta_{-+}} \tag{2.6}$$

One must determine the posterior distribution of the RR, which, of course, is induced by the posterior distributions of θ_{++} and θ_{-+}, both of which have beta posterior distributions as described earlier.

Our first example of a cohort study is described by Kahn and Sempos[5] and is based on the work of Abbott et al[6] on stroke among smokers. The information is provided in Table 2.3.

Thus, among the sample of 3435 smokers, 171 had a stroke, but among the 4437 nonsmokers, 117 had a stroke. Is there an association between smoking and suffering a stroke? If one uses an improper prior density (Equation 2.4), the posterior distribution of the incidence rate for the smokers is beta with parameter (171, 3264), while the posterior distribution of the incidence rate of stroke among the nonsmokers is beta with parameter vector (117, 4320). Therefore, the posterior mean of θ_{++} is the "usual" estimator of the incidence rate of stroke for smokers, namely, 171(171 + 3264) = 0.04978 and that (the posterior mean) for the incidence rate θ_{-+} among nonsmokers is 117(117 + 4320) = 0.026369. These facts will be confirmed by the Bayesian analysis based on the following code.

TABLE 2.3

A 12-Year Study of Stroke among Males

Risk factor	Disease Status	
	+ Stroke	− No stroke
+ Smoker	$\theta_{++}, n_{++} = 171$	$\theta_{+-}, n_{+-} = 3264$
− Nonsmokers	$\theta_{-+}, n_{-+} = 117$	$\theta_{--}, n_{--} = 4320$

Source: Abbott, R.D., et al., *N. Engl. J. Med.*, 315, 717, 1986.

BUGS CODE 2.2

```
model;
# uniform prior
{
# binomial distribution of the number of new cases for those
exposed to risk factor
nepp~dbin(thetapp,3435)
# binomial distribution for the number of new cases among
those not exposed to the risk factor
nemp~dbin(thetanp,4437)
# prior distribution of incidence rate thetapp is uniform
thetapp~dbeta(1,1)
thetanp~dbeta(1,1)
# RR is the relative risk
RR<- thetapp/thetanp}
# the number of new cases nepp = 171
# the number of new cases for nenp = 117
list(nepp = 171,nenp = 117)
# initial value of incidence thetapp and thetanp
list(thetapp =.5, thetanp = 117).
```

The Bayesian analysis is executed with a uniform prior and 55,000 observations generated from the joint posterior distribution of thetapp = θ_{++} and thetanp = θ_{-+}, with a burn-in of 5000 and a refresh of 100, and the results appear in Table 2.4.

To compare the incidence rate of stroke among smokers to that of nonsmokers, the RR, θ_{RR} is the relevant parameter, which has a posterior mean of 1.896 and a 95% credible interval (1.493, 2.376). This implies that smokers have an 88% greater chance of stroke compared to nonsmokers, and the credible interval gives us a good idea of the uncertainty in estimating the RR, which is also reflected by the posterior SD 0.2251. With regard to the incidence rate of stroke for smokers, the posterior mean is 0.05001, which implies a rate of approximately 5001 per 100,000 smokers, compared to a rate of 2660 per 100,000 nonsmokers. The "small" MCMC errors imply that 55,000 observations are sufficient for the WinBUGS simulation.

When the Bayesian analysis is performed with an improper distribution, the posterior analysis is shown in Table 2.5; Bayesian procedures are executed with BUGS CODE 2.3.

TABLE 2.4

Posterior Analysis of Stroke Cohort Study: Uniform Prior

Parameter	Mean	SD	MCMC Error	2½	Median	97½
θ_{++}	0.05001	0.003714	<0.00001	0.04295	0.04993	0.05756
θ_{-+}	0.0266	0.002426	<0.00001	0.02207	0.02654	0.03255
θ_{RR}	1.896	0.2251	0.001007	1.493	1.881	2.376

TABLE 2.5

Posterior Analysis of Stroke Cohort Study: Improper Prior

Parameter	Mean	SD	MCMC Error	2½	Median	97½
θ_{++}	0.04977	0.003707	<0.00001	0.04275	0.04968	0.05728
θ_{-+}	0.02636	0.002409	<0.000001	0.02186	0.0263	18
θ_{RR}	1.904	0.2253	<0.0001	1.505	1.89	2.387

By comparing Tables 2.4 and 2.5, one concludes that the posterior analysis with a uniform prior gives virtually the same results as that with an improper prior. Why? The solution to this question is left as an exercise. When the sample size is "small," the posterior analysis with a uniform prior will differ from that with an improper prior.

BUGS CODE 2.3

```
model;
{
# assume improper prior (2.5)
# posterior distribution of thetapp
thetapp~dbeta(npp,npm)
# posterior distribution of thetanp
thetamp~dbeta(nmp,nmm)
# RR is the relative risk
RR<-thetapp/thetamp
# AR is the attributable risk
AR<-(RR-1)/RR
}
# below is the data with an improper prior
list(npp = 171,nmp = 117,npm = 3264, nmm = 4320)
# initial value of incidence thetapp and thetanp
list(thetapp =.5, thetamp =.5).
```

2.4 Retrospective Studies: Association between Risk and Disease in Case–Control Studies

In a case–control scenario, a random sample of cases (those with disease) is selected, and separately, a random sample of controls (subjects without disease) is selected, and a good example appears in the study by Johnson and Johnson,[7] which was analyzed by Clayton and Hills,[8] where the association between having a tonsillectomy and acquiring Hodgkin's disease is analyzed. A sample of 174 subjects who have Hodgkin's disease

TABLE 2.6

Tonsillectomy and Hodgkin's Disease: A Case–Control Study

Tonsillectomy	+ Cases	− Controls
+ Positive	$\theta_{++}, n_{++} = 90$	$\theta_{+-}, n_{+-} = 165$
− Negative	$\theta_{-+}, n_{-+} = 84$	$\theta_{--}, n_{--} = 307$

Source: Johnson, S.K. and Johnson, R.E., *N. Engl. J. Med.*, 287, 1122, 1972.

was selected and among those 90 had a tonsillectomy, whereas among the 472 subjects without Hodgkin's disease, 165 had a tonsillectomy. See Table 2.6 for the study data. For this study, various patient characteristics such as age and gender are not taken into account in the Bayesian analysis. Is there an association between having a tonsillectomy and Hodgkin's disease? How is the association estimated? It is traditional to use the OR, which is defined as

$$\theta_{OR} = \frac{\theta_{++}/\theta_{-+}}{\theta_{+-}/\theta_{--}} \tag{2.7}$$

where the numerator is the odds of having a tonsillectomy among those with the disease and the denominator is the odds of having a tonsillectomy among the controls.

From a Bayesian viewpoint, determining the posterior distribution of the OR, θ_{OR} is similar to that of determining the posterior distribution of the RR. Assuming a uniform prior, the posterior distribution of θ_{++} is beta with parameter vector (91, 85), and independently, the posterior distribution of θ_{+-} is beta with parameter vector (166, 308). On the other hand, if one assumes an improper prior for θ_{++} and θ_{+-}, the posterior distribution of θ_{++} is beta with parameter vector (90, 84) and that for θ_{+-} is beta with parameter vector (165, 307). Of course, the posterior distribution of the OR is induced by the posterior distributions of θ_{++} and θ_{+-}. Suppose the analysis is based on the following code.

BUGS CODE 2.4

```
model;
# posterior distribution of the odds ratio
{
# assume improper prior (2.5)
# posterior distribution of thetapp
thetapp~dbeta(npp,nmp)
# posterior distribution of thetapm
thetapm~dbeta(npm,nmm)
# postrior distribution of thetamp
thetamp~dbeta(nmp,npp)
```

```
thetamm~dbeta (nmm, npm)
# OR is the odds ratio
OR<- (thetapp*thetamm) / (thetapm*thetamp)
# AR is the attributable risk
AR<- (OR-1) /OR
}
# below is the data for tonsillectomy
list(npp = 90, nmp = 84, npm = 165, nmm = 307)
# data for cataract study
list(npp = 55, nmp = 552, npm = 84, nmm = 1927)
# data for Israeli Heart Study for age
list(npp = 15, nmp = 41, npm = 188, nmm = 1767)
# data for Israeli Heart Study for blood pressure
list(npp = 29, nmp = 27, npm = 711, nmm = 1244)
# Israeli Heart Study age and sbp
list(npp = 124, nmp = 79, npm = 616, nmm = 1192)
# initial value of thetapp and thetapm
list(thetapp =.5, thetapm =.5) .
```

As shown in Table 2.7, the posterior of OR is approximately 2.015, and the odds that a case will have a tonsillectomy is 1.072, whereas that for a control is 0.5375; thus, the odds of having a tonsillectomy for a case is about twice that of the odds of having a tonsillectomy for a control. Note the 95% credible interval for the OR is (1.551, 2.573) and does not include 1, implying that the odds for a case is actually not the same as that for a control. It appears that having a tonsillectomy is a risk factor for Hodgkin's disease.

Of course, the ratio

$$\varphi_R = \frac{\theta_{++}/(\theta_{++} + \theta_{-+})}{\theta_{+-}/(\theta_{+-} + \theta_{--})} \tag{2.8}$$

could be used to compare the prevalence of the risk factor of the cases compared to the controls, in fact it is equivalent to using the OR. The OR is a one-to-one function of φ_R, the ratio of the fraction of cases with the risk factor relative to the fraction of controls with the risk factor.

TABLE 2.7

Posterior Analysis for Case–Control Study: Tonsillectomy and Hodgkin's Disease

Parameter	Mean	SD	MCMC Error	2½	Median	97½
θ_{OR}	2.015	0.2615	<0.0001	1.551	1.998	2.573
θ_{++}	0.5174	0.03773	<0.00001	0.4432	0.5173	0.591
θ_{+-}	0.3496	0.0219	<0.00001	0.3074	0.3495	0.3928
θ_{-+}	0.4826	0.0377	<0.00001	0.4088	0.4827	0.5568
θ_{--}	0.6504	0.0218	<0.00001	0.607	0.6506	0.6927

Can a case–control study estimate the RR (Equation 2.6)? Because of the sampling procedure for a case–control study, the observations do not represent random samples from the population of those exposed to the risk factor, and the observations do not represent a random sample from the population of those unexposed to the risk factor; thus, it appears that it is not possible to estimate the RR. Instead, as has been mentioned, for the case–control retrospective study, we have a random sample from the population of controls and a random sample from the population of non-cases or controls, and it is possible to estimate the OR, with a numerator that is the odds of being exposed for the cases versus a denominator that is the odds of being exposed to the risk factor for the controls.

See Kahn and Sempos (p. 73),[5] for additional information. This presentation on this subject will to some extent be based on their description of the equivalence between the OR and RR, when the disease is "rare." Recall that the OR for a case–control study is

$$\theta_{OR} = \frac{\theta_{++}/\theta_{-+}}{\theta_{+-}/\theta_{--}} \tag{2.9}$$

but for rare diseases it can be expressed as

$$\frac{\theta_{++}/\theta_{+-}}{\theta_{-+}/\theta_{--}} \approx \frac{\theta_{++}/(\theta_{++}+\theta_{+-})}{\theta_{-+}/(\theta_{-+}+\theta_{--})} = \theta_{RR} \tag{2.10}$$

Because the disease is rare,

$$\theta_{+-} \approx \theta_{++} + \theta_{+-}$$

and

$$\theta_{--} \approx \theta_{-+} + \theta_{--}$$

and the OR is approximately the same as the RR. Note that this is only an approximation; thus, for the tonsillectomy study, let us investigate the discrepancy between the value of the OR and the RR as computed by Equation 2.6.

It is easy to verify the posterior distribution of the RR for the tonsillectomy information as given in Table 2.8. I use an improper prior and BUGS CODE 2.3

TABLE 2.8

Posterior RR for Tonsillectomy Study

Parameter	Mean	SD	MCMC Error	2½	Median	97½
θ_{RR}	1.66	0.216	<0.0001	1.278	1.645	2.126
θ_{-+}	0.2147	0.0297	<0.00001	0.1753	0.2142	0.2564
θ_{++}	0.353	0.0300	<0.0001	0.2958	0.3527	0.4131

with 55,000 observations generated from the joint posterior distribution of the three parameters in Table 2.8, and the burn-in is 5000 with a refresh of 100.

It is seen that the RR posterior mean of 1.66 is smaller than the mean 2.077 for the posterior distribution of the OR as revealed by Table 2.7, implying that the RR is only an approximation to the OR! Thus, one should, in general, be careful in using the RR as a measure of the association between disease and risk in a case–control study. For a case–control study, the OR is a valid measure of the association between risk factor and disease, as is the RR for a cohort study.

2.5 Cross-Sectional Studies

The general format of a cross-sectional study appears in Table 2.9, where a random sample of size $n = n_{++} + n_{+-} + n_{-+} + n_{--}$ subjects is taken from a well-defined population and where the disease status and exposure status of each subject is known. Consider the ++ cell. θ_{++} is the probability that a subject will have the disease and will be exposed to the risk factor. It is assumed that n is fixed and that the cell frequencies follow a multinomial distribution with mass function

$$f(n_{++}, n_{+-}, n_{-+}, n_{--} \mid \theta_{++}, \theta_{+-}, \theta_{-+}, \theta_{--}) \propto \theta_{++}^{n_{++}} \theta_{+-}^{n_{+-}} \theta_{-+}^{n_{-+}} \theta_{--}^{n_{--}} \tag{2.11}$$

where the thetas are between zero and one and their sum is one and the n's are nonnegative integers with a sum equal to the sample size n.

As a function of the thetas, Equation 2.11 is recognized as a Dirichlet density. If the likelihood for the thetas is combined, via Bayes' theorem, with a prior distribution for the thetas, the result is the posterior density for thetas. For example, if a uniform prior is used for the thetas, the posterior distribution of the thetas is Dirichlet with parameter vector $(n_{++} + 1, n_{+-} + 1, n_{-+} + 1, n_{--} + 1)$, but on the other hand, if the improper prior density

$$f(\theta_{++}, \theta_{+-}, \theta_{-+}, \theta_{--}) \propto \left(\theta_{++}^{n_{++}} \theta_{+-}^{n_{+-}} \theta_{-+}^{n_{-+}} \theta_{--}^{n_{--}}\right) - 1 \tag{2.12}$$

is used, the posterior density of the thetas is Dirichlet with parameter vector $(n_{++}, n_{+-}, n_{-+}, n_{--})$. When the latter prior is used, the posterior means of the unknown parameters will be the same as the "usual" estimators. For

TABLE 2.9

A Cross-Sectional Study

Risk Factor	+ Disease	− Nondisease
+ Positive	θ_{++}, n_{++}	θ_{+-}, n_{+-}
− Negative	θ_{-+}, n_{-+}	θ_{--}, n_{--}

example, the usual estimator of θ_{++} is n_{++}/n, but the posterior distribution of θ_{++} is beta with parameter vector $(n_{++}, n - n_{++})$. Consequently, the posterior mean of θ_{++} is indeed the usual estimator n_{++}/n.

The sampling scheme for a cross-sectional study allows one to estimate the RR of disease (the incidence rate of those diseased among those exposed to the risk factor divided by the disease incidence rate among those unexposed to the risk) and the odds of exposure among those diseased versus the odds of exposure among the nondiseased.

A good example of a cross-sectional study is the Shields Heart Study carried out in Spokane, Washington from 1993 to 2001 at the Shields Coronary Artery Center where the disease status of coronary artery disease and various risk factors (coronary artery calcium, diabetes, smoking status, blood pressure, and cholesterol) were measured on each of about 4300 patients. The average age (SD) of 4386 patients was 55.14 (10.757) years, with an average age (SD) of 55.42 (10.78) years for 2712 males and 56.31 (10.61) years for 1674 females. The main emphasis of this study was to investigate the ability of coronary artery calcium (as measured by computed tomography) to diagnose coronary artery disease. A typical approach to this study investigates the association between having a heart attack and the fraction of subjects with a positive reading for coronary calcium. For additional information about the Shield Heart Study see Mielke et al.[9]

Thus, of the total of 4389 patients, 4259 did not have an infarction, while 130 did in fact have an infarction, and of those that had a heart attack, 119 had a positive reading for coronary artery calcium. Of the total of 4389 patients, 2580 had a positive reading for coronary artery calcium while 1809 did not. In fact, a negative reading for coronary artery calcium means that computed tomography was unable to detect any calcium and records a value of zero for the measured level. The reading for coronary artery calcium can vary from 0 to some positive number, and the larger the reading the more the risk for heart disease.

Of those 130 who did have a heart attack, 119 had a positive reading for calcium, while the remaining 11 had a zero reading, and among the 4259 who did not have an infarction, 2461 had a zero reading. Figure 2.2 depicts the distribution of the coronary artery calcium values by those who had an infarction versus those who did not. Figure 2.2 shows the distribution of coronary artery calcium by disease status. For those with no infarction, the median level of coronary artery calcium is 4, but for those with an infarction, the median level is 187.5.

To investigate the association between infarction and coronary artery calcium, the RR and OR are estimated using the information given in Table 2.10. From a Bayesian viewpoint, assuming an improper prior distribution for the four parameters shown in Equation 2.10, the posterior distribution of the four cell parameters is Dirichlet with parameter vector (119, 2461, 11, 1798), and the RR for heart disease is estimated from the parameter

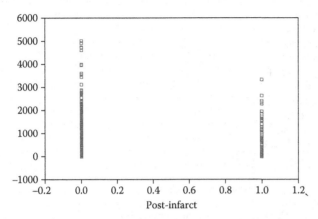

FIGURE 2.2
Calcium score versus infarction.

TABLE 2.10

Shields Heart Study

Coronary Calcium	+ Disease (Heart Attack)	Nondisease (No Infarction)
+ Positive	$\theta_{++}, n_{++} = 119$	$\theta_{+-}, n_{+-} = 2461$
− Negative	$\theta_{-+}, n_{-+} = 11$	$\theta_{--}, n_{--} = 1798$

$$\theta_{RR} = \frac{\theta_{++}/(\theta_{++} + \theta_{+-})}{\theta_{-+}/(\theta_{-+} + \theta_{--})} \qquad (2.13)$$

Note that the formula for RR has a numerator that is the probability of disease among those exposed (have a positive calcium score) to the risk factor, whereas the denominator is the probability of disease among those unexposed (negative coronary artery calcium). As for the OR, the posterior distribution of

$$\theta_{OR} = \frac{\theta_{pp}\theta_{mm}}{\theta_{mp}\theta_{pm}} \qquad (2.14)$$

will also be determined, and the Bayesian analysis executed with BUGS CODE 2.5 using 55,000 observations generated from the posterior distribution with a burn-in of 5000 and a refresh of 100.

BUGS CODE 2.5

```
model;
# the cross-sectional study
{
# below generates observations from the Dirichlet distribution
```

```
gpp~ dgamma(npp,2)
gpm~dgamma(npm,2)
gmp~dgamma(nmp,2)
gmm~dgamma(nmm,2)
sg<-gpp+gpm+gmp+gmm
thetapp<- gpp/sg
thetapm<-gpm/sg
thetamp<-gmp/sg
thetamm<-gmm/sg
# numerator of RR
nRR<-(thetapp)/(thetapp+thetapm)
# denominator of RR
dRR<-(thetamp)/(thetamp+thetamm)
# the odds ratio
OR<-(thetapp*thetamm)/(thetamp*thetapm)
# the relative rsk
RR<-nRR/dRR
# attributable risk based on the relative risk
ARRR<-(RR-1)/RR
# attributable risk based on the odds ratio
AROR<-(OR-1)/RR
}
# data with improper prior for Shields Heart Study
list(npp = 119,npm = 2461,nmp = 11,nmm = 1798)
# initial values generated from the specification tool
```

It is interesting that the posterior means of the RR and OR are quite similar, as are the posterior means, which implies that the disease (coronary infarction) is quite rare, which, of course, is obvious from Table 2.10, which indicates a disease rate of 2.96%. If the 4389 patients are actually a random sample, then one is confident that 2.9% is an accurate estimate of the "true" disease rate.

From Table 2.11, the posterior distribution of the RR is skewed to the right, which is shown in Figure 2.3. The 55,000 observations generated from the joint posterior distribution of the parameters in Table 2.11 appear

TABLE 2.11

Posterior Distribution for the Shields Heart Study

Parameter	Mean	SD	MCMC Error	2½	Median	97½
θ_{RR}	8.35	2.887	0.0133	4.378	7.805	15.49
θ_{OR}	8.709	3.308	0.01397	4.529	8.138	16.21
θ_{pp}	0.02711	0.00245	<0.00001	0.0225	0.02704	0.03215
θ_{mm}	0.4096	0.007413	<0.00001	0.395	0.4096	0.4243
θ_{mp}	0.002504	0.007553	<0.000001	0.001246	0.002427	0.004201
θ_{pm}	0.5608	0.0075	<0.00001	0.5461	0.5608	0.5754

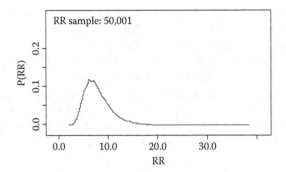

FIGURE 2.3
Posterior density of the relative risk.

to be sufficient for computing accurate estimates of the RR and OR. Also, note that the mean of the cell probabilities agree with the "usual" estimators easily computed from Table 2.11. For example, the usual estimate of θ_{pp} is $119/4389 = 0.0271132$ compared to its posterior mean of 0.02711, and the same is true for the other cell probabilities. This is true because an improper prior distribution (Equation 2.10) for the cell probabilities is used for the Bayesian analysis.

2.6 Attributable Risk

Recall that for case–control, cohort, and cross-sectional studies, for those exposed to the risk factor, not all subjects will develop disease, although we do expect more to develop disease, compared to those unexposed. For example, among those who smoke, not all develop lung cancer, however, we expect a larger fraction to develop the disease, compared to nonsmokers. How does one measure the risk of developing the disease, among those exposed to the risk factor? This question is answered by measuring the attributable risk (AR), which in essence is the risk of disease among those exposed to the risk factor in excess of the risk for those unexposed.

For a cohort study, where there are two groups, the exposed and unexposed, the AR is defined as either

$$\theta_{AR} = I_r - I_0 \tag{2.15}$$

or

$$\theta_{AR} = \frac{\left(I_r - I_0\right)}{I_r} \tag{2.16}$$

where I_0 and I_r are the incidence rates among the unexposed and exposed populations, respectively. For the first definition given in Equation 2.15, θ_{AR} is the excess risk attributable to the risk factor, as is the second definition, but measured as a fraction relative to the incidence rate of the exposed group. Leviton[10] proposed the second definition and explains various definitions of AR, but the second definition is usually used in epidemiological studies.

Following Kahn and Sempos (p. 73),[5] the AR can be thought of as the proportion of the incidence rate among those with the risk factor due to the association with the risk factor, and can be expressed as

$$\theta_{AR} = \frac{(\theta_{RR} - 1)}{\theta_{RR}} \tag{2.17}$$

where θ_{RR} (>1) is the population RR. Expression 2.17 is derived by dividing the numerator and denominator of Equation 2.16 by I_0. When Equation 2.17 is multiplied by I_r,

$$\theta_{IRRF} = I_r \left(\frac{(\theta_{RR} - 1)}{\theta_{RR}} \right) \tag{2.18}$$

which is referred to as the incidence rate due to the risk factor among those exposed to the risk factor. From a statistical point of view, the AR and the incidence rate (Equation 2.18) are parameters to be estimated by the information (the cell frequencies of Table 2.2 in a cohort study). Recall for a cohort study, the RR is defined as $\theta_{RR} = \theta_{++}/\theta_{-+}$. Thus, a Bayesian analysis for estimating the AR is easily executed using BUGS CODE 2.6.

Consider the Abbott et al.[6] cohort study of Table 2.3 for the association between smoking and stroke among 7872 males, where a random sample of 3435 smokers is selected from a population of smokers and the remaining 4437 nonsmokers are selected at random from a population of nonsmokers. Among the smokers, 171 had a stroke and 3264 did not, while among the nonsmokers 117 had a stroke and 4320 did not. The following code is executed for the Bayesian analysis, where 55,000 observations are generated from the posterior distribution with a burn-in of 1000 and a refresh of 100.

BUGS CODE 2.6

```
model;
# for a cohort study
{
# assume improper prior (2.5)
# posterior distribution of thetapp
thetapp~dbeta(npp,npm)
# posterior distribution of thetanp
thetamp~dbeta(nmp,nmm)
```

```
# RR is the relative risk
RR<-thetapp/thetamp
# AR is the attributable risk
AR<-(RR-1)/RR
}
# below is the data
list(npp = 171,nmp = 117,npm = 3264, nmm = 4320)
# below is the data for tonsillectomy study
list(npp = 90,nmp = 84,npm = 165, nmm = 307)
# initial value of incidence thetapp and thetanp
list(thetapp =.5, thetamp =.5).
```

Assuming an improper prior distribution for the cell probabilities, the Bayesian analysis for AR is reported in Table 2.12.

The analysis is focused on estimating the AR θ_{AR}, which has a posterior mean (SD) of 0.4667 (0.06337) and a 95% credible interval of (0.3319, 0.5807). Thus, the incidence rate of stroke among smokers has a posterior mean of 0.04975, and the AR is much larger than the raw incidence rate θ_{++} of smokers. Why? Should one be concerned that smoking is a serious risk factor for developing stroke? Does AR give us valuable additional information beyond that of the raw incidence rate θ_{++}?

For a case–control study, how does one estimate the AR? Recall that two samples are taken, one for cases and the other for non-cases (controls), and for each sample, the number of people exposed to the risk factor is recorded. When the disease is relatively rare, the OR should be a good approximation to the RR of disease, in which case, the AR can be estimated by substituting OR for RR in Equation 2.17.

A good example of this is the Hiller and Kahn[11] case–control study exhibited below, which investigates the association between cataracts and diabetes, where 607 subjects are selected at random from a population of diabetics, and 2011 nondiabetics are selected at random from the population of nondiabetics. The medical record of each subject was examined to see if they had experienced cataracts in the past. An accurate assessment of exposure in this case can be difficult, because it is often a problem to ascertain if a person has had cataracts.

TABLE 2.12

AR for Smoking and Stroke

Parameter	Mean	SD	MCMC Error	2½	Median	97½
θ_{AR}	0.4667	0.06337	<0.0001	0.3319	0.4702	0.5807
θ_{RR}	1.902	0.2265	<0.0001	1.497	1.888	2.385
θ_{-+}	0.02638	0.002417	<0.00001	0.02187	0.02631	0.0312
θ_{++}	0.04975	0.0037	<0.0000	0.04271	0.04968	0.05725

Based on BUGS CODE 2.4, the information from Table 2.13, and assuming an improper prior, the posterior analysis is performed with 65,000 observations for the simulation, with 5000 for the burn-in and a refresh of 100. The results are reported in Table 2.14.

Thus, one sees that the AR is estimated as 0.5612 with the posterior median and can vary over (0.3876, 0.6839), with a confidence of 95%. Also, the OR has a posterior mean of 2.31 implying an association between cataracts and diabetes. It appears that 65,000 observations are sufficient for the simulation, because the MCMC errors are quite small. To see the effect of the simulation sample size, the student should increase the sample size to see the change in the MCMC errors.

Our last topic in this chapter is to perform a Bayesian analysis that estimates the AR for a cross-sectional study. Recall for a cross-sectional study that one sample of n subjects is selected at random, where for each subject the disease status and exposure status are recorded. Refer to the general layout of such a design given by Table 2.9, where the cell frequencies follow a multinomial distribution and (when an improper prior distribution is assumed) the posterior distribution of the cell probabilities have a Dirichlet distribution. The details of the analysis are illustrated with the information from the Shields Heart Study exhibited by Table 2.10, and the analysis is executed with BUGS CODE 2.5 and the results reported in Table 2.11.

On the basis of the information for the Shields Heart Study, BUGS CODE 2.5 is executed, and the posterior analysis is performed with 75,000 observations generated for the simulation, with a burn-in of 5000 and a refresh of 100. The AR based on the OR and RR is defined by the parameters

$$\theta_{AROR} = \frac{(\theta_{OR} - 1)}{\theta_{OR}} \tag{2.19}$$

TABLE 2.13

Cataracts and Diabetes: A Case–Control Study

Cataracts	+ Cases (Diabetes)	− Controls (No Diabetes)
+ Positive	$\theta_{++}, n_{++} = 55$	$\theta_{+-}, n_{+-} = 84$
− Negative	$\theta_{-+}, n_{-+} = 552$	$\theta_{--}, n_{--} = 1927$

Source: Hiller, R.A. and Kahn, H.A. *Br. J. Opthalmol.*, 60, 283, 1976.

TABLE 2.14

Posterior Analysis for Cataract–Diabetes Study

Parameter	Mean	SD	MCMC Error	2½	Median	97½
θ_{AR}	0.5547	0.0757	<0.0001	0.3876	0.5612	0.6839
θ_{OR}	2.31	0.3913	0.00156	1.633	2.279	3.164

TABLE 2.15

ARs for Shields Heart Study: Association between Calcium and Coronary Infarction

Parameter	Mean	SD	MCMC Error	2½	Mean	97½
θ_{AROR}	0.9092	0.0451	<0.0001	0.8067	0.9142	0.9834
θ_{ARRR}	0.8672	0.04208	<0.0001	0.7716	0.8719	0.9354
θ_{OR}	8.709	3.038	0.01397	4.529	8.138	16.21
θ_{RR}	8.35	2.887	0.0133	4.378	7.805	15.49

and

$$\theta_{ARRR} = \frac{(\theta_{RR} - 1)}{\theta_{RR}} \tag{2.20}$$

respectively. For the posterior analysis, refer to Table 2.15.

It is seen that the two ARs in Equations 2.19 and 2.20 are quite similar because the RR and the OR are approximately the same. The RR and the OR both imply a strong association between coronary artery calcium (as measured by computed tomography) and infarction. Thus using the posterior mean of 0.9092 of θ_{AROR} as the measure of AR, the AR of 0.9092 of an infarction due to a positive calcium score is the excess risk due to a positive calcium score (in excess of the risk of a zero calcium score) relative to the risk of a positive calcium score. A similar interpretation for the AR θ_{ARRR} (based on the RR) is also appropriate.

2.7 Comments and Conclusions

To summarize, Chapter 2 lays the foundation for the Bayesian approach to estimating the association between a risk factor and disease, and three designs are considered for investigating that association: the cohort study, the case–control study, and the cross-sectional study. For the cohort study, two random samples are selected, one from the population of subjects exposed to the risk factor and the other selected from the population of those unexposed to the risk factor. The sampling scheme is reversed for the case–control study, where one random sample is selected from the population of cases (those with disease), and the other sample is selected from the population of controls (those without disease). In a cross-sectional study, one random sample is taken from a population where the exposure and disease status on each subject is available. It is important to remember that the populations for these designs must be carefully described because all Bayesian inferences are applicable only to those populations. For example, in investigating the association between a positive calcium score (measured by computed tomography) in the cross-sectional study of the Shields Heart Study, it is very import to define accurately the population from which the patients are selected (they were in fact patients referred to the Shields Coronary Artery Imaging Center by other physicians, and a small number were self-referred).

For each of the three designs, the measure of association is estimated via Bayesian techniques. For example, for the cohort design, the RR is estimated by its posterior distribution, but for the case–control design, the OR parameter is estimated by its posterior distribution. The association between disease and risk for the cross-sectional study is measured by both the RR and OR parameters.

For the three designs, the appropriate Bayesian methods are derived via Bayes' theorem and the analysis is executed with the corresponding BUGS CODE. For example, for a cohort study, the RR is estimated by its posterior distribution and the properties of that distribution are reported with the posterior mean, median, SD, and 95% credible interval. The analysis also includes the MCMC error for each parameter and a plot of each marginal posterior density. Recall the MCMC error tells one how close the computed posterior mean is to the "true" posterior mean, and the plots of the marginal densities reveal at a glance posterior characteristics such as multiple modality, kurtosis, and skewness.

Bayesian procedures are illustrated with realistic examples. Consider the cohort design used by Abbott et al.[6] for estimating the association between stroke and smoking. The study data are reported in Table 2.3, and the results of the Bayesian analysis are portrayed in Table 2.5. Using 55,000 observations for the simulation with a burn-in of 5000, the analysis is executed with BUGS CODE 2.3. The major focus of the analysis is the posterior distribution of the RR of a stroke for smokers versus nonsmokers.

For a case–control design, the Johnson and Johnson[7] study of the association between tonsillectomy and Hodgkin's disease is examined. The information from the study is reported in Table 2.6 and the analysis is executed with BUGS CODE 2.4. The posterior analysis is reported in Table 2.7. Association between tonsillectomy and Hodgkin's disease is estimated with the posterior mean of the OR, which measures the odds of smoking for people with Hodgkin's disease versus the odds of smoking for those without the disease.

Finally, for the cross-sectional study, the Shields Heart Study explores the association between coronary artery calcium and infarction with the Bayesian approach, Table 2.10 reports the data of the study of 4389 subjects, and the Bayesian analysis is executed with BUGS CODE 2.5; the posterior distribution of the OR and RR are given in Table 2.11.

Section 2.7 of this chapter presents the estimation of AR for the three major designs. To repeat, the chapter lays the foundation for the Bayesian approach to studying the association between disease and risk. Following chapters will deal with the same subject but in a more realistic manner. For example, more information than disease status and risk are usually available to the principal investigator and statistician. Patient demographics such as age and sex are usually available, as well as other patient information, and should be included in the Bayesian analysis. As will be seen, including such information improves the efficiency for the analysis, and provides more accurate estimates of the association between disease and various risk factors. Also to be presented are generalizations to when there are more than two disease levels, and when there are more than two risk factor levels.

There are many references to introductory epidemiology including Rothman[12] and Mausner and Bahn.[13] This book on Bayesian methods for epidemiology can be viewed as companions to statistically oriented books pertaining to epidemiology such as Kahn and Sempos[5] and to Clayton and Hills.[8] This chapter is the first concerning Bayesian methods for epidemiology, but for Bayesian books directed toward biostatistics, the reader is referred to the works by Broemeling.[14–16] The general area of statistics that examines the analysis for case–control, cohort, and cross-sectional designs is usually referred to as categorical data analysis, and two good references are Agresti[17] and Woolson.[18]

Exercises

1. Refer to Table 2.1 and using the information for the black race, use a Bayesian analysis to find a 95% credible interval for the incidence rate. To execute the analysis, assume a uniform prior, and using a sample of 100,000 with BUGS CODE 2.1, perform a Bayesian analysis and report the posterior mean, SD, median, and 95% credible interval for the incidence rate of breast cancer for blacks over the period 2004–2008. Generate 55,000 observations for the simulation, with a burn-in of 5000 and a refresh of 100.

    a. What is the posterior mean and median of the breast cancer incidence rate?

    b. What is the MCMC error for the incidence rate?

    c. Are 55,000 observations sufficient for estimating the posterior mean of the incidence rate?

2. Refer to Equation 2.5, the improper prior distribution for the odds of a cohort study, and show that with this prior the posterior distribution of θ_{++}, the incidence rate of disease for those exposed, is beta with parameter vector (n_{++}, n_{+-}).

3. Based on the experimental information in Table 2.3 for the cohort study for stroke and smoking and BUGS CODE 2.2, verify the posterior analysis of Table 2.4. Use 55,000 observations for the simulation, with a burn-in of 5000 and a refresh of 100.

    a. The posterior mean for the incidence of smokers is 0.05001. State the incidence rate per 100,000 smokers.

    b. The posterior mean of the RR is 1.896. What is the interpretation of this value and what is the risk of stroke for smokers versus nonsmokers?

    c. A 95% credible interval for the RR is (1.493, 2.376). Find a 99% credible interval for the RR.

    d. Are the MCMC errors sufficiently small so that one has confidence in the posterior mean value of 1.896 for estimating the true RR?

4. Compare Table 2.4 with Table 2.5. The former table reports the Bayesian analysis for the RR using a uniform prior, while the latter table reports the estimate for the same parameter, but using an improper prior (Equation 2.5). The posterior mean of the RR with a uniform prior is 1.896 and that with the improper prior is 1.904.

    a. Why the difference in the two estimates?

    b. Which prior is the most appropriate?

    c. Are the two estimates essentially the same?

5. For a case–control study

    a. What is the definition of the OR (Equation 2.7)?

    b. What is the interpretation of the numerator of the OR (Equation 2.7)?

    c. What is the interpretation of the denominator of Equation 2.7?

6. Consider the data of the case–control study of Johnson and Johnson.[7] Assuming a uniform prior for the cell probabilities, perform a Bayesian analysis using 65,000 observations for the simulation, a burn-in of 5000, and a refresh of 100.

    a. Plot the posterior density of the OR.

    b. What is the 95% credible interval for the OR?

    c. What is the MCMC error for the posterior mean of the OR? Are 65,000 observations for the simulation sufficient for estimating the posterior mean of the OR?

    d. The posterior mean of the OR is 2.015. Interpret this value. Is there an association between tonsillectomy and Hodgkin's disease?

7. a. What is the sampling scheme for a case–control study?

    b. Contrast the sampling scheme for a case–control study with the sampling scheme for a cohort study.

    c. Case–control studies are referred to as retrospective studies. Why?

    d. There are two populations for a case–control study. What are they?

    e. Why are cohort studies prospective?

8. a. Under what conditions is the OR a good approximation to the RR?
   b. For the tonsillectomy-Hodgkin's study, see Tables 2.7 and 2.8, is the OR a good approximation to the RR?
   c. The OR and the RR do not measure the association between disease and risk in the same way. Explain this statement.

9. Show that expression 2.8 can be expressed as a function of the OR (Equation 2.7). Thus, either measure can be used for the association between risk and disease in a case–control study.

10. Consider the cross-sectional study in Table 2.9.
    a. Describe the sampling scheme for a cross-sectional study given in Table 2.9.
    b. How does the sampling scheme for a cross-sectional study differ from those for case–control and cohort designs?
    c. What is the population for a cross-sectional design?
    d. What is the posterior distribution of the four cell probabilities assuming a uniform prior?

11. See the Shields Heart Study portrayed in Table 2.10. Assuming an improper prior for the cell probabilities and using BUGS CODE 2.5, perform a Bayesian analysis for estimating the OR and RR for the Shields Heart Study.
    a. What is the posterior mean of the RR?
    b. What is the posterior median of the OR?
    c. Is the OR a good approximation to the RR? Why?

12. Refer to Shields Heart Study of problem 11, and suppose that a previous study revealed the following association between coronary artery calcium.
    Of the 27 subjects who had a heart attack, 24 tested positive for coronary artery calcium, while among those that did not have an infarction, 566 had a positive reading for calcium. Suppose one uses the information from Table 2.16 as prior information that is to be combined with the data of the Shields Heart Study portrayed in Table 2.10. Using BUGS CODE 2.4, execute a Bayesian analysis with 55,000 observations for the simulation, with a burn-in of 5000 and a refresh of 100.
    a. Determine the posterior distribution of the RR, that is, find the posterior mean, median, SD, and 95% credible interval for the RR.

TABLE 2.16

Prior Information for the Shields Heart Study

Coronary Calcium	+ Disease (Heart Attack)	Nondisease (No Infarction)
+ Positive	$\theta_{++}, n_{++} = 24$	$\theta_{+-}, n_{+-} = 566$
− Negative	$\theta_{-+}, n_{-+} = 3$	$\theta_{--}, n_{--} = 342$

    b.   What is the posterior distribution of the OR? Find the posterior mean, median, SD, and 95% credible interval for the OR.

    c.   Compare the results of the posterior distribution for the OR and RR to Table 2.11. Recall that Table 2.11 reports the results for the Bayesian analysis of the Shields Heart Study, assuming a uniform prior. What is the difference in the posterior means? Which posterior mean is more accurate?

    d.   What are the MCMC errors for the posterior mean of the RR?

13. Refer to the AR for the smoking and stroke study reported in Table 2.3. The Bayesian analysis is based on an improper prior where 55,000 observations are generated for the simulation with a burn-in of 1000 and a refresh of 100.

    a.   Verify the posterior analysis of Table 2.12.

    b.   The posterior mean of the AR is 0.4667. Explain what this means for estimating the risk of smoking for developing a stroke.

    c.   The RR measures the association between smoking and stroke. Does the AR do the same thing? Explain.

14. Refer to the case–control study for estimating the risk between cataracts and diabetes. See Table 2.13. Answer the following based on BUGS CODE 2.4 and using 65,000 observations for the simulation with an improper prior.

    a.   Verify Table 2.14 for the posterior analysis.

    b.   What is the 95% credible interval for the AR?

    c.   Is having a cataract a risk factor for diabetes? Why?

    d.   Explain the significance of the posterior mean of 0.5547 for AR.

    e.   Is the AR for this example, the excess incidence of diabetes among those with cataracts, in excess of the incidence of diabetes among those without cataracts, relative to the incidence rate of diabetes among those with cataracts? Is this an accurate statement?

15. Refer to problem 14 for the case–control study for the association between cataracts and diabetes. Previous related studies are summarized below with Table 2.17. Using the reported values from Table 2.17 as prior information, conduct a Bayesian analysis with the results of Table 2.13 regarded as the "present" study. Use BUGS CODE 2.4 with 65,000 observations, a burn-in of 1000, and a refresh of 100.

TABLE 2.17

Cataracts and Diabetes: A Case–Control Study

Cataracts	+ Cases (Diabetes)	− Controls (No Diabetes)
+ Positive	$\theta_{++}, n_{++} = 28$	$\theta_{+-}, n_{+-} = 46$
− Negative	$\theta_{-+}, n_{-+} = 248$	$\theta_{--}, n_{--} = 886$

a. Perform a Bayesian analysis for estimating the odds ratio, and report the posterior mean, median, SD, and 95% credible interval for the OR.

b. Using the results from part a, compare the Bayesian analysis with the analysis reported in Table 2.14. What is the difference in the two analyses? In particular compare the posterior mean of the two analyses.

c. Plot the posterior density of the OR.

d. What is the MCMC error for estimating the posterior mean of the OR?

16. Consider the results of a case–control study (see Table 2.18) that examines the association between smoking and respiratory problems.

A random sample of 205 was selected from a population of smokers and after reviewing the medical history of each, it was found that 108 had respiratory problems while 97 did not. A random sample of 540 nonsmokers revealed 190 had respiratory problems but 350 did not. Answer the following using BUGS CODE 2.6 with 55,000 observations for the simulation, with a burn-in of 5000 and a refresh of 100.

a. Compute the OR and report the characteristics of posterior distribution of the OR. What is the posterior mean, posterior median, and 95% credible interval?

b. Plot the posterior distribution of the OR.

c. Is the posterior distribution of the OR skewed?

d. What is the MCMC error for estimating the posterior mean of the odds ratio? Are 55,000 observations sufficient for estimating the OR?

e. Is the OR a good approximation to the RR?

f. What is your overall conclusion about the association between smoking and respiratory problems?

g. What is the AR of smoking for the development of respiratory problems?

h. Describe how you would design a cohort study to examine the association between smoking and respiratory problems.

TABLE 2.18

Smoking and Respiratory Problems: A Case–Control Study

Respiratory Problems	+ Cases (Smokers)	− Controls (Nonsmokers)
+ Positive (Yes)	$\theta_{++}, n_{++} = 108$	$\theta_{+-}, n_{+-} = 190$
− Negative (No)	$\theta_{-+}, n_{-+} = 97$	$\theta_{--}, n_{--} = 350$

_____

References

1. Howlader, N., Noone, A.M., Krapcho, M., et al. (eds.). *SEER Cancer Statistics Review, 1975–2008*, National Cancer Institute, Bethesda, MD, http://seer.cancer. gov/csr/1975_2008/, based on November 2010 SEER data submission, posted to the SEER Web site, 2011.
2. Methodology for calculating incidence at National Cancer Institute: http://seer .cancer.gov/statfacts/html/breast.html#incidence-.
3. Methodology for calculating prevalence from the National Cancer Institute: http://surveillance.cancer.gov/prevalence.
4. Population estimates from the US census: http://www.census.gov/popest/ data/state/totals/2008/tables/NST-EST2008-01.csv.
5. Kahn, H.A. and Sempos, C.T. *Statistical Methods in Epidemiology*, Oxford University Press, 1989, New York.
6. Abbott, R.D., Yin, Y., Reed, D.M. and Yano, K. Risk of stroke in male cigarette smokers, *N. Engl. J. Med.*, 315, 717, 1986.
7. Johnson, S.K. and Johnson, R.E. Tonsillectomy in Hodgkin's disease, *N. Engl. J. Med.*, 287, 1122, 1972.
8. Clayton, D. and Hills, M. *Statistical Models in Epidemiology*, Oxford University Press, 2002, New York.
9. Mielke, C.H., Shields, J.P., and Broemeling, L.D. Coronary artery calcium scores for men and women of a large asymptomatic population, *CVD Prevention*, 2, 194–198, 1999.
10. Leviton, A. Definitions of attributable risk, *Am. J. Epidemiol.*, 98(3), 231, 1973.
11. Hiller, R.A. and Kahn, H.A. Senile cataract extraction and diabetes, *Br. J. Opthalmol.*, 60, 283–286, 1976.
12. Rothman, K.J. *Modern Epidemiology*, Little Brown and Company, 1986, Boston, Toronto.
13. Mausner, J.S. and Kramer, S.K. *Epidemiology, An Introductory Text*, W.B. Saunders Company, 1983, London, UK.
14. Broemeling, L.D. *Bayesian Biostatistics and Diagnostic Medicine*, CRC Press, Taylor & Francis Group, 2007, Boca Raton, FL.
15. Broemeling, L.D. *Bayesian Methods for Measures of Agreement*, CRC Press, Taylor & Francis Group, 2009, Boca Raton, FL.
16. Broemeling, L.D. *Advanced Bayesian Methods for Medical Test Accuracy*, CRC Press, Taylor & Francis Group, 2012, Boca Raton, FL.
17. Agresti, A.A. *Categorical Data Analysis*, John Wiley & Sons, 1990, New York.
18. Woolson, R.F. *Statistical Methods for the Analysis of Biomedical Data*, John Wiley & Sons, 1987, New York.

3

Bayesian Methods of Adjustment of Data

3.1 Introduction

A standard epidemiologic technique is the adjustment of data. This is done so that a fair comparison can be made between the mortality of one state versus that of another state when the age distribution of the two states is quite different. That is to say, the mortality of the states are adjusted for age, by assuming both states have the same age distribution. Under this assumption, the mortality of the two states can be computed and then compared with statistical techniques. Of course, in addition to mortality, one can adjust other variables, such as incidence rates of disease and prevalence rates of risk factors. There are many ways to adjust data, including the direct method, the indirect method, and regression techniques. Regression methods of adjustment are the subject of Chapter 4.

Direct adjustment of data (mortality) is to weight the mortality of the various age groups by the number of subjects in each age group, assuming a standard age distribution. The mortality of each age group is based on the actual number of subjects in each age group. On the contrary, the indirect method involves computing the mortality of each age group, where the mortality of a given age group is the mortality of that age group, assuming a standard age distribution. Such mortality (the probability of death) is weighted by the actual number of observations in each age category. Both methods of adjustment are given a Bayesian interpretation, in that the posterior distribution of the mortality rates for a state is based on the posterior distribution of the mortality rates of each age category. For a given category, the number of subjects who die has a binomial distribution; consequently, the unknown mortality rates (of each age category) have a beta distribution (assuming a uniform or improper prior distribution).

Chapter 3 is continued with the investigation of the possible confounding factors on the association between risk and disease. According to Kahn and Sempos,[1] a variable is a confounding factor if it is related to the disease and to the risk factor. For example, in the study of the association between coronary artery disease and systolic blood pressure (SBP), age is considered a confounding variable; thus, the association between disease and risk should be adjusted for age. To adjust for age, the data is partitioned into various age strata, and the interaction between age and the association (between coronary artery disease and SBP) is determined. Interaction exists if the association between risk and disease depends on the confounding factor age. If no interaction exists, then the data can be combined over strata, and the measure of association is estimated. For example, with a case–control study, if interaction exists, one would expect the odds ratio for two strata to be different statistically. For the Bayesian, one would compute the posterior distribution of the two odds ratios (corresponding to two strata) and test to see if they are the same. If interaction is present, there are ways to combine the two odds ratios to estimate an overall measure of association. If interaction does not exist, one would determine the posterior distribution of the one odds ratio.

For a case–control study, it is possible to control the effects of confounding variables by matching cases and controls in such a way that the pair share common values with regard to the confounding variable. For example, the case–control study of coronary artery disease and systolic blood for each of the four categories of the two levels of SBP, the case and control could share a common (or near common) age.

Each scenario (direct, indirect, and matching) described in the preceding discussion will be illustrated with a Bayesian analysis based on WinBUGS.

3.2 Direct Adjustment of Data

Consider two areas 1 and 2 and let r denote the reference area used for the standard age distribution to be used for the direct adjustment of mortality. The following notation is used for this scenario:

1: area 1

2: area 2

r: refers to the area of the reference standard

m_{i1}: the number of individuals in the i-th age category of area 1

m_{i2}: the number of individuals of area 2 in the i-th age category

m_{ir}: the number of subjects of the reference area r in the i-th age category

y_{i1}: the number of deaths in area 1 of the i-th age category

y_{i2}: the number of deaths in area 2 of the i-th age category

p_{i1}: the mortality rate for the i-th age category of area 1

The observed mortality rate for area 1 of the i-th age category is

$$p_{i1} = \frac{y_{i1}}{m_{i1}}, \quad \text{for } i = 1, 2, \ldots, k \tag{3.1}$$

with similar expressions for the death rates of area 2.

The directly standardized mortality rate for area 1 is

$$DSR_1 = \frac{\sum_{i=1}^{i=k} \theta_{i1} m_{ir}}{\sum_{i=1}^{i=k} m_{ir}} \tag{3.2}$$

where θ_{i1} is the unknown "true" mortality of the i-th category for area 1. If it is assumed that the number of individuals is fixed, Equation 3.2 is an unknown parameter that can be estimated by Bayesian techniques. To employ Bayesian methods, it is assumed that the number of deaths occurring in the area 1 for category i has a binomial distribution, that is,

$$y_{i1} \sim \text{binomial} \, (\theta_{i1}, m_{i1}) \tag{3.3}$$

for $i = 1, 2, \ldots, k$.

Note the standardized mortality for area 1, DSR_1, is a weighted average of the unknown "true" mortalities θ_{i1}, which is weighted by the corresponding number of subjects m_{ir} in the reference standard of area r. In a similar way, the standardized mortality rate for area 2 is

$$DSR_2 = \frac{\sum_{i=1}^{i=k} \theta_{i2} m_{ir}}{\sum_{i=1}^{i=k} m_{ir}} \tag{3.4}$$

and it is assumed that

$$y_{i2} \sim \text{binomial} \, (\theta_{i2}, m_{i2}) \tag{3.5}$$

for $i = 1, 2, \ldots, k$.

By assuming the number of deaths in a given age group have a binomial distribution, one is also assuming the probability of death of each individual in that age category is the same.

To illustrate Bayesian inferences for comparing the standardized mortality rates for two areas, the mortality rates for California will be compared to that of Florida, using the U.S. census data for 2010 as the reference population.

Consider the populations of California, Florida, and the United States as shown in Table 3.1. Also shown are the number of deaths by age for the United States, California, and Florida.

TABLE 3.1

U.S., California, and Florida Population by Age

	United States (Area r)		California (Area 1)		Florida (Area 2)	
Age	Number	Percent	Number	Percent	Number	Percent
Total	308,745,538	100	37,253,956	100	18,801,310	100
<5	20,201,362	6.5430	2,531,333	6.7948	1,073,506	5.7097
5–24	84,652,193	33.9611	10,686,658	28.6859	4,668,242	24.8293
25–44	82,134,554	26.6026	10,500,587	28.1864	4,720,799	25.1088
45–64	81,489,445	26.3937	9,288,864	24.9338	5,079,161	27.0149
65–84	34,774,551	11.2631	3,645,546	9.7856	2,825,477	15.0280
>85	5,493,433	1.7792	600,968	1.6131	434,125	2.3090

Source: U.S. Census Bureau 2010 Census Summary. http://factfinder2.census.gov/faces/tableservices/jsf/pages/productview.xhtml?fpt=table.

TABLE 3.2

Number of Deaths for United States, California, and Florida by Age

	United States (Area r)	California (Area 1)	Florida (Area 2)
Age	Mortality	Mortality	Mortality
<5	28,869	3,446	1,987
5–24	34,790	4,137	1,940
25–44	112,178	10,038	3,993
45–64	493,566	29,511	9,823
65–84	1,031,816	31,918	12,092
>85	764,582	16,332	5,225
Total	2,465,801	95,383	35,060

Source: Worktable 3 of National Center of Health Statistics, 2, http://www.cdc.gov/nchs/data/dvs/MortFinal2007_Worktable308.pdf

What are the standardized mortality rates for California and Florida, using the 2010 census population figures (Table 3.1)? Using the notation defined earlier, the vector of deaths for the reference standard is

$$y_r = (y_{1r}, y_{2r}, ..., y_{6r})$$
$$= (28869, 34790, 112178, 493566, 1031816, 764,582)$$

and the vector of deaths for area 1 (California) is

$$y_1 = (y_{11}, y_{21}, ..., y_{61})$$
$$= (3446, 4137, 10038, 29511, 31918, 16332)$$

As for Florida, the vector of deaths is

$$y_2 = (y_{12}, y_{22}, ..., y_{62})$$
$$= (1987, 1940, 3993, 9823, 12092, 5225)$$

In a similar fashion, the population for the reference standard by the six age categories is

$$m_r = (m_{1r}, m_{2r}, ..., m_{6r})$$
$$= (20201362, 84652193, 82134554, 81489445, 34774551, 5493433)$$

For California the vector is

$$m_1 = (m_{11}, m_{21}, ..., m_{61})$$
$$= (2531333, 10686658, 10500587, 9288864, 3645546, 600968)$$

and for Florida the vector is

$$m_2 = (m_{12}, m_{22}, ..., m_{62})$$
$$= (1073506, 4668242, 4720799, 5079161, 2825477, 434125)$$

Recall that for California, it is assumed $y_{i1} \sim$ binomial (θ_{i1}, m_{i1}), for i = 1, 2, ... , 6; therefore, in particular for the under 5-year age, the number of deaths is distributed as a binomial with parameters θ_{11} and m_{11}, that is,

$$y_{11} \sim \text{binomial}(\theta_{11}, m_{11})$$
$$\sim \text{binomial}(\theta_{11}, 2531333) \tag{3.6}$$

where the number of observed deaths is 3446 (Table 3.2). Thus, one is assuming that 3466 is a sample from a binomial distribution with parameter vector $(\theta_{11}, 2,531,333)$. It follows that if one assumes an improper prior distribution for θ_{11}, the posterior distribution of θ_{11} is beta with parameter vector (3466, 2527867), and the posterior mean of θ_{11} is .001369239, the estimated probability of death for a person who is less than 5 years old. This is equivalent to a mortality rate of 137 per 100,000 in that age group. Our goal is to determine the posterior distribution of the directly standardized mortality rate for California,

$$\text{DSR}_1 = \frac{\sum_{i=1}^{i=k} \theta_{i1} m_{ir}}{\sum_{i=1}^{i=k} m_{ir}}$$

which depends on the posterior distribution of the probabilities θ_{i1} for i = 1, 2, ... , 6.

On the other hand, if an improper prior distribution is used, the posterior distribution of θ_{i1} is beta with vector $(y_{i1}, m_{i1} - y_{i1})$ for i = 1, 2, ... , 6. The improper prior for θ_{i1} is expressed as

$$f(\theta_{i1}) \propto [\theta_{i1}(1 - \theta_{i1})]^{-1}$$
$$\text{for } 0 < \theta_{i1} < 1$$

(3.7)

To determine the posterior distribution of the directly standardized mortality rates for California and Florida, BUGS CODE 3.1 will be executed.

BUGS CODE 3.1

```
model;
{
for(i in 1:6){y1[i]~dbin(theta1[i],m1[i])}
for(i in 1:6){y2[i]~dbin(theta2[i],m2[i])}
numds1<-theta1[1]*mr[1]+theta1[2]*mr[2]+theta1[3]*mr[3]+
theta1[4]*mr[4]+theta1[5]*mr[5]+theta1[6]*mr[6]
denoms1<-sum(mr[])
# directly standardized mortality for area 1
s1<-numds1/denoms1
numds2<-theta2[1]*mr[1]+theta2[2]*mr[2]+theta2[3]*mr[3]+
theta2[4]*mr[4]+theta2[5]*mr[5]+theta2[6]*mr[6]
denoms2<-sum(mr[])
#standardized mortality for area 2
s2<-numds2/denoms2
r<-s1/s2
d<-s1-s2
}
# deaths for area 2
list(y2 = c(1987,1940,3993,9823,12092,5225),
# deaths for area 1
y1 = c(3446,4137,10038,29511,31918,16332),
# population for area 1
m1 = c(2531333,10686658,10500587,9288864,3645546,600968),
# population for area 2
m2 = c(1073506,4668242,4720799,5079161,2825477,434125),
# population for reference standard
mr = c(20201362,84652193,82134554,81489445,34774551,5493433))
# initial values for the thetas
list(theta1 = c(.5,.5,.5,.5,.5,.5),theta2 = c(.5,.5,.5,.5,.5,.5))
```

The code closely follows the notation used in this section and is executed with 55,000 observations for the simulation, a burn in of 5000, and a refresh of 100. BUGS statements give a binomial distribution to the age-specific number of deaths with the first parameter θ_{ij} (i = 1, 2, ... , 6, j = 1, 2) unknown. The posterior distribution of the θ_{ij} is induced by the binomial distribution of the age-specific rates.

Therefore, the mortality rate for California is 276 per 100,000 and 167 per 100,000 for Florida, and their ratio (California to Florida) has a posterior mean of 1.654 with a 95% credible interval of (.634, 1.675). Note that the posterior mean of the probability of death (directly standardized mortality) for California is .002758, which implies an approximate mortality rate of 276 per 100,000. Since the Markov chain Monte Carlo (MCMC) errors are extremely small, one has confidence 55,000 observations are sufficient to estimate the true posterior mean of the five parameters given in Table 3.3. Of interest is the 95% credible interval of r, the ratio of the directly standardized mortality of California to that of Florida. Since the interval does not include unity, one is inclined to believe that the two mortality rates are indeed not the same.

Choosing a standard reference population can be a problem. For example, the U.S. population is used as a standard for comparing the directly standardized mortalities for California and Florida, but others could have been chosen. For example, the population of either state could have been used, but another interesting choice is described by Kahn and Sempos,[1] which is to weight the mortality probabilities by the inverse of the variance of the differences of the two estimated mortality probabilities for each age category. This seems reasonable if one is interested in estimating the difference in the two standardized mortalities.

Consider the directly standardized mortality for area 1:

$$DSR_1 = \frac{\sum_{i=1}^{i=k} \theta_{i1} m_{ir}}{\sum_{i=1}^{i=k} m_{ir}}$$

TABLE 3.3

Bayesian Analysis for Directly Standardized Mortality Rates of California and Florida

Parameter	Mean	SD	Error	2½	Median	97½
d	.001091	.0000126	<.00000001	.001066	.001091	.001116
r	1.654	.01030	<.00001	1.634	1.654	1.675
DSR_1	.002758	<.000001	<.00000001	.00274	.002758	.002775
DSR_2	.001667	<.000001	<.00000001	.001649	.001667	.001685
θ_{11}	.001362	<.000001	<.0000001	.001317	.001362	.001407

where each age-specific mortality probability θ_{i1} is weighted by m_{ir}, the population size for that age category of the standard population. Note the difference d between the two is

$$\text{DSR}_1 - \text{DSR}_2 = \frac{\sum\limits_{i=1}^{i=k}(\theta_{i1} - \theta_{i2})m_{ir}}{\sum\limits_{i=1}^{i=k}m_{ir}} \tag{3.8}$$

Thus, the Kahn and Sempos (p. 89)[1] choice is to replace the weights m_{ir} by the inverse of the variance of $\theta_{i1} - \theta_{i2}$. For a non-Bayesian, this is an optimal choice because it minimizes the variance of the estimated $\text{DSR}_1 - \text{DSR}_2$. A reasonable analogy for the Bayesian would be to choose the inverse of the variance of the posterior distribution of $\theta_{i1} - \theta_{i2}$. What is the variance of the posterior distribution of $\theta_{i1} - \theta_{i2}$? Using an improper prior for the mortality probabilities, see Equation 3.7, it is known that θ_{i1} has a beta posterior distribution with vector $(y_{i1}, m_{i1} - y_{i1})$; therefore, using the formulas of Degroot (p. 40),[3] the variance of the age-specific difference is

$$\text{Var}(\theta_{i1} - \theta_{i2} \mid \text{data}) = \frac{[y_{i1}(m_{i1} - y_{i1})]}{[m_{i1}^2(m_{i1} + 1)]} + \frac{[y_{i2}(m_{i2} - y_{i2})]}{[m_{i2}^2(m_{i2} + 1)]} \tag{3.9}$$

for i = 1, 2, ... , k, where k is the number of age categories.

Code for the Bayesian analysis based on weighting the mortality probabilities by the inverse of the variance of the difference in the mortality probabilities is portrayed in BUGS CODE 3.2.

BUGS CODE 3.2

```
model;
# estimated standardized rates based on the inverse of the
variance of the difference
{
for(i in 1:6){y1[i]~dbin(theta1[i],m1[i])}
for(i in 1:6){y2[i]~dbin(theta2[i],m2[i])}
for(i in 1:6){nw1[i]<-y1[i]*(m1[i]-y1[i])/
m1[i]*m1[i]*(m1[i]+1)}
for(i in 1:6){nw2[i]<-y2[i]*(m2[i]-y2[i])/
m2[i]*m2[i]*(m2[i]+1)}
# the variance of the difference is the next statement
for(i in 1:6){v[i]<-nw1[i]+nw2[i]}
# below are the computed weights
for(i in 1:6){w[i]<-1/v[i]}
numds1<-theta1[1]*w[1]+theta1[2]*w[2]+theta1[3]*w[3]+
```

```
theta1[4]*w[4]+theta1[5]*w[5]+theta1[6]*w[6]
denoms1<-sum(w[])
# directly standardized mortality for area 1
s1<-numds1/denoms1
numds2<-theta2[1]*w[1]+theta2[2]*w[2]+theta2[3]*w[3]+
theta2[4]*w[4]+theta2[5]*w[5]+theta2[6]*w[6]
denoms2<-sum(w[])
# standardized mortality for area 2
s2<-numds2/denoms2
# uniform prior for thetas
for(i in 1:6){theta1[i]~dbeta(1,1)}
for(i in 1:6){theta2[i]~dbeta(1,1)}
r<-s1/s2
d<-s1-s2
}
# deaths for area 2
list(y2 = c(1987,1940,3993,9823,12092,5225),
# deaths for area 1
y1 = c(3446,4137,10038,29511,31918,16332),
# population for area 1
m1 = c(2531333,10686658,10500587,9288864,3645546,600968),
# population for area 2
m2 = c(1073506,4668242,4720799,5079161,2825477,434125))
# initial values for the thetas
list(theta1 = c(.5,.5,.5,.5,.5,.5),theta2 = c(.5,.5,.5,.5,.5,.5))
```

The Bayesian analysis is based on BUGS CODE 3.2 and executed with 55,000 observations, a burn-in of 5000, and a refresh of 100. The results are revealed in Table 3.4.

The conclusions based on Table 3.4 are similar to those implied by Table 3.3 in that one would conclude that the two standardized mortality rates are not the same because the 95% credible interval for r is (2.13, 2.26). However, the rate for California is .02114 (based on the posterior mean) compared to 1.654 of Table 3.3. Of course, this demonstrates that the standard populations are not the same. It is also observed that the standard deviation of the posterior

TABLE 3.4

Bayesian Analysis for Directly Standardized Mortality for California and Florida

Parameter	Mean	SD	Error	2½	Median	97½
d	.0115	.0002038	<.0000001	.01111	.0115	.0119
r	2.194	.03318	<.0001	2.13	2.194	2.26
DSR$_1$	.02114	.00016	<.0000001	.02082	.02114	.02145
DSR$_2$	.009634	.0001262	<.0000001	.009388	.009634	.009883
θ_{11}	.001362	.0000231	<.0000001	.001317	.001362	.001407

Note: Standard population is the inverse of the variance of the posterior distribution of the difference.

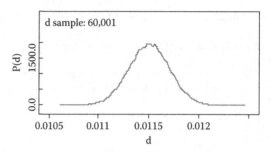

FIGURE 3.1
Posterior density of difference in standardized rates for California and Florida.

density of the difference d is extremely small and that the 95% credible interval does not include zero, as depicted in Figure 3.1.

Recall that for the analysis of Table 3.3, the standard population is the U.S. population of 2010, but for the second analysis, the standard population has weights proportional to the variance of the posterior distributions of the age-specific differences in the mortality probabilities. Note the directly standardized mortality rates for California and Florida are 211 and 96, respectively, per 100,000 people.

I prefer to compare the directly standardized mortality rates of two states based on some mutually agreed real population, such as the U.S. population. The second choice is based on a different idea, namely, that of estimating the difference in the two rates with minimum variance.

3.3 Indirect Standardization Adjustment

3.3.1 Introduction

There are two methods to adjust mortality data of two states when the two states have different age distributions, and direct standardization was employed in Section 3.2. The two approaches are direct and indirect, but what is the difference between the two? Recall for the direct way, one must know the total population and the population for each age category for both areas and the standard population. With the indirect approach, one need not know the age-specific mortality and population for the areas to be standardized, but one does need to know the age distribution for each group to be standardized, the total number of deaths in each group, and the age-specific mortality and population for the standard population.

In Section 3.2, the groups were the states of Florida and California, and the reference population was the U.S. population; however, it is important to know that the methods of direct and indirect standardization apply to other entities besides geographical units. For example, the groups could be two groups of patients receiving therapy, and the reference standard population could be a historical control, where the main focus is to compare the mortality between the two groups of patients, standardized with respect to the patient characteristic of the historical control. Another example is in a cohort study, where the incidence rates of disease are to be compared among the two groups. One would want to adjust the two groups (those exposed to the risk factor and the group not exposed to the risk factor) relative to some standard population of patients. For this scenario, the standard population could be one of the two groups or it could be the one where each age-specific mortality is weighted by the variance of the difference in the two mortality probabilities.

3.3.2 Indirect Standardization

Consider two areas 1 and 2 and a standard reference population r; the indirect standardized mortality of area 1 is

$$ISR_1 = \left(\frac{\sum\limits_{i=1}^{i=k} \theta_{i1} m_{i1}}{\sum\limits_{i=1}^{i=k} \theta_{ir} m_{i1}} \right) \left(\frac{\sum\limits_{i=1}^{i=k} \theta_{ir} m_{ir}}{\sum\limits_{i=1}^{i=k} m_{ir}} \right) \tag{3.10}$$

where $\sum\limits_{i=1}^{i=k} \theta_{i1} m_{i1}$ is the size of the population of the number of deaths for area 1 and $\sum\limits_{i=1}^{i=k} \theta_{ir} m_{i1}$ is the expected number of deaths for area 1 if the standard rates are used. The number of age categories is k, and the crude mortality rate for the standard group is $\sum\limits_{i=1}^{i=k} \theta_{ir} m_{ir} / \sum\limits_{i=1}^{i=k} m_{ir}$.

Thus for the first factor of Equation 3.10, the numerator is the number of deaths for area 1 and the denominator is the expected number of deaths, if the age-specific mortality rates of the standard population apply. Thus, one does not need to know the mortality and population size for the age categories of area 1. For the Bayesian approach, one would estimate the expected number of death in area 1 as if standard rates apply by the posterior mean of the probability of death in the standard population by the population size of area 1. One does not need to know the age-specific population sizes. One would assume the number of deaths in the standard population to have a binomial distribution with parameter vector θ_r and $\sum\limits_{i=1}^{i=k} m_{ir}$, the size of the standard population. For the crude rate of the standard population $\sum\limits_{i=1}^{i=k} \theta_{ir} m_{ir} / \sum\limits_{i=1}^{i=k} m_{ir}$, the Bayesian approach would assume the number of deaths

y_{ir} for age category i to have a binomial distribution with parameters θ_{ir} and m_{ir}, and the posterior distribution of the standardized rate (Equation 3.10) is easily determined.

Of course, the definition of the indirectly standardized rate for area 2 is

$$ISR_2 = \left(\frac{\sum\limits_{i=1}^{i=k} \theta_{i2} m_{i2}}{\sum\limits_{i=1}^{i=k} \theta_{ir} m_{i2}} \right) \left(\frac{\sum\limits_{i=1}^{i=k} \theta_{ir} m_{ir}}{\sum\limits_{i=1}^{i=k} m_{ir}} \right) \tag{3.11}$$

and the Bayesian approach is to determine its posterior distribution. Recall that one need not know the age-specific mortalities and population sizes for areas 1 and 2, the two areas to be standardized, but one does need to know the age-specific mortality and population sizes for the standard population and the total number of deaths and population size for the two areas.

The first factor of the indirectly standardized rates (Equations 3.10 and 3.11) is referred to as the standard mortality ratio (SMR), and, in fact, the ratio of the standardized ratios is simply the ratio of the standard mortality rates. As emphasized by Kahn and Sempos (p. 97),[1] the weighting factors of the two indirect standard rates do not have the same age-specific weights; thus, the two cannot be compared. However, $\sum\limits_{i=1}^{i=k} \theta_{i1} m_{i1}$ can be compared to $\sum\limits_{i=1}^{i=k} \theta_{ir} m_{i1}$, and in a similar fashion $\sum\limits_{i=1}^{i=k} \theta_{i2} m_{i2}$ can be compared to $\sum\limits_{i=1}^{i=k} \theta_{ir} m_{i2}$. The latter comparisons are valid because they have the same age-specific weights.

3.3.3 Bayesian Inferences for Indirect Adjustment

Our initial inferences will be for estimating the indirectly standardized rate (Equation 3.10) for area 1. Consider the numerator of the first factor, which is the number of deaths for area 1, and the Bayesian approach is to assume that the number of deaths y of area 1 has a binomial distribution with parameters (θ_1, m_1), where m_1 is the total population of area 1; therefore, the posterior distribution of θ_1 is beta with parameter $(y_1, m_1 - y_1)$, assuming an improper prior for θ_1. The number of deaths for area 1 would be estimated by multiplying the posterior mean of θ_1 by the population of area 1, namely, m_1.

As for the denominator of the first factor of Equation 3.10, it is the expected number of deaths assuming the standard mortality rates apply. How does the Bayesian approach estimate this quantity? For the Bayesian approach, one could assume the number of deaths y_r in the standard population has a binomial distribution with parameter vector (θ_r, m_r), where θ_r is the probability of death for the standard population and m_r is the size of the population for the standard. Again, assuming an improper prior distribution for θ_r implies that its posterior distribution is beta with parameter vector $(y_r, m_r - y_r)$, with a posterior mean of $y_r \, / \, m_r$. If the posterior mean of θ_r is multiplied by the population size m_1, the result is the average number of deaths for area 1, assuming the standard mortality rates apply to area 1. The posterior distribution of the standard mortality rate for area 1

$$
SMR_1 = \frac{\sum\limits_{i=1}^{i=k} \theta_{i1} m_{i1}}{\sum\limits_{i=1}^{i=k} \theta_{ir} m_{i1}}
$$

is induced by the posterior distributions of the numerator and denominator as described earlier. Of course, a similar course of action is easily taken for the standard mortality rate for area 2.

The second factor of Equation 3.10 is the crude rate for the standard population.

A Bayesian approach would assume the age-specific deaths y_{ir} to have a binomial distribution with parameter (θ_{ir}, m_{ir}), which induces a beta posterior distribution with parameter (y_{ir}, m_{ir}); thus, the crude mortality for the standard population is easily determined. Note this crude rate is the common second factor of the indirectly standardized rates for both areas.

3.3.4 Example of Indirect Standardization

Returning to the example of comparing the mortality of California to that of Florida, using the U.S. population as a standard, the information is portrayed in Tables 3.1 and 3.2. First, consider the indirect standardization of California. Actually, one only needs the total mortality (95,383) and total population size (37,253,956) of California. Consider the numerator of the standard mortality rate for California. The Bayesian approach would assume the number of deaths $y_1 \sim \text{beta}(\theta_1, m_1)$, where $m_1 = 37,253,956$ and $y_1 = 95,383$.

Assuming an improper prior distribution for θ_1 implies that posterior distribution of $\theta_1 \sim \text{beta}(95383, 37158573)$ and the posterior distribution of the total number of deaths is the posterior distribution of $m_1 \theta_1$.

As for the denominator of the standard mortality rate for California, one needs the posterior distribution of the number of deaths y_r of the U.S. population and the population size of California m_1; therefore, y_r has a binomial distribution with parameter (θ_r, m_r), where $m_r = 308,745,538$ is the population of the United States, and the posterior distribution of θ_r is beta with parameter $(y_r, m_r) = (2465801, 308745538)$. The denominator of the SMR for California is then given by the posterior mean of the posterior distribution of $\theta_r m_1$. Thus, all the information necessary to induce the posterior distribution of the SMR is at hand, and the Bayesian analysis is easily performed.

Finally, the second factor $\sum\limits_{i=1}^{i=k} \theta_{ir} m_{ir} / \sum\limits_{i=1}^{i=k} m_{ir}$ of Equation 3.10 is the crude mortality rate for the U.S. population. For that population, the age-specific mortality rates and population sizes are known. Recall that for the Bayesian approach, the age-specific number of deaths $y_{ir} \sim$ binomial(θ_{ir}, m_{ir}) and the posterior distribution of θ_{ir} is beta ($y_{ir}, m_{ir} - y_{ir}$). Thus, all the information required to induce the posterior distribution of the second factor of Equation 3.10 is at hand, and the posterior distribution of ISR_1 can be determined.

To execute the Bayesian analysis for the indirectly standardized mortality rates for California and Florida, consider BUGS CODE 3.3.

BUGS CODE 3.3

```
model;
{
for (i in 1:6){yr[i]~dbin(thetar[i],mr[i])}
# prior distribution of thetar
for (i in 1:6){thetar[i]~dbeta(.5,.5)}
y1~dbin(theta1,m1)
y2~dbin(theta2,m2)
ys~dbin(thetas, ms)
# prior distribution of theta1, theta2, and thetas
theta1~dbeta(.5,.5)
theta2~dbeta(.5,.5)
thetas~dbeta(.5,.5)
# numerator of SMR1
nums1<-theta1*m1
denoms1<- thetas*m1
# numerator of SMR2
nums2<-theta2*m2
denoms2<-thetas*m2
# crude rate for the standard
# Numerator of crude rate for standard
numst<-thetar[1]*mr[1]+thetar[2]*mr[2]+thetar[3]*mr[3]+thetar
  [4]*mr[4]+
thetar[5]*mr[5]+thetar[6]*mr[6]
CRST<-numst/ms
ISR1<-(nums1/denoms1)*CRST
ISR2<-(nums2/denoms2)*CRST
# difference in the indirectly standardized rates
disr<-ISR1-ISR2
SMR1<-nums1/denoms1
SMR2<-nums2/denoms2
# difference in the standard mortality rates
dsmr<-SMR1-SMR2
# ratio of the standard mortality rates
rsmr<-SMR1/SMR2
# ratio of the indirectly standardized rates
risr<-ISR1/ISR2
}
```

```
# data from Tables 3.1 and 3.2
list(mr = c(20201362,84652193,82134554,81489445,34774551,5493433),
yr = c(28869,34790,112178,493566,1031816,764582), m1 = 37253956, m2 =
18801310,y1 = 95383, y2 = 35060, ms = 308745538,ys = 2465801))
# initial values
list(theta1 =.5,theta2 = .5)
```

The BUGS CODE is well documented by the comment statements denoted with a #. For example, the comment "# difference in the standard mortality rates" indicates that dsmr is the appropriate parameter. A Bayesian analysis is executed with 65,000 observations, a burn-in of 5000, and a refresh of 100, and the results are reported in Table 3.5.

The Bayesian analysis provides extremely accurate posterior means for estimating the various characteristics of the posterior distribution of the various parameters appearing in Table 3.5. Considering the crude rate for the standard population in California, the MCMC error is $< 10^{-7}$ implying that the posterior mean of .007926 is extremely close to the "true" posterior mean and that 65,000 observations are indeed sufficient for the simulation. All distributions appear to be symmetric about their posterior mean.

For the standard mortality rate SMR_1 for California, the posterior mean is .3206 implying that the observed number of deaths in California is 32% of the expected number of deaths in California, assuming the mortality rates of the U.S. population apply to California. On the other hand, the observed number of deaths for Florida is approximately 23% of the expected number of deaths for Florida, assuming the mortality rates of the U.S. population apply to Florida.

The main focus of the analysis is the indirectly standardized mortally rates for the two states, using the U.S. population as the standard. For California, the rate is ISR_1 with a posterior mean of .00256. Thus, the probability of death for California is estimated as .00256 with a 95% credible interval of

TABLE 3.5

Bayesian Analysis for Indirect Standardization for California and Florida

Parameter	Mean	SD	Error	2½	Median	97½
CRST	.007926	<.000001	<.00000001	.007977	.007987	.007996
ISR_1	.00256	<.000001	<.00000001	.002544	.00256	.002577
ISR_2	.001865	<.00001	<.00000001	.001845	.001865	.001885
SMR_1	.3206	.00106	<.000001	.3185	.3206	.3226
SMR_2	.2335	.001257	<.00001	.231	.2335	.236
risr	1.373	.008567	<.00001	1.356	1.373	1.39
rsmr	1.373	.008567	<.00001	1.356	1.373	1.39

(.002544, .002577), while the probability of death in Florida is .001865 (via the posterior mean) with a 95% credible interval of (.001845, .001885). Note the posterior distribution of the ratio of the two standardized rates has a mean of 1.373 and a 95% credible interval (1.356, 1.39) implying that there is indeed a difference in the two rates and that the rate of California is 1.37 times that of Florida. The reader is invited to compare the results for the directly standardized rate of California depicted in Table 3.3, and the corresponding indirectly standardized rate portrayed in Table 3.5. See problem 7.

3.4 Stratification and Association between Disease and Risk Exposure

3.4.1 Introduction

Factors other than the risk factor may have an effect on the association between risk and disease. Consider for example the Israeli Ischemic Heart Disease Study, which is analyzed by Kahn and Sempos (p. 105).[1] The data is reported in Table 3.6.

I used BUGS CODE 2.2 with 55,000 observations for the simulation, a burn-in of 5000, and a refresh of 100 and found the posterior mean of the odds ratio for the association between age and myocardial infarction (MI) to be 3.48 with a 95% credible interval of (2.004, 5.355). As for the association between SBP and MI, the odds ratio has posterior mean 1.918 with a standard deviation of .3773 and a 95% credible interval of (1.284, 2.764). Because an improper prior is assumed for the prior distribution of the cell probabilities, the Bayesian posterior means of the odds ratio agree with the odds ratios computed the conventional way by Kahn and Sempos (p. 105).[1]

For this study, the main focus is on the association between SBP and the occurrence of MI. How does one adjust for the other factor, age? Age appears

TABLE 3.6

Association of Age and Systolic Blood Pressure on Myocardial Infarction (MI)

	MI	No MI
Age ≥60	15	188
Age <60	41	1767
SBP ≥140	29	711
SBP <140	27	1244
Total	56	1955

Source: Kahn, H.A., Sempos, C.T. *Statistical Methods in Epidemiology*, Oxford University Press, New York, 1989, p. 105.

TABLE 3.7

Association between Systolic Blood Pressure and Age
(Israeli Heart Disease Study)

	Age ≥60	Age <60	Total
SBP ≥140	124	616	740
SBP <140	79	1192	1271
Total	203	1808	2011

Source: Kahn, H.A., Sempos, C.T. *Statistical Methods in Epidemiology*, Oxford University Press, New York, 1989, p. 106.

to be a confounding factor because it is related to the disease. It also appears that age and SBP are related, thus, consider Table 3.6.

A Bayesian analysis reveals the posterior mean of the odds ratio to be 3.066 with a 95% credible interval (2.461, 3.807), which agrees with the value of 3.44 of Kahn and Sempos. Refer to the list statements of BUGS CODE 2.3 for the information from Tables 3.6 and 3.7. Thus, age is related to the risk factor, SBP, and disease (MI).

3.4.2 Interaction and Stratification

If a factor affects both the disease (or morbidity) and the risk factor, the factor is referred to as a confounder, and one would expect a different association between the risk factor and disease for different levels of the possible confounder. If this is true, interaction between the risk factor and confounder is said to exist. This should be the case for the Israeli Heart Disease Study. To determine the status of the interaction between age and SBP, consider Table 3.8, which depicts the association between MI and SBP stratified by two age classes.

TABLE 3.8

Myocardial Infarction (MI) versus Systolic Blood
Pressure by Age (Israeli Heart Disease Study)

	MI	No MI	Total
Age ≥60			
SBP ≥140	9	115	124
SBP <140	6	73	79
Total	15	188	203
Age <60			
SBP ≥140	20	596	616
SBP <140	21	1171	1192
Total	41	1767	1808

Source: Kahn, H.A., Sempos, C.T. *Statistical Methods in Epidemiology*, Oxford University Press, New York, 1989, p. 106.

TABLE 3.9

Bayesian Analysis for Interaction (Israeli Heart Disease Study)

Parameter	Mean	SD	Error	2½	Median	97½
AR1	−.1118	.4568	.002087	−1.226	−.0394	.5757
AR2	.4511	.1276	.000524	.1585	.4655	.6582
OR1	1.071	.5082	.002217	.4493	.9621	2.357
OR2	1.919	.4442	.001834	1.188	1.871	2.925
RAR	−.273	7.746	.03438	−3.314	−.08838	1.588
ROR	.588	.316	.001425	.2108	.517	1.382

For the Bayesian analysis, the posterior distribution of the odds ratio between SBP and MI will be estimated by the posterior mean for each stratum, then compared by the posterior distribution of the ratio for the stratum age ≥60 versus age <60. The results are showed in Table 3.9, which use BUGS CODE 3.4 and 65,000 observations for the simulation, a burn-in of 5000, and a refresh of 100.

BUGS CODE 3.4

```
model;
# posterior distribution of the odds ratio for two strata
{
# stratum 1
# assume improper prior (2.5)
# posterior distribution of thetapp1
thetapp1~dbeta(npp1,nmp1)
# posterior distribution of thetapm1
thetapm1~dbeta(npm1,nmm1)
# postrior distribution of thetamp1
thetamp1~dbeta(nmp1,npp1)
thetamm1~dbeta(nmm1,npm1)
# OR1 is the odds ratio for stratum 1
OR1<-(thetapp1*thetamm1)/(thetapm1*thetamp1)
# AR1 is the attributable risk for stratum 1
AR1<-(OR1-1)/OR1

# stratum 2
# assume improper prior (2.5)
# posterior distribution of thetapp2
thetapp2~dbeta(npp2,nmp2)
# posterior distribution of thetapm2
thetapm2~dbeta(npm2,nmm2)
# posterior distribution of thetamp2
thetamp2~dbeta(nmp2,npp2)
thetamm2~dbeta(nmm2,npm2)
# OR2 is the odds ratio for stratum 2
```

```
OR2<-(thetapp2*thetamm2)/(thetapm2*thetamp2)
# AR2 is the attributable risk
AR2<-(OR2-1)/OR2
DOR<-OR1-OR2
ROR<-OR1/OR2
RAR<-AR1/AR2
DAR<-AR1-AR2

}
# Israeli Heart Study Strata 1 and 2 list(npp1 = 9,nmp1 = 6,npm1
= 115,nmm1 = 73,npp2 = 20,nmp2 = 21,npm2 = 596,nmm2 = 1171)
# initial value of thetas
list(thetapp1 =.5, thetapm1 =.5, thetamp1 =.5,thetamm1 =.5,
thetapp2 =.5, thetapm2 =.5, thetamp2 =.5,thetamm2 =.5)
```

For the first stratum, the attributable risk has a posterior mean of −.1118 with a 95% credible interval (−1.226, .5757); however, the two parameters of interest are the two odds ratio. The posterior median of the odds ratio is .9621 for stratum 1 and 1.871 for stratum 2 (age <60). Note the posterior distribution for the two odds ratio is skewed; therefore, the posterior medians are more appropriate as estimators of the odds ratio. Figure 3.2 clearly shows the skewness of the posterior distribution of the odds ratio for stratum 1, and it can be shown that the posterior distribution of the odds ratio for the second stratum is also skewed. Of special interest is the posterior distribution of the ratio of the odds ratio, which has a posterior median with a 95% credible interval of (.2108, 1.382). Thus, there is not sufficient evidence to declare a difference between the odds ratios of the two strata; however, it appears that there is sufficient evidence to be suspicious that an interaction might exist.

The formal definition of interaction between the risk factor and the confounder is based on Table 3.10.

Note that θ_{pq} is the probability that the confounder is present, the risk factor is positive, and that each of the n_{pq} subjects does not have the disease, where the first subscript p denotes the risk factor is present and the second

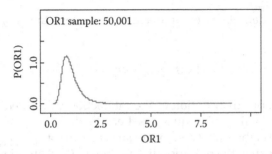

FIGURE 3.2
Posterior density of the odds ratio for stratum 2.

TABLE 3.10

Cell Frequencies and Probabilities for Two Levels of the Confounder,
Two Levels of the Risk Factor, and Two Levels of Disease

	Disease	No Disease	Total
Confounder present			
RF present	n_{pp}, θ_{pp}	n_{pq}, θ_{pq}	n_p
RF not negative	n_{qp}, θ_{qp}	n_{qq}, θ_{qq}	n_q
Confounder negative			
RF present	m_{pp}, ϕ_{pp}	m_{pq}, ϕ_{pq}	m_p
RF not present	m_{qp}, ϕ_{qp}	m_{qq}, ϕ_{qq}	m_q
Total			n

subscript q denotes the disease is not present. When the confounder is not present, φ_{pp} is the probability that the risk factor is present and that the disease is present. The cell frequencies have a multinomial distribution with parameter vector

$$\theta = (\theta_{pp}, \theta_{pq}, \theta_{qp}, \theta_{qq}; \phi_{pp}, \phi_{pq}, \phi_{qp}, \phi_{qq})$$

where the thetas are the cell probabilities when the confounder is present, while the phis are the cell probabilities when the confounder is not present.

Assuming an improper prior distribution implies that the cell probabilities have a Dirichlet with parameter vector

$$(n_{pp}, n_{pq}, n_{qp}, n_{qq}; m_{pp}, m_{pq}, m_{qp}, m_{qq})$$

and the marginal distribution of the cell frequencies are distributed as a beta. For example, the posterior distribution of θ_{pp} is beta with parameter vector $(n_{pp}, n - n_{pp})$.

The formal definition of interaction between the risk factor and confounder is

$$I = \theta_{pp} - \phi_{pp} - \theta_{qp} - \phi_{qp} \tag{3.12}$$

where the posterior distribution of the interaction is induced by the posterior distribution of the cell frequencies. Let n be the total number of subjects totaled over the eight cells. Then the usual estimator of θ_{pp} is n_{pp} / n, which is the posterior mean of the posterior distribution of θ_{pp}. Thus, the interaction is the probability the disease is present when the confounder and risk factor are present, minus the probability the disease is present when the confounder is not present and the risk factor is present, minus the probability of disease when the confounder is present and the risk factor is not present, plus the probability of disease when the confounder is not present and the risk factor is absent.

3.4.3 An Example of Stratification

Consider the Israeli Heart Disease Study, where the information is portrayed in Tables 3.8 and 3.11, where the confounder is age and considered to be present when age ≥ 60 and not present when age <60. The risk factor is systolic blood pressure (SBP) and is considered to be present when the SBP ≥ 140 and not present when the SBP <140. The "disease" in this case is MI and is present when a subject has had an infarction and not present when the subject has not had a heart attack. For the cell frequencies, I used the notation given in Table 3.10. Recall that the results of the Bayesian analysis for this study are given in Table 3.9, where the odds ratio for the association between SBP and heart attack are computed for both strata (age ≥ 60 and age <60). The analysis is executed with BUGS CODE 3.4 using 65,000 observations for the simulation and showed that the odds ratios could not be considered different. Our objective is to estimate the interaction as defined by I.

Assuming an improper prior distribution for the cell frequencies,

$$\theta = (\theta_{pp}, \theta_{pq}, \theta_{qp}, \theta_{qq} ; \phi_{pp}, \phi_{pq}, \phi_{qp}, \phi_{qq})$$

θ has a Dirichlet posterior distribution with parameter vector (9,115,6, 73;20,596,21,1171). For example, this implies the posterior distribution of θ_{pp} (the probability a subject with the risk factor [SBP >140] and with the

TABLE 3.11

Myocardial Infarction versus Systolic Blood Pressure by Age (Israeli Heart Disease Study)

	MI	No MI	Total
Age ≥60			
SBP ≥140	$9 = n_{pp}$	$115 = n_{pq}$	$124 = n_p$
SBP <140	$6 = n_{qp}s$	$73 = n_{qq}$	$79 = n_q$
Total	15	188	203
Age <60			
SBP ≥140	$20 = m_{pp}$	$596 = m_{pq}$	$616 = m_p$
SBP <140	$21 = m_{qp}$	$1171 = m_{qq}$	$1192 = m_q$
Total	41	1767	1808

Source: Kahn, H.A., Sempos, C.T. *Statistical Methods in Epidemiology*, Oxford University Press, New York, 1989, p. 106.

confounder present [age >60] will have a heart attack) is beta (9, 1799). I have assumed a random sample of 1808 individuals are selected and that the individual is classified as to status of age, SBP, and the occurrence of a heart attack.

To do the Bayesian analysis, refer to BUGS CODE 3.5.

BUGS CODE 3.5

```
Model;
{
gpp~dgamma(npp,2)
gpq~dgamma(npq,2)
gqp~dgamma(nqp,2)
gqq~dgamma(nqq,2)
hpp~dgamma(mpp,2)
hpq~dgamma(mpq,2)
hqp~dgamma(mqp,2)
hqq~dgamma(mqq,2)
sgh<- gpp+gpq+gqp+gqq+hpp+hpq+hqp+hqq

# Dirichlet distribution for the thetas and phis
thetapp<-gpp/sgh
thetapq<-gpq/sgh
thetaqp<-gqp/sgh
thetaqq<-gqq/sgh
phipp<-hpp/sgh
phipq<-hpq/sgh
phiqp<-hqp/sgh
phiqq<-hqq/sgh
# the interaction
I<-thetapp-phipp-thetaqp+phiqp
MH<- MHn/MHd
MHn<- (thetapp*thetaqq/n1)+(phipp*phiqq/n2)
MHd<- (thetapq*thetaqp/n1)+(phipq*phiqp/n2)

}
# data from Table 3.11
list(npp = 9,npq = 115,nqp = 6,nqq = 73,mpp = 20,mpq = 596,mqp
    = 21,mqq = 1171,
n1 = 203,n2 = 1808)
# initial values
list(gpp = 1,gpq = 1,gqp = 1,gqq = 1,hpp = 1,hpq = 1,hqp =
    1,hqq = 1)
```

Based on BUGS CODE 3.5, 55,000 observations are generated for the simulation, with a burn-in of 5000 and a refresh of 100, and the results are reported in Table 3.12.

Since the 95% credible interval is (−.005308, .009324), it appears there is no interaction between age and SBP. The posterior mean of θ_{pp} is .0044880,

TABLE 3.12

Bayesian Analysis for Interaction (Israeli Heart Disease Study)

Parameter	Mean	SD	Error	2½	Median	97½
I	.0019950	.00372	<.00001	−.005308	.001985	.009324
ϕ_{pp}	.009942	.0022099	<.000001	.006091	.009788	.01468
θ_{pp}	.0044880	.001507	<.000001	.002037	.004328	.007919

which agrees with the "usual" estimate of 9/1808 = .004977876. This calculation serves as a check on the validity of the Bayesian analysis. The analysis for the interaction agrees with the previous Bayesian analysis reported in Table 3.9, which implied that the odds ratios for the two strata are not different.

3.5 Mantel–Haenszel Estimator of Association

When analyzing the association between disease and a risk factor in the presence of a confounding factor, one should determine if interaction is present. If interaction is present, the odds ratios of the various strata vary, while on the contrary if no interaction is present, the various odds ratios do not differ. It is important to know the status of interaction in a stratified study.

One way to estimate the association between disease and risk factor in a stratified study is with the Mantel–Haenszel (MH)[4] estimator. If interaction is present, the MH estimator is a weighted average of the odds ratios of the various strata, while if interaction is not present, the estimator estimates the common association.

Consider the general format of a stratified study given in Table 3.10, then the MH estimator is defined for two strata as

$$MH = \frac{\left(\dfrac{\theta_{pp}\theta_{qq}}{n_1} + \dfrac{\varphi_{pp}\varphi_{qq}}{n_2} \right)}{\left(\dfrac{\theta_{pq}\theta_{qp}}{n_1} + \dfrac{\varphi_{pq}\varphi_{qp}}{n_2} \right)} \qquad (3.13)$$

where n_i is the number of subjects in stratum i, i = 1, 2.

Note that the θ_{ij} are the cell probabilities for the first stratum, while the φ_{ij} are the cell probabilities for the second stratum, for i, j = p, q. Note that for the Bayesian approach, an estimator estimates a parameter such as MH with the posterior distribution of the parameter.

Recall the Israeli Heart Disease Study, where in stratified form, the results are reported in Table 3.12. BUGS CODE 3.5 is executed with 55,000 observations for the simulation, a burn-in of 5000, and a refresh of 100. The code identifies the Mantel-Haenszel as MH as the ratio of its numerator, labeled as MHn and denominator, labeled as MHd. It is interesting to observe that the posterior density of MH is highly skewed. Figure 3.3 shows that skewness.

Skewness is also demonstrated by the characteristics of the posterior distribution of MH reported in Table 3.13.

The median of the posterior distribution of MH is 1.573, which is the estimate of the overall association between heart attack and SBP. Note the small values for the numerator and denominator of MH, which probably accounts for the skewness of the posterior distribution of MH. Recall from Table 3.9 that the odds ratios for the two strata are .9621 and 1.871, respectively, and the MH estimate is 1.573 with more weight given to the second stratum (because there are many more observations for the second stratum, with 203 in the first and 1808 for the second). See exercise 15 for additional information about using the MH estimator for determining the association between disease and a risk factor. Woolson (p. 418)[5] presents very valuable information about the MH estimator and the reader is encouraged to read that material, and in addition, Rothman (p. 195)[6] reports interesting examples of using the estimator.

We follow up on an alternative estimator explained by Woolson (p. 422),[5] namely, a weighted estimator of the individual odds ratios of the various strata.

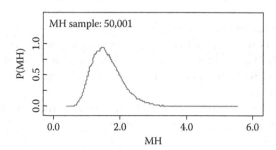

FIGURE 3.3
Posterior density of the Mantel–Haenszel estimator.

TABLE 3.13

Posterior Distribution of Mantel–Haenszel Estimator

Parameter	Mean	SD	Error	2½	Median	97½
MH	1.636	.4728	.001372	.9049	1.573	2.735
MHn	3.998×10^{-6}	7.541×10^{-7}	3.229×10^{-9}	2.66×10^{-6}	3.95×10^{-6}	5.6×10^{-6}
MHd	2.54×10^{-6}	5.038×10^{-7}	2.201×10^{-9}	1.666×10^{-6}	2.512×10^{-6}	3.636×10^{-6}

Recall the MH estimator provides an overall estimate of the association between disease and risk. If there is no interaction, the MH estimator provides an estimate with no need for stratification; however, if interaction is present, the MH estimator is a weighted average of the individual odds ratios. An alternative to this is to use a weighted average of the logs of the individual odds ratios, where the weights are the inverse of the variance of the log odds.

Consider the general situation of two strata depicted in Table 3.10. Then define the logs odds of the weighted estimator as

$$\log(\theta_{OR}) = \frac{(w_1 \log(\theta_{OR1}) + w_2 \log(\theta_{OR2}))}{(w_1 + w_2)} \tag{3.14}$$

where the weights are the inverse of the variance of the log odds, namely,

$$w_1 = \left(\frac{1}{n_{pp}} + \frac{1}{n_{pq}} + \frac{1}{n_{qp}} + \frac{1}{n_{qq}} \right)^{-1} \tag{3.15}$$

and

$$w_2 = \left(\frac{1}{m_{pp}} + \frac{1}{m_{pq}} + \frac{1}{m_{qp}} + \frac{1}{m_{qq}} \right)^{-1} \tag{3.16}$$

Therefore, the overall odds ratio for a stratified study is expressed as

$$\theta_{OR} = \exp(\log(OR)) \tag{3.17}$$

BUGS CODE 3.6 provides the posterior distribution of the individual odds ratio for each stratum, the estimator of the weighted log odds ratios, and finally the overall odds ratio, expressed as the anti-log of the log odds.

BUGS CODE 3.6

```
model;
{
gpp~dgamma(npp,2)
gpq~dgamma(npq,2)
gqp~dgamma(nqp,2)
gqq~dgamma(nqq,2)
hpp~dgamma(mpp,2)
hpq~dgamma(mpq,2)
hqp~dgamma(mqp,2)
hqq~dgamma(mqq,2)
sgh<- gpp+gpq+gqp+gqq+hpp+hpq+hqp+hqq
```

```
# Dirichlet distribution for the thetas and phis
thetapp<-gpp/sgh
thetapq<-gpq/sgh
thetaqp<-gqp/sgh
thetaqq<-gqq/sgh
phipp<-hpp/sgh
phipq<-hpq/sgh
phiqp<-hqp/sgh
phiqq<-hqq/sgh
# odds ratio first stratum
OR1<-(thetapp*thetaqq)/(thetapq*thetaqp)
# odds ratio second stratum
OR2<-(phipp*phiqq)/(phipq*phiqp)
V1<-(1/npp)+(1/npq)+(1/npq)+(1/nqq)
V2<-(1/mpp)+(1/mpq)+(1/mpq)+(1/mqq)
# weight for first stratum
w1<-1/V1
w2<-1/V2
# weighted average of log odds ratios
lodds<-(w1*log(OR1)+w2*log(OR2))/(w1+w2)
# overall odds ratio
OR<-exp(lodds)
}
# data from Table 3.11
list(npp = 9,npq = 115,nqp = 6,nqq = 73,mpp = 20,mpq = 596,
mqp = 21,mqq = 1171,
n1 = 203,n2 = 1808)
# initial values
list(gpp = 1,gpq = 1,gqp = 1,gqq = 1,hpp = 1,hpq = 1,hqp = 1
,hqq = 1)
```

Bayesian analysis for the overall odds ratio and the individual odds ratios are portrayed in Table 3.14. An amount of 55,000 observations are generated for the simulation, with a burn-in of 5000 and a refresh of 100. The list statement of the code contains the information from the Israeli Heart Disease study as reported in Table 3.11.

The results are interesting, because the standard deviation of the posterior distribution of the weighted odds ratio θ_{OR} is less than that of the other two. Also, the posterior distribution of the three odds ratios is slightly skewed,

TABLE 3.14

Weighted Estimator of the Odds Ratio for Israeli Heart Disease Study

Parameter	Mean	SD	Error	2½	Median	97½
θ_{OR}	1.627	.4685	.001941	.9094	1.565	2.706
θ_{OR1}	1.155	.7519	.003323	.3289	.9735	3.059
θ_{OR2}	1.969	.6465	.002758	.9992	1.872	3.5

and the posterior median of the weighted estimator is weighted toward the odds ratio of the second stratum. Note that the weighted estimator θ_{OR} has a posterior standard deviation quite similar to that of the MH estimator MH and a posterior median that is almost the same. The student should compare Table 3.13 with Table 3.14. Thus, it appears that there is very little difference between the two estimators of the measure of association between disease and risk (at least for the association between MI and SBP).

3.6 Matching to Adjust Data in Case–Control Studies

Consider a series of 2 by 2 tables, where each table corresponds to a case of a case–control study. For example, in the Shields Heart Study, each patient with an infarction could be matched with a patient without an infarction, with age as a confounder. The two matched patients would have the same age or be within, say, 1 year of each other in age. This approach can result in adjusting the effect of the confounder on the association between risk and disease. A layout for a matched study is given in Table 3.15.

Cell entries are the cell probabilities and cell frequencies. For example, for the cell where the case has the risk factor present and for the corresponding control the risk factor is absent, there are n_{+-} such pairs and the probability of each pair is θ_{+-}. To analyze such a situation, the study is stratified so that each case corresponds to a stratum. It can be shown that the MH estimator MH is given by

$$\theta_{MH} = \frac{\theta_{+-}}{\theta_{-+}} \tag{3.18}$$

Note that the two cells where the status of the risk factor is the same play no part in the calculation of MH (Equation 3.18). Consider a hypothetical scenario with 114 pairs of subjects, with 34 pairs where both case and control have been exposed to the risk factor (Table 3.16). On the contrary, there are 21 pairs where the case exposed to the risk factor, but the control is not. The confounder is age, where each case–control pair have an age that is within

TABLE 3.15

Matched Case–Control Study

Cases	Controls		
	RF present	RF absent	Total
RF present	θ_{++}, n_{++}	θ_{+-}, n_{+-}	$n_{+.}$
RF absent	θ_{-+}, n_{-+}	θ_{--}, n_{--}	$n_{-.}$
Total	$n_{.+}$	$n_{.-}$	$n_{..}$

TABLE 3.16

Hypothetical Example of a Matched Case–Control Study

Cases	Controls		Total
	RF present	RF absent	
RF present	$\theta_{++}, n_{++} = 34$	$\theta_{+-}, n_{+-} = 21$	$n_{+.} = 55$
RF absent	$\theta_{-+}, n_{-+} = 17$	$\theta_{--}, n_{--} = 42$	$n_{-.} = 59$
Total	$n_{.+} = 51$	$n_{.-} = 63$	$n_{..} = 114$

1 year of each other. The risk factor might be SBP and the disease might be coronary artery disease.

I assume the 114 pairs are selected at random and that the cell frequencies follow a multinomial distribution, and the cell probabilities follow a Dirichlet posterior distribution with parameter (34, 21, 17, 42).

This assumes an improper prior distribution for the cell probabilities. Even though MH depends explicitly on the two cell frequencies θ_{+-} and θ_{-+}, it should be recognized that MH depends on all four cell probabilities, because of the restriction $\theta_{++} + \theta_{+-} + \theta_{-+} + \theta_{--} = 1$.

BUGS CODE 3.7

```
# Analysis for matched case-control study
  model;
{
gpp~dgamma(npp,2)
gpm~dgamma(npm,2)
gmp~dgamma(nmp,2)
gmm~dgamma(nmm,2)
sgh<- gpp+gpm+gmp+gmm

# Dirichlet distribution for the thetas and phis
thetapp<-gpp/sgh
thetapm<-gpm/sgh
thetamp<-gmp/sgh
thetamm<-gmm/sgh

# Mantel-Haenszel for a matched case-control study

MH<- thetapm/thetamp
}
# data from Table 3.16
list(npp = 34,npm = 21,nmp = 17,nmm = 59)
# initial values
list(gpp = 1,gpm = 1,gmp = 1,gmm = 1)
```

MH posterior distribution is skewed to the right with a mean of 1.31 and a median of 1.236; thus, it is appropriate to use 1.236 as one's estimate of the association between risk and disease (Table 3.17). The student should observe

TABLE 3.17

Bayesian Analysis for a Matched Case–Control Example

Parameter	Mean	SD	Error	2½	Median	97½
θ_{MH}	1.31	.4496	.002063	.6522	1.236	2.385
θ_{mm}	.4504	.0432	<.0001	.3661	.4502	.5363
θ_{mp}	.1298	.0295	<.0001	.0783	.128	.1921
θ_{pm}	.1601	.0317	<.0001	.1029	.1584	.227
θ_{mm}	.2597	.0389	<.0001	.188	.2583	.3382

that that posterior means of the four cell probabilities add to one, and these calculations serve as a check on the code. An MCMC error of .002063 implies the estimate of 1.31 is within .0021 units of the "true" posterior mean.

3.7 Comments and Conclusions

Chapter 3 introduces the reader to a basic topic in epidemiology, namely, the adjustment for the effects of a confounder on the estimation of the association between risk and disease. The chapter began with a description of the Bayesian approach to direct adjustment of mortality for two states, California and Florida. The ages of each state and the U.S. population are partitioned into six categories, then, using the U.S. population as the standard, the adjusted mortality rate for both states is determined by a Bayesian approach. The direct standardization mortality rate for both states was compared by finding the posterior distribution of the ratio of two rates.

Another way to adjust for a difference in the age distribution of the two states is by indirect standardization. With this method, one only needs to know the total population size and the total mortality for the two states, but one does need age-specific mortality and population of the standard population. Recall that for direct standardization, one must know the age-specific mortality and population for the two states. A by-product of indirect standardization is the standard mortality rate for the states, which for a particular state is the ratio of the number of deaths of that state to the number expected for the state, assuming standard mortality rates for the state.

The chapter continues with a presentation of stratified studies, where the sample results are partitioned into two subsets: (1) the first stratum corresponds to subjects where the confounder is not present, and (2) the second stratum corresponds to patient results when the confounder is present. A good example of this is the Israeli Heart Disease Study, analyzed by Kahn and Sempos,[1] where age is a confounder and the association between SBP and occurrence of a heart attack is examined. The study should be stratified if the confounder is related to the disease and the association between risk and disease is different for the two levels of the confounder. That is, the interaction between the confounder and the risk factor can be examined by finding the posterior distribution of the interaction. Interaction is defined and determined for the Israeli Heart Disease Study.

A Bayesian approach for the measure of the association between disease and risk in the presence of confounder is to estimate the MH odds ratio. If interaction is present, the MH estimator is a weighted average of the separate odds ratios of each stratum and gives an estimate of the overall association between disease and risk. A different estimator of the overall association is also presented and is based on a weighted average of the log odds ratio for each stratum, where the weights are the inverse of the variance of the log odds.

For each concept for this chapter, an example is provided and analyzed with the Bayesian posterior distribution of the relevant parameters. Exercises at the end of the chapter present an opportunity for the reader to expand their understanding of the concepts introduced in this chapter. Data adjustment is the main theme of this chapter, which will be expanded in Chapter 4 on regression techniques.

Exercises

1. Refer to BUGS CODE 3.1 and Table 3.3, the Bayesian analysis that compare the standardized mortality rates between California and Florida. Now consider BUGS CODE 3.1 Alternative portrayed in the following and execute a Bayesian analysis using 55,000 observations, a burn-in of 5000, and a refresh of 100. Find the posterior distribution of r, s, DSR_1, and DSR_2. Refer to the first two statements of BUGS CODE 3.1 Alternative, where beta distributions are given to the θ_{ij}, and note that the first two statements of BUGS CODE 3.1 give a binomial distribution to the age-specific number of deaths.

   a. Show that the posterior characteristics (mean, median, standard deviation, MCMC error, and 2 ½ and 97 ½ percentiles) are exactly the same as those given in Table 3.3.

   b. Plot the posterior density of r and d.

BUGS CODE 3.1 Alternative

```
model;
{
for(i in 1:6){p1[i]<-m1[i]-y1[i]}
for(i in 1:6){p2[i]<-m2[i]-y2[i]}
# beta distributions for the thetas assuming improper prior
for(i in 1:6){theta1[i]~dbeta(y1[i],p1[i])}
for(i in 1:6){theta2[i]~dbeta(y2[i],p2[i])}
numds1<-theta1[1]*mr[1]+theta1[2]*mr[2]+theta1[3]*mr[3]+
```

```
theta1[4]*mr[4]+theta1[5]*mr[5]+theta1[6]*mr[6]
denoms1<-sum(mr[])
# directly standardized mortality for area 1
s1<-numds1/denoms1
numds2<-theta2[1]*mr[1]+theta2[2]*mr[2]+theta2[3]*mr[3]+
theta2[4]*mr[4]+theta2[5]*mr[5]+theta2[6]*mr[6]
denoms2<-sum(mr[])
# standardized mortality for area 2
s2<-numds2/denoms2
r<-s1/s2
d<-s1-s2
}
# deaths for area 2
list(y2 = c(1987,1940,3993,9823,12092,5225),
# deaths for area 1
y1 = c(3446,4137,10038,29511,31918,16332),
# population for area 1
m1 = c(2531333,10686658,10500587,9288864,3645546,600968),
# population for area 2
m2 = c(1073506,4668242,4720799,5079161,2825477,434125),
# population for reference standard
  mr = c(20201362,84652193,82134554,81489445,34774551,5493433))
# initial values for the thetas
list(theta1 = c(.5,.5,.5,.5,.5,.5),theta2 =
  c(.5,.5,.5,.5,.5,.5))
```

2. Consider the number of deaths and the population (by age category) for Maine given in Table 3.18 and refer to Tables 3.1 and 3.2 for the population and the number of deaths by age for California.

TABLE 3.18

Population and Mortality by Age

	Maine (Area 2)	
Age	Mortality	Population
<5	104	68,570
5–24	129	321,956
25–44	406	316,376
45–64	2,255	409,856
65–84	6,382	183,045
>85	4,235	27,797
Total	13,511	1,327,600

Source: U.S. Census Bureau 2010 Census Summary. http://factfinder2.census.gov/faces/tableservices/jsf/pages/productview.xhtml?fpt=table. and National Center of Health Statistics, http://www.cdc.gov/nchs/data/dvs/MortFinal2007_Worktable308.pdf.

Using BUGS CODE 3.1, compute the posterior distribution of the directly standardized mortality for California and for Maine using 55,000 observations for the simulation with a burn-in of 5000 and a refresh of 100. Use the U.S. population for the standard reference. Note that this assumes an improper prior for the unknown mortality probabilities θ_{ij} for i = 1, 2, ... , 6, and j = 1, 2!

a. What are the posterior means of DSR_1 (California) and DSR_2 (Maine)?

b. What is the posterior mean for the difference d in the two?

c. What is the 95% credible interval for the ratio r of the two?

d. Is there a difference in the two directly standardized mortalities?

e. Repeat the above with BUGS CODE 3.1 Alternative and show the Bayesian analysis is the same as with using BUGS CODE 3.1. Use 55,000 observations for the simulation, a burn-in of 5000, and a refresh?

f. Plot the posterior density of the ratio r.

3. The crude rates for California and Florida are 95,383/37,253,956 = .00175506 and 35,060/18,801,310 = .001864763, respectively. See Tables 3.1 and 3.2 for the information required to calculate the crude rates.

a. Compare the crude rate of California to its directly standardized mortality.

See Table 3.3. Is there a difference? If not, why?

b. Compare the crude morality of Florida with its directly standardized mortality. Is there a difference? If so, why?

4. Refer to Tables 3.3 and 3.4.

a. What is the standard population for the analysis of Table 3.3?

b. What is the standard population for the analysis of Table 3.4?

c. Why does one standardize the mortality rates of two states?

d. Why does one standardize using the inverse of the variance of the difference of the two mortality probabilities?

e. To compare the standardized rates for California and Florida, which population do you prefer as the standard reference?

5. Kahn and Sempos (p. 90)[1] illustrate the non-Bayesian approach to estimating the directly standardized mortality rates for California and Maine, using data from 1970, and the information they used is given in Table 3.19.

Thus, the population for the United States in 1970 was 203,212,000 and for Maine it was 992,000. Using the United States as the standard

TABLE 3.19

Population and Mortality for California and Maine

Age	United States		California		Maine	
	Pop./1000	Deaths	Pop./1000	Deaths	Pop./1000	Deaths
<15	57,900	103,062	5,524	8,751	286	535
15–24	35,411	45,261	3,558	4,747	168	192
25–34	24,907	39,193	2,677	4,036	110	152
35–44	23,088	72,617	2,359	6,701	109	313
45–54	23,220	169,517	2,330	15,675	110	759
55–64	18,590	308,373	1,704	26,276	94	1,622
65–74	12,436	445,531	1,105	36,259	69	2,690
>75	7,630	736,758	696	63,840	46	4,788
Total	203,212	1,920,312	19,953	166,285	992	11,051

Source: Bureau of the Census 1970 and Center for Health Statistics 1970.

population, perform a Bayesian analysis with BUGS CODE 3.1 with 65,000 observations, a burn-in of 5000, and a refresh of 100. Use the results of your analysis for parts a and b that follow.

a. What is the posterior distribution of the difference d between the directly standardized mortality rates of California and Maine?

b. What is the posterior distribution of the ratio r of the directly standardized mortality rates of California versus that for Maine?

c. Compare the results of your analysis in parts a and b with Table 3.3.

d. Do the two standardized rates appear to be different? Explain your answer.

e. Suppose one uses the variance of the posterior distribution of the age-specific difference in the mortality probabilities. See Equation 3.7 and execute the analysis with 55,000 observations for the simulation, a burn-in of 5000, and a refresh of 100.

f. Compare the results of your analysis of parts d and e with Table 3.3.

6. Verify the Bayesian analysis of Table 3.5 using BUGS CODE 3.3 with 55,000 observations, a burn-in of 5000, and a refresh of 100.

Why do the two ratios disr and dsmr have the same posterior characteristics? For example, the posterior means are identical.

7. Compare the results of Table 3.3 with those of Table 3.5. Table 3.3 reports the Bayesian analysis for the direct adjustment of the

mortality rates for California and Florida, while Table 3.5 reports the analysis for the indirectly standardized rates of the two states.

   a. What is the difference in the two posterior means for California?

   b. What is the difference in the two posterior means for Florida?

   c. Since they are different, should one be concerned?

8. Refer to the Table 3.19 of problem 6, which reports the age-specific deaths and population for the United States, California, and Florida. To estimate the directly standardized rates in problem 5, you executed the Bayesian analysis based on BUGS CODE 3.1. Perform a Bayesian analysis for estimating the indirect adjustment of mortality for 1970 California and Florida, using 55,000 observations for the simulation, a burn-in of 5000, and a refresh of 100. Assume the U.S. population is the standard.

   a. What is the posterior mean of the crude rate for the U.S. population?

   b. What is the posterior mean of the indirectly standardized rate for California?

   c. What is the posterior mean of the indirectly standardized rate for Florida?

   d. What is the mean of the posterior distribution of the ratio of the two indirectly standardized mortality rates?

   e. What is the 95% credible interval for the standard mortality rate for Florida?

   f. What is the posterior mean of the ratio of the indirect rate for California relative to that of Florida? Do you believe the two indirect rates are not the same? Why?

9. Consider the population size and mortality by age group for the United States, Alaska, and Texas in Tables 3.20 and 3.21.

TABLE 3.20

U.S., Alaska, and Texas Population by Age

Age	United States (Area r)		Alaska (Area 1)		Texas (Area 2)	
	Number	Percent	Number	Percent	Number	Percent
Total	308,745,538	100	710,231	100	25,145,561	100
<5	20,201,362	6.5430	53,996	7.6	1,928,473	7.66
5–24	84,652,193	33.9611	208,263	29.3	7,510,329	29.86
25–44	82,134,554	26.6026	196,099	27.61	7,071,855	28.12
45–64	81,489,445	26.3937	196,935	27.72	6,033,027	23.99
65–84	34,774,551	11.2631	50,227	7.1	2,296,707	9.13
>85	5,493,433	1.7792	4,711	.663	305,179	1.213

Source: U.S. Census Bureau 2010 Census Summary. http://factfinder2.census.gov/faces/tableservices/jsf/pages/productview.xhtml?fpt=table.[2]

TABLE 3.21

Number of Deaths for the United States, Alaska, and Texas by Age

Age	United States (Area r) Mortality	Alaska (Area 1) Mortality	Texas (Area 2) Mortality
<5	28,869	640	25,668
5–24	34,790	311	7,375
25–44	112,178	678	20,296
45–64	493,566	2,401	82,411
65–84	1,031,816	3,874	167,384
>85	764,582	554	39,036
Total	2,465,801	8,458	342,170

Source: Worktable 3 of National Center of Health Statistics http://www.cdc .gov/nchs/data/dvs/MortFinal2007_Worktable308.pdf

Using BUGS CODE 3.1 and the information in Tables 3.20 and 3.21, perform a Bayesian analysis that will estimate the directly standardized mortality rates for Alaska and Texas. Execute the analysis with 65,000 observations for the simulation, a burn-in of 5000, and a refresh of 100.

a. What is the posterior mean of the directly standardized mortality rate for Alaska?

b. What is the posterior mean of the directly standardized mortality rate for Texas?

c. What is the 95% credible interval for the ratio of the directly standardized mortality for Alaska to that of Texas?

d. Are you confident the two rates are the same? Why?

e. Based on the six age groups, does the age distribution for Alaska differ from that of the U.S. population?

10. Answer the following using the results of the Bayesian analysis performed in problem 9:

a. What is the posterior mean of the indirectly standardized mortality rate for Alaska?

b. What is the posterior mean of the indirectly standardized mortality rate for Texas?

c. What is the 95% credible interval for the ratio of the indirectly standardized mortality for Alaska to that of Texas?

d. Are you confident the two rates are the same? Why?

e. What is the posterior median of the standard mortality rate for Alaska?

f. What is the 95% credible interval for the standard mortality rate for Texas?

g. What is the crude mortality rate for Texas?

11. a. Compare the directly standardized mortality rate for Texas with the indirectly standardized mortality rate for Texas.

   b. Compare the directly standardized mortality rate for Alaska with the indirectly standardized mortality rate for Alaska.

   c. For Alaska, is the directly standardized rate the same as the indirectly standardized mortality? Why?

12. Show the validity of Table 3.12 by performing a Bayesian analysis based on BUGS CODE 3.5. Use 55,000 observations for the simulation with a burn-in of 5000 and a refresh of 100. Plot the posterior density of the interaction I.

13. Table 3.22 is information from the Shields Heart Study.

The confounder is age, where the first stratum corresponds to age <55 and the second to age ≥55 years. Blood pressure is the risk factor where high blood pressure indicates the risk factor is present, while low blood pressure indicates the risk factor is not present. The "disease" is the presence of coronary artery calcium; thus, θ_{qq} is the probability that a subject under age 55 and who does not have high blood pressure will have a zero reading for coronary artery calcium (as determined by computed tomography).

Based on BUGS CODE 3.4 and the information from the Shields Heart Study, execute a Bayesian analysis with 65,000 observations for the simulation, a burn-in of 5000, and a refresh of 100.

   a. Find the posterior mean of the two odds ratios for the two strata.

   b. Are the two odds ratios the same? Why?

   c. Plot the posterior density of the odds ratio for stratum 1 (age <55).

TABLE 3.22

Stratified Version of the Shields Heart Study

	Positive Coronary Calcium	Zero Coronary Calcium	
Stratum 2 (age <55)			
RF present High BP	$n_{pp} = 239, \theta_{pp}$	$n_{pq} = 212, \theta_{pq}$	$n_{p.} = 451$
RF not present Low BP	$n_{qp} = 631, \theta_{qp}$	$n_{qq} = 1005, \theta_{qq}$	$n_{q.} = 1636$
Total			2087
Stratum 2 (age ≥55)			
RF present High BP	$m_{pp} = 625, \phi_{pp}$	$m_{pq} = 154, \phi_{pq}$	$m_{p.} = 779$
RF not present Low BP	$m_{qp} = 1032, \phi_{qp}$	$m_{qq} = 446, \phi_{qq}$	$m_{q.} = 1478$
Total			2257

14. Based on BUGS CODE 3.5 and the information for the Shields Heart Study, execute a Bayesian analysis with 65,000 observations for the simulation, a burn-in of 5000 and a refresh of 100.

    a. What is the posterior mean for the interaction I between age and blood pressure?

    b. What is the 95% credible interval for the interaction?

    c. Is there interaction between age and blood pressure?

    d. Plot the posterior distribution of the interaction I.

    e. Find the posterior mean of θ_{pp} and show its posterior mean is the usual estimate of $234/4334 = .055145362$. This serves as a check on the Bayesian analysis.

15. Table 3.23 is information from the Shields Heart Study and portrays information about the association between diabetes and the presence (or lack of presence) of coronary artery calcium. This represents a stratified study where age is the possible confounding factor.

    The confounder is age, where the first stratum corresponds to age <55 and the second to age ≥55 years. Diabetes is the risk factor and indicates the risk factor is present, while no diabetes indicates the risk factor is not present. The "disease" is the presence of coronary artery calcium; thus, θ_{pp} is the probability that a subject under age 55 and who has diabetes will have a positive reading for coronary artery calcium (as determined by computed tomography).

TABLE 3.23

Stratified Study for the Association between Diabetes and Coronary Artery Calcium, Shields Heart Study

	Positive Coronary Calcium	Zero Coronary Calcium	
Stratum 1 (age <55)			
RF present Diabetes	$n_{pp} = 44, \theta_{pp}$	$n_{pq} = 34, \theta_{pq}$	$n_{p.} = 78$
RF not present No diabetes	$n_{qp} = 845, \theta_{qp}$	$n_{qq} = 1195, \theta_{qq}$	$n_{q.} = 2040$
Total			2118
Stratum 2 (age ≥55)			
RF present Diabetes	$m_{pp} = 115, \phi_{pp}$	$m_{pq} = 18, \phi_{pq}$	$m_{p.} = 133$
RF not present No diabetes	$m_{qp} = 1576, \phi_{qp}$	$m_{qq} = 562, \phi_{qq}$	$m_{q.} = 2138$
Total			2271

Based on BUGS CODE 3.4 and the information from the Shields Heart Study, execute a Bayesian analysis with 65,000 observations for the simulation, a burn-in of 5000, and a refresh of 100.

a. Find the posterior mean of the two odds ratios for the two strata.

b. Are the two odds ratios the same? Why?

c. Plot the posterior density of the odds ratio for stratum 1 (age <55).

d. Is diabetes a risk factor for coronary calcium?

Based on BUGS CODE 3.5 and the information for the Shields Heart Study, execute a Bayesian analysis with 65,000 observations for the simulation, a burn-in of 5000, and a refresh of 100:

a. What is the posterior mean for the interaction I between age and diabetes?

b. What is the 95% credible interval for the interaction?

c. Is there interaction between age and diabetes?

d. Plot the posterior distribution of the interaction I.

e. Find the posterior mean of θ_{pp} and show its posterior mean is the "usual" estimate of $44/4389 = .010025062$. This serves as a check on the Bayesian analysis.

16. Refer to problem 15 and the information from the Shields Heart Study about the association between diabetes and coronary artery calcium, where age is a confounder. Using BUGS CODE 3.5, execute a Bayesian analysis to estimate the overall association through the MH estimator MH (2.13). Generate 55,000 observations for the simulation, with a burn-in of 5000 and refresh of 100.

a. What are the posterior mean and median of MH?

b. Is the posterior distribution of MH skewed?

c. What is the 95% credible interval for MH?

d. Does the posterior distribution of MH imply an association between diabetes and coronary artery calcium?

e. Plot the posterior density of MH.

17. Refer to problem 15 about the Shields Heart Study. Table 3.23 gives the results of studying the association between coronary artery calcium and diabetes with age as a possible confounder. Suppose prior information from a previous related study is available and reported in Table 3.24.

Combine the prior information contained in Table 3.24 of problem 17 with the information from the table for the Shields Heart Study reported in Table 3.23 of problem 5, and using BUG CODE 3.5, execute a Bayesian analysis with 65,000 observations for the simulation, a burn-in of 5000, and a refresh of 100. Note the list statement of BUG

TABLE 3.24

Prior Information for Shields Heart Study

	Positive Coronary Calcium	Zero Coronary Calcium	
Stratum 1 (age <55)			
RF present Diabetes	$n_{pp} = 3, \theta_{pp}$	$n_{pq} = 3, \theta_{pq}$	$n_{p.} = 6$
RF not present No diabetes	$n_{qp} = 59, \theta_{qp}$	$n_{qq} = 72, \theta_{qq}$	$n_{q.} = 131$
Total			137
Stratum 2 (age ≥55)			
RF present Diabetes	$m_{pp} = 6, \phi_{pp}$	$m_{pq} = 3, \phi_{pq}$	$m_{p.} = 9$
RF not present No diabetes	$m_{qp} = 126, \phi_{qp}$	$m_{qq} = 39, \phi_{qq}$	$m_{q.} = 165$
Total			2271

CODE 3.5 for the data must be revised to accommodate the prior information from Table 3.24 of problem 17.

a. Is interaction present for this study?

b. Determine the posterior mean and median of the MH estimator MH.

c. Is the posterior distribution of MH skewed about the posterior mean?

18. Compare Table 3.13 with Table 3.15 and explain why the posterior mean of the MH estimate is almost the same as the posterior mean of the weighted odds ratio θ_{OR}.

19. Use BUGS CODE 3.6 and validate the posterior analysis reported in Table 3.14. I used 55,000 observations for the simulation, with a burn-in of 5000 and a refresh of 100.

a. What is the MCMC error for estimating the posterior mean of θ_{OR}?

b. Increase the number of observations by 55000 and report the MCMC error for estimating the posterior mean of θ_{OR}.

c. Compare the MCMC error for estimating the posterior mean of θ_{OR} with 55000 observations to that using 110,000 observations.

20. Use BUGS CODE 3.7 to validate Table 3.17, the Bayesian analysis that determines the posterior distribution of the MH estimator of the association between risk and disease, adjusted for the confounder.

21. Show that for the matched case–control study depicted by Table 3.15, the odds ratio is as follows:

$$\theta_{MH} = \frac{\theta_{+-}}{\theta_{-+}}$$

TABLE 3.25

Prior Information for a Matched Case–Control Study

Cases	Controls		
	RF present	RF absent	Total
RF present	$\theta_{++}, n_{++} = 10$	$\theta_{+-}, n_{+-} = 8$	$n_{+.} = 18$
RF absent	$\theta_{-+}, n_{-+} = 9$	$\theta_{--}, n_{--} = 17$	$n_{-.} = 26$
Total	$n_{.+} = 19$	$n_{.-} = 25$	$n_{..} = 44$

Treat each case as a stratum and refer to Kahn and Sempos (pp. 109–113).[1]

22. Refer to Table 3.16 and suppose a previous related study for a matched case–control study is reported as in Table 3.25.

   Using Table 3.25 as prior information for the study results of Table 3.16, execute a Bayesian analysis for estimating the MH parameter θ_{MH} for a matched case–control study. Based on BUGS CODE 3.7, generate 55,000 observations for the simulation, with a burn-in of 5000, and a refresh of 100.

   a. What is a 95% credible interval for θ_{MH}?

   b. Is the posterior distribution of θ_{MH} skewed? Why?

   c. Plot the posterior density of θ_{MH}.

   d. Compare the posterior density of θ_{MH} (using the prior information of Table 3.25 and the study results of Table 3.16) with Table 3.17.

References

1. Kahn, H.A. and Sempos, C.T. *Statistical Methods in Epidemiology*, Oxford University Press, 1989, New York.
2. U.S. Census Bureau 2010 Census Summary. Table QP-Q1. http://factfinder2.census.gov/faces/tableservices/jsf/pages/productview.xhtml?fpt=table
3. DeGroot, M.H. *Optimal Statistical Decisions*, McGraw-Hill Book Company, 1970, New York.
4. Mantel, N. and Haenszel, W.J. Statistical aspects of the analysis of data from retrospective studies of disease, *J. Natl. Cancer Inst.*, 22, 719, 1959.
5. Woolson, R.R. *Statistical Methods for the Analysis of Biomedical Data*, John Wiley & Sons Inc., 1987, New York.
6. Rothman, K.J. *Modern Epidemiology*, Little Brown & Company, 1986, Boston, MA.
7. Work Table 28R from The National Center of Health Statistics. http://www.cdc.gov/nchs/data/dvs/MortFinal2007_Worktable308.pdf.

4

Regression Methods for Adjustment

4.1 Introduction

Data adjustment by regression provides the epidemiologist a powerful method to study the association between various exposures to various risk factors and disease status. Recall Chapter 3 where introductory adjustment techniques are described and are confined to one confounder and one risk factor. When several risk factors and confounders are available, regression models are ideal and are a standard tool for the analysis of data. Regression analysis is an important topic in statistics, and many textbooks are accessible. For a good introduction see the study by Chatterjee and Price,[1] and for a Bayesian approach see the study by Ntzoufras.[2]

Several types of regression models are presented, including logistic regression models, simple and multiple linear regression models, categorical or ordinal regression models, and nonlinear models. The types of models differ with respect to the type of dependent variables. A regression model has a dependent variable and at least one independent variable. When the dependent variable is binary (two values), the association between disease and other variables can be modeled by a logistic regression. If the dependent variable is continuous or quantitative normal-theory simple linear and multiple regression models are applicable, but if the dependent variable is categorical (several values) multinomial regression models are appropriate. It is of interest to the epidemiologist that the dependent variable is an indicator of disease status or morbidity and the independent variables are measurements of various risk factors or confounders.

Logistic models are appropriate for the type of adjustment techniques (direct and indirect standardization, matched pair designs, etc.) encountered in Chapter 3. For example, for the Israeli Heart Disease Study, the dependent variable is disease status (myocardial infarction) and the independent variables are age and systolic blood pressure (SBP), where age is a possible confounder and SBP can be considered a risk factor.

In this case the independent variables are binary (take two values), but they could be regarded as continuous if one uses the actual age and blood pressure values. Using the logistic regression model, odds ratios are used for measuring the association between disease and risk factors and are computed with the logistic regression model, which allows one to reanalyze the examples given in Chapter 3. For a good introduction to logistic regression, and many examples, see the work of Hosmer and Lemeshow.[3]

Simple linear and multiple regression models are appropriate when the dependent variable (an indicator of disease status) is quantitative (often called continuous) and considered to have a normal distribution. A simple linear regression model has one independent variable, whereas a multiple linear regression model has more than one independent variable. For example, for a population of diabetics one might want to study the association between blood glucose values and age. One could use the simple linear regression model in which blood glucose values constitute the dependent variable and age the independent variable; but if another variable such as gender is included as an independent variable, one would have a multiple linear regression model with two independent variables in which age and gender are risk factors. The word linear in regression refers to the fact that the average value of the dependent variable is a linear function of the regression coefficients. Obviously, a model does not have to be linear, but can be nonlinear, a topic that is explained in more detail. For additional information about normal-theory regression models, see the book by Chatterjee and Price,[1] a book that is frequently referred throughout this chapter.

For categorical regression models, the dependent variable can have a small number of values, for example, the disease status for cancer can be labeled as local and confined to the primary tumor, metastasized to the lymph nodes, or metastasized beyond the lymph nodes, in which case the dependent variable assumes three values. Special models are used in this situation, which are also referred to as multinomial regression models. Independent variables can be either continuous or categorical. As mentioned earlier, the dependent variable is an indicator of disease status, whereas the independent variables are risk factors or confounders. For example, for cancer, the possible independent variables are age, sex, type of treatment, and so on, and one wants to study their effect on the stage of disease.

Additional references are works by Agresti,[4] who describes categorical regression models, and Congdon,[5] who presents a Bayesian approach to the general subject of regression analysis.

When the dependent variable is survival time, special regression methods using the Cox proportional hazards model are described in Chapter 6 on life tables.

4.2 Logistic Regression

4.2.1 Introduction

For n subjects, suppose for the ith subject the response y_i is binary, that is, $y_i = 0$ or 1, and y_i is distributed as a Bernoulli with parameter θ_i, that is, $\theta_i = P(y_i = 1)$ and $1 - \theta_i = P(y_i = 0)$. In addition, if there is one independent variable x that assumes the value x_i for subject i, then the logistic model is defined as follows:

$$\text{logit}(\theta_i) = \alpha + \beta x_i \tag{4.1}$$

where

$$\text{logit}(\theta_i) = \ln\left[\frac{\theta_i}{(1-\theta_i)}\right] \tag{4.2}$$

The name logistic comes from the fact that Equation 4.2 is equivalent to the following logistic transformation:

$$\theta_i = \frac{[\exp(\alpha + \beta x_i)]}{[1 + \exp(\alpha + \beta x_i)]} \tag{4.3}$$

Of course, an additional $p - 1$ independent variables can be added to Equation 4.1 in which case

$$\text{logit}(\theta_i) = \alpha + \beta_1 x_{1i} + \beta_2 x_{2i} + \dots + \beta_p x_{ip} \tag{4.4}$$

Unknown regression coefficients in the model are $\beta_1, \beta_2, \dots, \beta_p$, which are estimated from the data (x_i, y_i), and for the Bayesian approach they are estimated from their joint posterior distribution. The model is linear for the regression coefficients on the logit scale (Equation 4.4) but not for the probability parameter (Equation 4.3). Note also that the odds that $y_i = 1$ is the antilog of Equation 4.1 or

$$\left[\frac{\theta_i}{(1-\theta_i)}\right] = \exp(\alpha + \beta x_i) \tag{4.5}$$

If x is a categorical variable with $x_i = 0$ or 1, then the odds that $y_i = 1$ when $x_i = 1$ divided by the odds that $y_i = 1$ when $x_i = 0$ gives the odds ratio, $\exp(\beta)$.

Formally, the odds ratio is expressed as

$$\frac{\left[\dfrac{\theta_i}{(1-\theta_i)\,|\,x_i=1}\right]}{\left[\dfrac{\theta_i}{(1-\theta_i)\,|\,x_i=0}\right]} = \exp(\beta) \tag{4.6}$$

or the logarithm of the odds ratio is β. On the other hand, if x is continuous then $\exp(\beta)$ is the odds ratio for a unit increase in x, that is, when x increases from a value of x to a value of $x + 1$.

The same interpretation also applies for the multiple linear logistic model (Equation 4.4). Note that for a unit increase in x_{1i}, the odds that $y_i = 1$ given $x_{1i} = 1$ divided by the odds that $y_i = 1$ given $x_{1i} = 0$ is $\exp(\beta_1)$, assuming the other $p - 1$ variables are constant, for all possible values of the remaining independent variables. It is important to remember the restriction that $\exp(\beta_1)$ is the odds ratio, assuming the other variables are constant for all possible values of those other variables. Bayesian inferences about the odds are based on the posterior distribution of β_1, which induces the posterior distribution of the odds $\exp(\beta_1)$.

4.2.2 An Example of Heart Disease

Recall the example of the Israeli Heart Disease Study from Chapter 3, in which there is an association between heart attack and two independent variables, age and SBP. Age is categorized as age <60 years and age ≥60 years, whereas SBP is categorized as SBP ≥140 and SBP <140. The occurrence of heart attack is modeled as $y = 0$ for no attack and $y = 1$ for heart attack. In a similar way, let $x_1 = 1$ for age ≥60 years and $x_1 = 0$ otherwise, and $x_2 = 1$ for SBP ≥140 mmHg and $x_2 = 0$ otherwise. For the logistic regression, we have a model with two independent variables, and both are binary; thus, β_1 is the odds of a heart attack for patients over 60 years of age divided by the odds of a heart attack for patients under the age of 60 years, for all values of blood pressure.

Consider Table 4.1, which portrays the results of a hypothetical study similar to the Israeli Heart Disease Study. There are a total of 450 subjects of which 9 had a heart attack and 391 did not. Logistic regression will be used to assess the association between SBP and heart attack, adjusting for age. There are two levels of age, two of blood pressure, and two of disease status. The Bayesian analysis is based on BUGS CODE 4.1 and is executed with 55,000 observations for the simulation, with a burn-in of 5000 and a refresh of 100. The code closely follows the various formulas for logistic regression given by Equations 4.1 through 4.6. Note that the data is given by the first list statement in the code, where y is the column

TABLE 4.1

Myocardial Infarction versus SBP by Age

Hypothetical Study

	MI	No MI	
Age ≥60			Total
SBP ≥140	1	15	16
SBP <140	0	8	8
Total	2	23	24
Age <60			
SBP ≥140	2	80	82
SBP <140	3	118	121
Total	5	198	203

Note: MI = Myocardial infarction.

of values for the occurrence of a heart attack, x_1 is the column of age indicators with a 1 indicating age >60, and x_2 is the column indicating low and high $(x_2 = 1)$ SBPs.

A glance at the study results shows that for age <60, the number of subjects who had a heart attack and a blood pressure of at least 140 is 2. It appears that age does have an effect on the association between infarction and blood pressure; however, the numbers are quite small and caution is required. Therefore, a Bayesian posterior analysis is executed on the basis of BUGS CODE 4.1. It is noted that 65,000 observations are generated for the simulation, with a burn-in of 5000 and a refresh of 100.

BUGS CODE 4.1

```
model;
{
# Bernoulli distribution for the observations
# theta is the probability of a heart attack
for (i in 1:400){y[i]~dbern(theta[i])}
for (i in 1:400){z[i]~dbern(theta[i])}
# logistic regression of theta on age and systolic
  blood pressure
for (i in 1:400){logit(theta[i])<-alpha+beta[1]*x1[i]+beta[2]*
  x2[i]}
# prior distributions for the regression coefficients
# uninformative priors
alpha~ dnorm(0.0000,.00001)
beta[1]~dnorm(0.0000,.00001)
beta[2]~dnorm(0.0000,.00001)
```

```
# the odds ratio for age
ORage<-exp(beta[1])
ORsbp<-exp(beta[2])
}
# y is the occurrence of a heart attack
# x1 is age
# x2 is systolic blood pressure
list(y = c(1,1,
           1,
           0,0,0,0,0,0,0,0,0,0,0,0,0,0,0,0,0,0,0,0,0,0,0,0,
            0,0,0,0,
           0,0,0,0,0,0,0,0,0,0,0,0,0,0,0,0,0,0,0,0,
           1,1,1,1,
           1,1,1,1,1,
           0,0,0,0,0,0,0,0,0,0,0,0,0,0,0,0,0,0,0,0,0,0,0,0,0,0,
           0,0,0,0,0,0,0,0,0,0,0,0,0,0,0,0,0,0,0,0,0,0,0,0,0,0,
           0,0,0,0,0,0,0,0,0,0,0,0,0,0,0,0,0,0,0,0,0,0,0,0,0,0,
           0,0,0,0,0,0,0,0,0,0,0,0,0,0,0,0,0,0,0,0,0,0,0,0,0,0,
           0,0,0,0,0,0,0,0,0,0,0,0,0,0,0,0,0,0,0,0,0,0,0,0,0,0,
           0,0,0,0,0,0,0,0,0,0,0,0,0,0,0,0,0,0,0,0,0,0,0,0,0,0,

           0,0,0,0,0,0,0,0,0,0,0,0,0,0,0,0,0,0,0,0,0,0,0,0,0,0,
           0,0,0,0,0,0,0,0,0,0,0,0,0,0,0,0,0,0,0,0,0,0,0,0,0,0,
           0,0,0,0,0,0,0,0,0,0,0,0,0,0,0,0,0,0,0,0,0,0,0,0,0,0,
           0,0,0,0,0,0,0,0,0,0,0,0,0,0,0,0,0,0,0,0,0,0,0,0,0,0,
           0,0,0,0,0,0,0,0,0,0,0,0,0,0,0,0,0,0,0,0,0,0,0,0,0,0,
           0,0,0,0,0,0,0,0,0,0,0,0,0,0,0,0,0,0,0,0,0,0,0,0,0,0,
           0,0,0,0,0,0,0,0,0,0,0,0,0,0,0,0,0,0,0,0,0,0,0,0,0,0,
           0,0,0,0,0,0,0,0,0,0,0,0,0,0,0,0,0,0,0,0,0,0,0,0,0,0,
           0,0,0,0,0,0,0,0,0,0,0,0,0,0,0,0,0,0,0,0,0,0,0,0,0,0,
           0,0,0,0,0,0,0,0,0,0,0,0,0,0,0,0,0,0,0),

x1 = c(1,1,
           1,
           1,1,1,1,1,1,1,1,1,1,1,1,1,1,1,1,1,1,1,1,1,1,1,1,1,
            1,1,1,1,
           1,1,1,1,1,1,1,1,1,1,1,1,1,1,1,1,1,1,1,1,
           0,0,0,0,
           0,0,0,0,0,
           0,0,0,0,0,0,0,0,0,0,0,0,0,0,0,0,0,0,0,0,0,0,0,0,0,0,
           0,0,0,0,0,0,0,0,0,0,0,0,0,0,0,0,0,0,0,0,0,0,0,0,0,0,
           0,0,0,0,0,0,0,0,0,0,0,0,0,0,0,0,0,0,0,0,0,0,0,0,0,0,
           0,0,0,0,0,0,0,0,0,0,0,0,0,0,0,0,0,0,0,0,0,0,0,0,0,0,
           0,0,0,0,0,0,0,0,0,0,0,0,0,0,0,0,0,0,0,0,0,0,0,0,0,0,
           0,0,0,0,0,0,0,0,0,0,0,0,0,0,0,0,0,0,0,0,0,0,0,0,0,0,

           0,0,0,0,0,0,0,0,0,0,0,0,0,0,0,0,0,0,0,0,0,0,0,0,0,0,
           0,0,0,0,0,0,0,0,0,0,0,0,0,0,0,0,0,0,0,0,0,0,0,0,0,0,
           0,0,0,0,0,0,0,0,0,0,0,0,0,0,0,0,0,0,0,0,0,0,0,0,0,0,
           0,0,0,0,0,0,0,0,0,0,0,0,0,0,0,0,0,0,0,0,0,0,0,0,0,0,
```

```
            0,0,0,0,0,0,0,0,0,0,0,0,0,0,0,0,0,0,0,0,0,0,0,0,0,
            0,0,0,0,0,0,0,0,0,0,0,0,0,0,0,0,0,0,0,0,0,0,0,0,0,
            0,0,0,0,0,0,0,0,0,0,0,0,0,0,0,0,0,0,0,0,0,0,0,0,0,
            0,0,0,0,0,0,0,0,0,0,0,0,0,0,0,0,0,0,0,0,0,0,0,0,0,
            0,0,0,0,0,0,0,0,0,0,0,0,0,0,0,0,0,0,0,0,0,0,0,0,0,
            0,0,0,0,0,0,0,0,0,0,0,0,0,0,0,0,0,0,0)),

x2 = c(1,1,
            0,
            1,1,1,1,1,1,1,1,1,1,1,1,1,1,1,1,1,1,1,1,1,1,1,1,1,
              1,1,1,1,
            0,0,0,0,0,0,0,0,0,0,0,0,0,0,0,0,0,0,0,
            1,1,1,1,
            0,0,0,0,0,
            1,1,1,1,1,1,1,1,1,1,1,1,1,1,1,1,1,1,1,1,1,1,1,1,1,
            1,1,1,1,1,1,1,1,1,1,1,1,1,1,1,1,1,1,1,1,1,1,1,1,1,
            1,1,1,1,1,1,1,1,1,1,1,1,1,1,1,1,1,1,1,1,1,1,1,1,1,
            1,1,1,1,1,1,1,1,1,1,1,1,1,1,1,1,1,1,1,1,1,1,1,1,1,
            1,1,1,1,1,1,1,1,1,1,1,1,1,1,1,1,1,1,1,1,1,1,1,1,1,
            1,1,1,1,1,1,1,1,1,1,1,1,1,1,1,1,1,1,1,1,1,1,1,1,

            0,0,0,0,0,0,0,0,0,0,0,0,0,0,0,0,0,0,0,0,0,0,0,0,0,
            0,0,0,0,0,0,0,0,0,0,0,0,0,0,0,0,0,0,0,0,0,0,0,0,0,
            0,0,0,0,0,0,0,0,0,0,0,0,0,0,0,0,0,0,0,0,0,0,0,0,0,
            0,0,0,0,0,0,0,0,0,0,0,0,0,0,0,0,0,0,0,0,0,0,0,0,0,
            0,0,0,0,0,0,0,0,0,0,0,0,0,0,0,0,0,0,0,0,0,0,0,0,0,
            0,0,0,0,0,0,0,0,0,0,0,0,0,0,0,0,0,0,0,0,0,0,0,0,0,
            0,0,0,0,0,0,0,0,0,0,0,0,0,0,0,0,0,0,0,0,0,0,0,0,0,
            0,0,0,0,0,0,0,0,0,0,0,0,0,0,0,0,0,0,0,0,0,0,0,0,0,
            0,0,0,0,0,0,0,0,0,0,0,0,0,0,0,0,0,0,0,0,0,0,0,0,0,
            0,0,0,0,0,0,0,0,0,0,0,0,0,0,0,0,0)))

# below is the data with the actual values of age and systolic
blood pressure
list( y = c(1,1,
            1,
            0,0,0,0,0,0,0,0,0,0,0,0,0,0,0,0,0,0,0,0,0,0,0,0,0,
              0,0,0,0,
            0,0,0,0,0,0,0,0,0,0,0,0,0,0,0,0,0,0,0,
            1,1,1,1,
            1,1,1,1,1,
            0,0,0,0,0,0,0,0,0,0,0,0,0,0,0,0,0,0,0,0,0,0,0,0,0,
            0,0,0,0,0,0,0,0,0,0,0,0,0,0,0,0,0,0,0,0,0,0,0,0,0,
            0,0,0,0,0,0,0,0,0,0,0,0,0,0,0,0,0,0,0,0,0,0,0,0,0,
            0,0,0,0,0,0,0,0,0,0,0,0,0,0,0,0,0,0,0,0,0,0,0,0,0,
            0,0,0,0,0,0,0,0,0,0,0,0,0,0,0,0,0,0,0,0,0,0,0,0,0,
            0,0,0,0,0,0,0,0,0,0,0,0,0,0,0,0,0,0,0,0,0,0,0,0,0,
            0,0,0,0,0,0,0,0,0,0,0,0,0,0,0,0,0,0,0,0,0,0,0,0,

            0,0,0,0,0,0,0,0,0,0,0,0,0,0,0,0,0,0,0,0,0,0,0,0,0,
            0,0,0,0,0,0,0,0,0,0,0,0,0,0,0,0,0,0,0,0,0,0,0,0,0,
```

```
0,0,0,0,0,0,0,0,0,0,0,0,0,0,0,0,0,0,0,0,0,0,0,0,0,0,0,
0,0,0,0,0,0,0,0,0,0,0,0,0,0,0,0,0,0,0,0,0,0,0,0,0,0,0,
0,0,0,0,0,0,0,0,0,0,0,0,0,0,0,0,0,0,0,0,0,0,0,0,0,0,0,
0,0,0,0,0,0,0,0,0,0,0,0,0,0,0,0,0,0,0,0,0,0,0,0,0,0,0,
0,0,0,0,0,0,0,0,0,0,0,0,0,0,0,0,0,0,0,0,0,0,0,0,0,0,0,
0,0,0,0,0,0,0,0,0,0,0,0,0,0,0,0,0,0,0,0,0,0,0,0,0,0,0,
0,0,0,0,0,0,0,0,0,0,0,0,0,0,0,0,0,0,0,0,0,0,0,0,0,0,0,
0,0,0,0,0,0,0,0,0,0,0,0,0,0,0,0,0,0,0,0,0,0,0,0,0,0,0,
0,0,0,0,0,0,0,0,0,0,0,0,0,0,0,0,0,0),

x1 = c(77,85,87,74,56,76,69,71,74,83,67,69,52,77,72,76,77,76,
70,67,70,74,72,74,57,88,79,64,67,67,69,77,68,86,58,77,79,86,
70,80,60,72,63,92,56,79,76,65,70,76,55,56,48,54,50,49,45,36,
54,47,52,47,47,49,46,39,50,51,39,59,46,44,52,53,49,53,52,46,
49,46,41,58,61,50,45,58,51,58,44,50,57,60,48,51,56,47,44,53,
41,39,50,36,55,47,59,43,51,43,40,43,39,38,62,49,46,45,42,50,
53,43,47,31,40,43,45,51,51,55,56,54,49,46,53,49,34,43,60,44,
56,50,46,44,49,46,53,45,46,56,43,53,45,49,49,41,46,44,43,48,
39,59,55,46,42,38,40,42,29,41,38,52,51,54,50,48,47,47,47,55,
44,50,55,50,43,54,42,55,42,38,52,49,50,56,50,39,51,49,51,54,
47,42,59,54,52,47,52,44,44,45,50,49,48,35,43,45,49,46,47,47,
59,48,52,44,48,50,58,47,49,43,56,50,42,48,41,52,41,43,47,43,
47,54,52,50,62,47,56,42,46,53,43,43,50,56,54,46,51,49,47,42,
42,54,53,45,35,48,52,45,47,41,48,40,42,50,41,56,56,49,53,34,
46,50,54,42,48,51,51,56,49,56,44,47,51,51,36,46,58,58,51,40,
53,53,44,44,51,47,48,61,40,50,49,39,47,36,45,47,57,44,44,51,
40,44,51,60,48,51,46,51,47,48,57,48,39,49,40,58,50,45,41,47,
49,41,64,36,47,49,58,59,45,45,53,37,43,43,50,44,49,43,53,48,
54,53,44,47,55,52,40,43,47,47,54,46,51,65,54,53,56,53,49,47,
38,59,55,44,52,41,55,43,47,46,55,38,45,42,54,52,42,56,48,44,
40,47,37,44,51,37,50,47,46,45,46,44,44,52,45,38,48,54,46,41,
41,43,44,44,49,48,41,57,45,46,46,35,50,43,41,52,43,54,46,40,
50,61,49,44,50,40,50,38,52,46,49,53),

x2 = c(158,164,150,157,161,165,166,147,169,159,150,166,161,161,
167,164,167,163,166,159,162,166,145,163,161,162,156,160,178,164,
154,115,114,128,117,123,122,114,116,114,119,112,123,119,122,130,
119,128,120,115,163,166,162,154,167,163,155,160,157,154,149,159,
167,162,159,160,162,157,158,168,168,164,157,171,150,158,150,163,
163,159,171,165,148,156,164,168,158,167,153,167,149,158,159,144,
167,151,156,150,154,164,158,147,151,167,151,151,160,161,160,160,
156,168,162,171,155,150,159,156,147,162,163,170,158,163,168,153,
151,150,156,162,164,164,159,152,164,159,158,163,162,152,162,157,
158,162,149,166,142,151,160,154,168,165,161,152,151,159,155,169,
169,168,166,163,158,150,151,150,152,167,155,150,157,152,158,168,
158,152,162,165,154,168,173,158,164,157,155,161,160,160,159,155,
162,164,160,158,164,157,168,159,169,159,155,161,150,119,114,120,
114,128,126,116,119,109,121,114,121,120,122,112,113,116,126,117,
109,120,111,123,113,119,110,112,120,127,116,124,111,134,137,113,
```

```
124,127,125,123,120,119,121,115,119,110,123,123,124,126,128,119,
122,118,115,121,114,125,117,125,124,133,115,121,120,127,120,118,
136,125,119,120,111,119,121,111,125,130,117,131,127,109,124,114,
120,112,130,127,118,119,121,119,117,117,117,123,122,121,119,125,
113,115,131,126,116,126,120,119,125,113,124,127,116,119,128,119,
129,125,122,121,113,118,113,130,119,117,125,130,120,115,118,131,
121,117,130,103,136,127,118,122,125,121,123,117,110,126,118,125,
123,128,129,116,127,109,118,127,134,123,124,115,113,126,120,116,
122,112,109,121,110,114,133,131,123,122,122,116,120,104,120,125,
117,121,125,121,121,116,115,121,126,121,113,126,128,112,112,126,
120,122,113,126,119,131,118,112,126,131,124,116,122,120,115,104,
115,127,122,123,134,120,124,104,113,110,118,117,118,128,121,120,
116,121,116,121,116,128,115,123,118,114,123,121,116,111,118,119,
127,129,114,122))
```

```
# Initial values
list(alpha = 0,beta = c(0,0))
```

The analysis of the association between myocardial infarction and SBP adjusted for age is reported in Table 4.2.

The interpretation of the parameters is as follows: for the odds ratio of age with a posterior median of 2.207, the implication is that the odds of a heart attack for a person older than 60 years is 2.207 times the odds of a heart attack for a person younger than 60. For the odds ratio of SBP, the median is 1.069, implying that there is very little difference in the odds of a heart attack for a person with a blood pressure greater than 140 mmHg and the odds of a heart attack for a person with a blood pressure less than 140 mmHg. The 95% credible intervals for the two odds ratios include the value 1, implying that age and SBP have very little effect on the occurrence of a heart attack. It is of interest to note that the Monte Carlo Markov chain (MCMC) errors are not as small here as in earlier examples. For example, an error of .02174 indicates that the posterior mean of 2.776 is within .02174 units of the "true" posterior

TABLE 4.2

Bayesian Analysis of Logistic Regression for a Heart Study

Binary Values for All Variables						
Parameter	Mean	SD	Error	2½	Median	97½
Odds ratio age	2.776	2.11	.02174	.4575	2.207	8.318
Odds ratio SBP	1.292	.8591	.01157	.316	1.069	3.58
α	−3.765	.4535	.005423	−4.729	−3.731	−2.963
β_1	.7673	.7367	.008129	−.7819	.7917	2.118
β_2	.06499	.6179	.006504	−1.152	.06713	1.275

Note: SD = Standard deviation.

mean. The posterior distribution of the two odds ratios is highly skewed; therefore, I use the posterior median (see Figure 4.1).

It is interesting to analyze the heart study data from Table 4.1 using the actual values for age and SBP; see the second list statement of BUGS CODE 4.1, which contains the actual values. I performed a Bayesian analysis with the actual data, using 55,000 observations for the simulation, with a burn-in of 5000 and a refresh of 100, and the analysis is reported in Table 4.3.

Table 4.2 should be compared with Table 4.3, because the differences are quite interesting. Recall from Table 4.2 that when age and blood pressure are coded with binary values, there is a small hint that age and SBP can impact the probability of a heart attack; but from Table 4.3 when the actual values of age and blood pressure are used, one sees that there is strong evidence that age and blood pressure have absolutely no effect on the probability of a heart attack. Of course, the analysis with the actual values of age and blood pressure should be preferred, because coding age and SBP with binary values involves a loss of information. For example, all age values greater than 60 are treated the same, and so on.

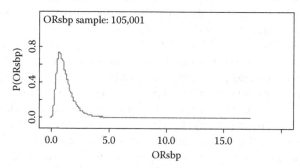

FIGURE 4.1
Skewed posterior density of the odds ratio: in the figure, ORsbp refers to odds ratio systolic blood pressure.

TABLE 4.3

Bayesian Analysis for the Heart Study with Actual Age and SBP Values

Parameter	Mean	SD	Error	2½	Median	97½
Odds ratio age	1.041	.02495	.0004006	.9908	1.041	1.089
Odds ratio blood pressure	1.087	.03363	.0009797	1.034	1.082	1.167
α	−18.29	5.089	.1478	−30.42	−17.65	−10.2
β_1	.03692	.02402	.0003863	−.009285	.04021	.08518
β_2	.08256	.03054	.000466	.03362	.0786	.1546

4.2.3 An Example with Several Independent Variables

An example exhibiting four independent variables is provided by Hosmer and Lemeshow (p. 48),[3] in which the disease is coronary heart disease and the four variables correspond to four races. The data appear in the first list statement of BUGS CODE 4.2, where the y column vector (5,20,15,10) gives the number of white, black, Hispanic, and other subjects, respectively, with the disease.

In addition, the vector n is (25,30,25,20) and consists of the total number of white, black, Hispanic, and other subjects, respectively, in the study.

Thus, among 25 whites 5 have coronary heart disease and among 20 of another race (not white, black, or Hispanic) 10 have heart disease; thus, the odds of having heart disease among whites relative to that among other races is estimated as (5/20)/(10/10) = .25. This information is analyzed with the logistic model

$$\text{logit}(\theta_i) = \alpha + \beta_1 x_{1i} + \beta_2 x_{2i} + \beta_3 x_i \tag{4.7}$$

where θ_i is the probability of disease for the ith race, $i = 1, 2, 3$. The four racial groups are coded as indicator variables $x_1 = (1,0,0,0)$, $x_2 = (0,1,0,0)$, and $x_3 = (0,0,1,0)$; thus, the group corresponding to other races is coded as $x_{14} = x_{24} = x_{34} = 0$. This coding makes group 4 (the one with all other races) the reference group. To perform a Bayesian analysis for this study, consider the following. Note that this information appears in the first list statement of BUGS CODE 4.2.

BUGS CODE 4.2

```
model;
{
# binomial distribution for the number of respondents
for(i in 1:4){y[i]~dbin(theta[i],n[i])}
# logistic regression of theta
for(i in 1:4){logit(theta[i])<-alpha+beta[1]*x1[i]+beta[2]*x2
[i]+beta[3]*x3[i]}
# prior distributions
alpha~dnorm(.0000,.0001)
for(i in 1:3){beta[i]~dnorm(.0000,.00001)}
# odds ratio of first group versus fourth
OR1<-exp(beta[1])
OR2<-exp(beta[2])
OR3<-exp(beta[3])
for( in 1:4){z[i]~dbin(theta[i],n[i])}
}
# data from Hosmer and Lemeshow page 48
# below is the number with CHD
list(y = c(5,20,15,10),
# n is the total number in each group
```

```
n = c(25,30,25,20),
# indictor for white
x1 = c(1,0,0,0),
# indicator for black
x2 = c(0,1,0,0),
# indicator for Hispanic
x3 = c(0,0,1,0))
# data below Table 4.1
list(y = c(2,1,4,5),
# n is the number in each group
n = c(31,19,153,247),
x1 = c(1,0,0,0),
x2 = c(0,1,0,0),
x3 = c(0,0,1,0))

# initial values
list(alpha = 0, beta = c(0,0,0))
```

Based on BUGS CODE 4.2, a posterior analysis is conducted with 55,000 observations generated for the simulation, with a burn-in of 5000 and a refresh of 100, and the calculations reported in Table 4.4.

This is an interesting analysis because the posterior distributions of the odds ratio are skewed. For example, the posterior mean of the odds ratio for group 1 (whites) is .2932, but the posterior median is .2346. Consider the "usual" estimate of (5/20)/(10/10) = .25 for the odds ratio for coronary artery disease of whites relative to the group consisting of other races: neither the posterior mean or median is equal to the usual estimate. I would use .2346 as the estimate, because of the skewness of the posterior distribution.

TABLE 4.4

Posterior Analysis for the Association between Race and Coronary Artery Disease

Parameter	Mean	SD	Error	2½	Median	97½
OR1	.2932	.2239	.001455	.05672	.2346	.8732
OR2	2.477	1.694	.01225	.6358	2.084	6.851
OR3	1.85	1.287	.009219	.4571	1.523	5.158
alpha	$<10^{-20}$	.458	.00399	−.9015	$<10^{-20}$	.9008
beta[1]	−1.464	.6965	.004732	−2.87	−1.45	−.1352
beta[2]	.7211	.6057	.004637	−.4529	.7171	1.924
beta[3]	.4225	.6184	.00469	−.783	.421	1.641
theta[1]	.2001	.07883	<.0001	.07162	.1916	.3752
theta[2]	.6671	.08462	<.0001	.4924	.6707	.8214
theta[3]	.5999	.09587	<.0001	.4065	.6025	.7775
theta[4]	.5	.1089	<.0001	.2887	.4998	.7111

Note: OR = Odds ratio.

4.2.4 Goodness of Fit

Goodness of fit of regression models is a vast topic and can read by referring to Ntzoufras[2] and Hosmer and Lemeshow[3]; thus, what is presented here on the topic is based on comparing the actual number with disease to the number predicted by the model. Recall that for the logistic model the number y_i is the number of diseased subjects for the *i*th group (or individual) and has a binomial distribution with parameters θ_i and n_i, and the Bayesian analysis involves determining the posterior distribution of θ_i.

The Bayesian predictive distribution is the conditional distribution of a future observation given the data y_i, and it will be used to generate future observations z_i corresponding to y_i. WinBUGS easily generates the future observations.

Consider the example of the effect of race on the probability of coronary heart disease, where BUGS CODE 4.2 is used for the Bayesian analysis of estimating the odds of the various racial groups relative to some reference group. Table 4.5 reports the Bayesian analysis with the posterior distribution of the odds ratios and other relevant parameters. The last statement

```
for(i in 1:4){z[i]~dbin(theta[i],n[i])}
```

before the first list statement of the code produces the future number of diseased subjects for each group. Returning to the Hosmer and Lemeshow example of four racial groups, the posterior distribution of the four predicted distributions is shown in Table 4.5.

The posterior means of future observations are almost equal to the exact number of people with disease; thus, the logistic model appears to give an excellent fit. Some uncertainty exists for the four predicted observations. For example, for the first observation the posterior standard deviation is 2.768, and this is demonstrated in Figure 4.2. A Bayesian analysis is performed via BUGS CODE 4.2 with 106,000 observations for the simulation, with a burn-in of 5000, and a refresh of 100.

In summary, to test how well the logistic model fits the data, generate a future observation for each group and compare it with the actual number with disease.

TABLE 4.5

Predicted Number with Coronary Artery Disease

Parameter	Mean	SD	Error	2½	Median	97½
z[1]	5.007	2.768	.008715	1	5	11
z[2]	20.0	3.596	.010081	13	20	26
z[3]	15.01	3.393	.01056	8	15	21
z[4]	9.99	3.093	.01906	4	10	16

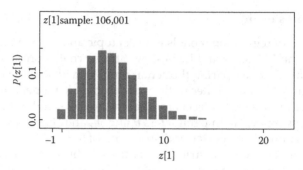

FIGURE 4.2
Predictive density of the number of Caucasians with coronary heart disease.

4.3 Linear Regression Models

4.3.1 Introduction

Simple and multiple linear regression models are used in epidemiology to determine associations between a dependent variable and several independent variables, and the subject is vast. If one refers to the latest issue of the *American Journal of Epidemiology*, one will mostly likely find a regression analysis that is employed to find some type of association between disease and exposure to various risk factors and confounders. The models considered in this chapter differ from the logistic model in that the dependent variable is quantitative. Quantitative variables assume values like what one would encounter in measuring blood glucose values, where in principle between any two values one can find another blood glucose value; thus, measurements such as age, weight, and SBP are examples of a quantitative variable.

The presentation of regression analysis is initiated with the definition of a simple linear regression model, which has one dependent variable and one independent variable, with the goal of establishing an association between the two. For example, the dependent variable might be SBP and the independent variable indicating two groups, where the subjects are with and without coronary artery disease. Simple linear regression has very strict assumptions, such as the dependent variable must be normally distributed and the variance of the dependent variable must be constant over all values of the independent variable. These assumptions will be relaxed to some extent by allowing for unequal variances, where a weighted regression is appropriate. Of course, the approach is Bayesian, where the posterior distribution of regression coefficients (intercept and slope) and the variance about the regression line are determined. Several examples relevant to epidemiologic studies are presented. One problem to be explained is that of interpreting the estimated value of the regression coefficient.

Simple linear regression models are generalized to multiple linear regression models, where the goal is to establish an association between one quantitative dependent variable and several (more than one) independent variables. For example, the dependent variables might be blood glucose values and the dependent variables might be age, weight, gender, and subjects with and without diabetes. The goal is to estimate the effect of age, weight, gender, and diabetes on the average blood glucose value. One challenge with such a regression analysis is the interpretation of the regression coefficients of the model, and this is carefully explained with many examples relevant to epidemiology.

One uses the term linear regression with the emphasis on linear. This means that the average value of the dependent variable is linear in the unknown regression coefficients, which are to be estimated from their posterior distribution. In practice, the dependent and independent variables are often transformed to achieve the linear assumption and the assumption of constant variance.

Often, one must use nonlinear regression models to assess the association between disease and various risk factors, and this is presented and explained in Chapter 8.

4.3.2 Simple Linear Regression

The definition of simple linear regression is as follows:

$$y_i = \alpha + \beta x_i + e_i \qquad (4.8)$$

where the values of the dependent variable y and the independent variable x are paired as (x_i, y_i) for the ith individual, for $i = 1, 2, \ldots, n$. If one plots the n pairs of observations, one would expect a linear association to develop; however, the relationship would not appear exactly linear because of the error term e with n values e_i, which are assumed to be independent and normally distributed with mean zero and unknown variance σ^2. This implies that the average value of the dependent variable y is as follows:

$$\text{Avg}(y_i) = \alpha + \beta x_i \qquad (4.9)$$

Thus, the average value of y is linear in x, namely, $\alpha + \beta x_i$ for $i = 1, 2, \ldots, n$.

With simple linear regression, there are three unknown parameters: the intercept term α; the slope β; and the variance of the error term σ^2, which is also the variance of the dependent variable y.

Consider Table 4.6, which gives 15 pairs of observations for the independent variable y and the dependent variable x. This is the first example of simple linear regression, and the values are hypothetical.

TABLE 4.6

(x,y) Pairs for Simple Linear
Regression

x	y
1	3.2
1	2.9
1	3.0
1	3.3
1	2.8
2	5.3
2	4.7
2	4.9
2	5.2
2	5.1
3	7.4
3	6.9
3	7.2
3	6.8
3	7.0

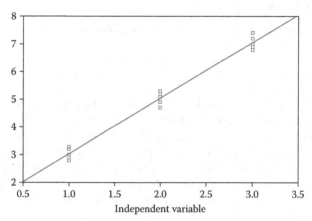

FIGURE 4.3
Simple linear regression.

These pairs are plotted in Figure 4.3, and the red line is the fitted simple linear regression line. This fitted line is the estimated average value of the dependent variable plotted against the three values 1, 2, 3 of the independent variable. The line is fitted by a procedure called least squares, and a comparable Bayesian technique is employed and explained in the following discussion. This is an example of replication, where the independent variable occurs with the same value five times, at $x = 1, 2,$ and 3. There are five corresponding y values for $x = 1, 2,$ and 3, and one can see the variability of the y values for each x value, and the variance σ^2 is assumed the same for $x = 1, 2,$ and 3. The term σ^2 is called the variance about the regression line, and its posterior distribution is determined.

To estimate the three unknown parameters, BUGS CODE 4.3 is executed with 55,000 observations for the simulation, with a burn-in of 5000 and a refresh of 100.

BUGS CODE 4.3

```
model;
{
for(i in 1:n){y[i]~dnorm(mu[i], tau)}
for(i in 1:n){z[i]~dnorm(mu[i], tau)}
for(i in 1:n){mu[i]<- alpha+beta*x[i]}
# alpha is intercept
# uninformative priors
alpha~dnorm(0,.0001)
# beta is slope
beta~dnorm(0,.0001)
tau~dgamma(.00001,.00001)
# sigsq is variance about regression line
sigsq<-1/tau
}
# hypothetical example Table 4.6
list(n = 15,x = c(1,1,1,1,1,2,2,2,2,2,3,3,3,3,3),
y = c(3.2,2.9,3.0,3.3,2.8,5.3,4.7,4.9,5.2,5.1,7.4,6.9,
7.2,6.8,7.0))
# data from Table 4.8
list(n = 30,
x = c(14,12,16,13,11,21,17,15,17,20,21,17,23,12,16,15,19,24,
24,12,23,23,20,13,18,18,19,24,25,22),
y = c(97,100,100,106,119,111,106,117,111,126,115,118,124,126,
116,131,125,125,128,130,138,131,129,127,131,126,142,165,
157, 163))

# initial values
list(alpha = 0,beta = 0,tau = 1)
```

The code is self-explanatory, where alpha is the intercept; beta is the slope; sigsq is the variance about the regression line; and tau is the precision about the regression line, that is, $\sigma^2 = 1/\tau$. Note the first list statement is the data from Table 4.7, and the first and second statements of the code make up the linear regression of y on x. See Table 4.7 for the Bayesian analysis of the data. I used uninformative prior distributions for the slope and intercept with a normal distribution with mean 0.0 and variance

TABLE 4.7

Posterior Analysis for Linear Regression

Parameter	Mean	SD	Error	2½	Median	97½
α	1.029	.1661	.002606	.7002	1.029	1.36
β	2.009	.07002	.001206	1.855	2.008	2.16
σ^2	.05801	.02769	<.0001	.02573	.05143	.1282
τ	20.42	8.02	.04845	7.8	19.44	38.87

1/.0001 = 10,000. As for the precision parameter tau, I used a gamma with mean 1 and variance 100,000.

Thus, the intercept is estimated as 1.029 with the posterior mean and median, and the slope as 2.009 with the posterior mean and median, and lastly the variance about the regression line is estimated as .05801 with the posterior mean. The interpretation of the slope is that for each unit increase in the independent variable, the average value of the dependent variable increases by 2.009 units.

Refer to Figure 4.3; the estimated intercept and slope appear to be very reasonable estimates. Of course, the hypothetical values are designed to give a slope of 2 and an intercept of 1 and one can conclude that the Bayesian analysis provides correct estimates of the unknown parameters. Simulation errors are small where the MCMC error for the intercept implies that the estimate of 1.029 is within .002606 of the true posterior mean. Also note that the estimated value of .05143 for the variance about the regression line is the estimate of the "scatter" of the y values when $x = 1, 2,$ or 3.

4.3.3 Another Example of Simple Linear Regression

Consider another example of simple linear regression that examines the effect of age x on SBP y; the data are given in Table 4.8. This example is taken from the study by Woolson (p. 298)[6] and is a subset of a larger study investigating the effect of weight on SBP adjusted for age.

The dependent variable is the SBP y; the independent variable is age x; and a plot of blood pressure versus age appears in Figure 4.4, which includes a plot of the regression line of SBP on age. It appears that there is a linear relationship between age and SBP, but it also seems as if there is a lot of variation in the y variable for each value of x. BUGS CODE 4.4 is employed to perform Bayesian regression analysis, where the slope, intercept, and variance about the regression line are calculated. The second list statement contains the information from Table 4.8, and I would expect the goodness of fit to be not as good as that of the previous example. Using 55,000 observations for the simulation with a burn-in of 5000 and a refresh of 100, the Bayesian analysis is reported in Table 4.9.

TABLE 4.8

Age and SBP for 30 Patients

Age	SBP	Weight	Subject
14	97	94	1
12	100	95	2
16	100	104	3
13	106	107	4
11	119	108	5
21	111	116	6
17	106	117	7
15	117	117	8
17	111	122	9
20	126	124	10
21	115	125	11
17	118	125	12
23	124	125	13
12	126	128	14
16	116	128	15
15	131	129	16
19	125	134	17
24	125	134	18
24	128	136	19
12	130	137	20
23	138	137	21
23	131	138	22
20	129	139	23
13	127	141	24
18	131	141	25
18	126	143	26
19	142	150	27
24	165	171	28
25	157	172	29
22	163	172	30

Source: Woolson, R.F., *Statistical Methods for the Analysis of Biomedical Data*, John Wiley & Sons Inc., New York, 1987.

Alpha, the intercept, is estimated as 84.94 with the posterior mean and a 95% credible interval (60.98, 108.7), and the slope is estimated as 2.19 with (.9094, 3.478) as the 95% credible interval for that parameter. Thus, for every year increase in age the average SBP increases by 2.19 mmHg, but the average increase in SBP can vary from .9094 mmHg to 3.478 mmHg with 95% confidence. Of course, the surprise is the estimate of 207.7 with

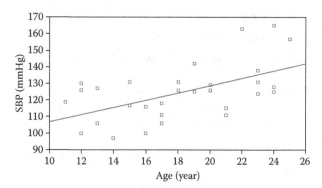

FIGURE 4.4
Simple linear regression: systolic blood pressure (SBP) versus age.

TABLE 4.9

Bayesian Analysis for Association between Blood Pressure and Age

Parameter	Mean	SD	Error	2½	Median	97½
α	84.94	12.09	.2181	60.98	84.02	108.7
β	2.19	.6497	.01172	.9094	2.191	3.478
σ^2	218.3	63.01	.2658	127.5	207.7	371.1
τ	.004936	.001324	<.000001	.002695	.004814	.007844

the posterior median for the variance about the regression line, which implies a less-than-good fit of the model with the data. This is somewhat confirmed by Figure 4.5, which is a graph of the predicted values of SBP versus the actual blood pressure values, y, in Table 4.9. Figure 4.5 is an informal way of assessing how well the model fits the data, but there are more formal approaches including calculating the correlation between the observed and predicted values of SBP. One may show that the correlation between the observed and predicted values is $R = .551$, corresponding to $R^2 = .3036$, where R^2 is the traditional way to assess the fit of a linear regression model.

There is a linear association between the two, but there is a lot of variation in the predicted values for each of the actual blood pressure values. One assumption that needs to be checked is that the variance of the dependent variables is the same for all values of the independent variable x. For this example, the residuals (SBP minus the predicted blood pressure) are plotted against the independent variable age.

The absolute value of the residual for a given age is an approximate estimate of the variance of the dependent variable SBP, and Figure 4.6 does not reveal any surprises. Note that the length of the residual is its absolute value.

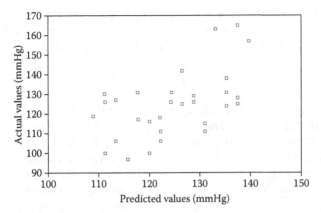

FIGURE 4.5
Predicted values of SBP versus actual values predicted.

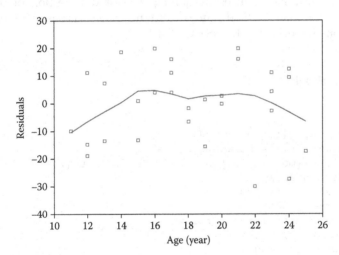

FIGURE 4.6
Residuals versus age.

The lowess curve portrays no discernible trend in the residuals; thus, there is no sufficient information to question the assumption of constant variance for all possible ages.

4.3.4 More on Multiple Linear Regression

The study results that relate SBP to age and weight are portrayed in Table 4.9, and a simple linear regression is employed to assess the association between age and SBP based on information from 30 subjects.

Age can be viewed as a confounder; thus, multiple linear regression is performed to assess the association between weight and blood pressure adjusted for age. Therefore, consider the following model:

$$y_i = \alpha + \beta_1 x_{1i} + \beta_2 x_{2i} + e_i \tag{4.10}$$

where the independent variable x_1 is the age of the ith individual and x_2 is the corresponding weight. The model (Equation 4.10) implies that the average value of the dependent variable y is a linear function of the regression coefficients and that for a unit increase of 1 lb in weight the average value of SBP increases by β_2, holding the value of age constant (and for all possible ages). As with simple linear regression, the assumptions are as follows: the n observations are independent random variables; for each value of x_1 and x_2, the variance of y is the same; and the errors are normally distributed with a mean of 0 and a variance of σ^2.

Bayesian analysis produces the joint posterior distribution of the four unknown parameters α, β_1, β_2, and σ^2, and the analysis is based on the information from Table 4.8 and BUGS CODE 4.4.

BUGS CODE 4.4

```
model;
{
for(i in 1:n){y[i]~dnorm(mu[i], tau)}
for(i in 1:n){z[i]~dnorm(mu[i], tau)}
for(i in 1:n){mu[i]<- alpha+beta[1]*x1[i]+beta[2]*x2[i]}
# alpha is intercept
# uninformative priors
alpha~dnorm(0,.0001)
# regression coefficients for x1 and x2
beta[1]~dnorm(0,.0001)
beta[2]~dnorm(0,.0001)
# prior for precision
tau~dgamma(.00001,.00001)
# sigsq is variance about regression line
sigsq<-1/tau
}
# data from Table 4.8
# y is systolic blood pressure
# x1 is age
# x2 is weight
list(n = 30,
x1 = c(14,12,16,13,11,21,17,15,17,20,21,17,23,12,16,15,19,24,
   24,12,23,23,20,13,18,18,19,24,25,22),
y = c(97,100,100,106,119,111,106,117,111,126,115,118,124,126,
   116,131,125,125,128,130,138,131,129,127,131,126,142,165,
   157, 163),
```

```
x2 = c(94,95,104,107,108,116,117,117,122,124,125,125,125,128,
    128,129,134,134,136,137,137,138,139,141,141,143,150,171,172,
    172))
# cigarette data below
# x1 is income
# x2 is price per pack
# y is consumption


list(n = 51, x1 = c(2948,4644,3665,2878,4493,3855,4917,4524,
    5079,3738,3354,4623,
3290,4507,3772,3751,3853,3112,3090,3302,4309,4340,4180,3859,
2626,3781,3500,3789,4563,3737,4701,3077,4712,3252,3086,4020,
3387,3719,3971,3959,2990,3123,3119,3606,3227,3468,3712,4053,
3061,3812,3815),


x2 = c(42.70,41.80,38.50,38.80,39.70,31.10,45.50,41.30,32.60,
    43.80,
35.80,36.70,33.60,41.40,32.20,38.50,38.90,30.10,39.30,38.80,
34.20,41.00,39.20,40.10,37.50,36.80,43.70,34.70,44.00,34.10,
41.70,41.70,41.70,29.40,38.90,38.10,39.80,29.00,44.70,40.20,
34.30,38.50,41.60,42.00,36.60,39.50,30.20,40.30,41.60,40.20,
34.40),


y = c(90,121,115,100,123,125,120,155,200,124,110,82,102,125,
    135,109,
114,156,116,129,124,124,129,104,93,121,111,108,190,266,
    121,90,
119,172,94,122,108,157,107,124,104,93,100,106,66,123,124,97,
115,106,132))
# inital values
list(alpha = 0,beta = c(0,0,0))
```

The BUGS statements in BUGS CODE 4.4 are self-explanatory and closely follow the notation of the defined model (Equation 4.10). Before analyzing the data, one should plot the data, as, for example, in Figure 4.7.

In order to show trends, a linear regression line is shown for each plot. All possible relations are shown: SBP versus age, blood pressure versus weight, and age versus weight. It can be seen that blood pressure versus weight is more linear than blood pressure versus age.

BUGS CODE 4.4 was executed with 55,000 observations for the simulation, a burn-in of 5000, and a refresh of 100, and it produced the posterior analysis shown in Table 4.10 for the multiple linear regression.

One is interested in assessing the association between SBP and weight adjusted for age. The effect of weight is the increase in average blood pressure by .8173 mmHg for an increase of 1 lb, for all ages. The 95% credible interval

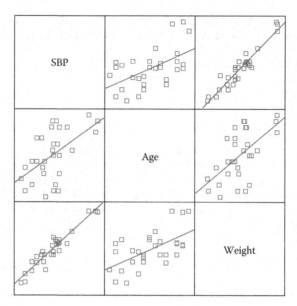

FIGURE 4.7
Two-dimensional plots of age, weight, and SBP.

TABLE 4.10

Bayesian Analysis for Multiple Linear Regression

SBP versus Weight and Age						
Parameter	Mean	SD	Error	2½	Median	97½
α	19.09	7.127	.2839	5.169	19.01	33.4
β_1	−.05125	.3119	.009947	−.6557	−.05085	.578
β_2	.8173	.06849	.002954	.68	.8181	.9498
σ^2	31.96	9.481	.08804	18.55	30.34	54.78
τ	.0338	.009188	<.00001	.01826	.03296	.05391

for β_2 is (.68,.9489), suggesting a real effect on blood pressure, although that for age is (−.6557,.578), suggesting that age can be eliminated as a predictor of blood pressure.

How well does the model (Equation 4.10) fit the data? To answer this question, one should plot the predicted blood pressure values versus the corresponding actual blood pressure values y_i, $i = 1, 2, \dots , 30$.

Figure 4.8 demonstrates that the multiple linear regression model is a good fit to the data and in fact appears to be a linear relationship. In addition, the R^2 value is .9025, giving further evidence of an excellent fit to the data.

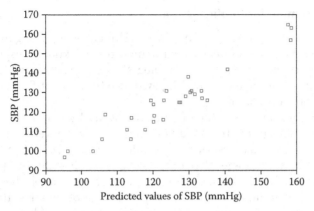

FIGURE 4.8
SBP versus predicted SBP.

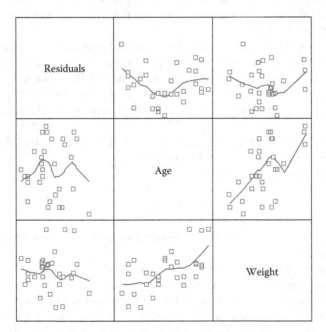

FIGURE 4.9
Residuals versus age and weight.

One should check to see if the variance of the dependent variable is constant for all pairs (x_{1i}, x_{2i}), $i = 1, 2, ..., n$. A plot of residuals versus age and weight is presented in Figure 4.9. Can you detect a trend for the residuals when plotted against age and plotted against weight?

The lowess curve is plotted, and no discernible trend is perceived for residuals versus age and residuals versus weight.

4.3.5 An Example for Public Health

One of the major health problems in the United States is cigarette smoking, and the dangerous effects of smoking are well known, especially the development of lung cancer. Epidemiologists took an active role in promoting smoking cessation and providing evidence from scientific studies of the deleterious effects of smoking.

The following dataset is taken from the study by Chatterjee and Price (p. 265),[1] which reports the 1970 consumption of cigarettes and certain social demographic factors for the 50 states and the District of Columbia, including the median age of people living in each state, the percentage of people over 25 years of age who completed high school, the per capita income, the percentage of blacks, the percentage of females, the average price of a pack of cigarettes, and the number of packs sold on a per capita basis (Table 4.11). Such information is valuable to public health officials, state attorneys general, and insurance companies. This is a good example of aggregate data where various state variables are averaged over the relevant population of a state.

For example, what is the effect of the level of education on cigarette consumption? What is the average price per capita? If one knew the effect of price per pack on consumption, one could adjust the price (via taxes) by increasing it and decreasing the consumption. It is well known that price per pack and income are the most important variables in predicting cigarette consumption (number of packs sold in a state per capita) and that the other factors, education, percentage of blacks, and percentage of females, do not play an important role in cigarette consumption.

A multiple linear regression model

$$y_i = \alpha + \beta_1 x_{1i} + \beta_2 x_{2i} + e_i$$

will be utilized to model the association between cigarette consumption and income and price.

One should plot consumption versus income and price and verify the linear association. Bayesian analysis is executed using the information in Table 4.11, which is also listed as the third list statement in BUGS CODE 4.4. I used 55,000 observations for the simulation, with a burn-in of 5000 and a refresh of 5000, and the posterior distribution for the parameters are reported in Table 4.12. I deleted the value 265.7 for the consumption of NH, the state of New Hampshire, which is considered an outlier and not used in the Bayesian analysis.

The effect of price (dollars/pack) is estimated by the posterior mean as −2.17, implying that for fixed income (regardless of the income value) the average annual decrease in consumption (number of packs sold per capita) is −2.17 with a 95% credible interval of (−3.528,−.7488). The relatively long interval is a reflection of the standard deviation.

TABLE 4.11

Cigarette Consumption by State in 1970

State	Age	Education	Income	Percentage of Blacks	Percentage of Females	Price	Consumption
AL	27.0	41.3	2948	26.2	51.7	42.7	89.8
AK	22.9	66.7	4644	3.0	45.7	41.8	121.3
AZ	26.3	58.1	3665	3.0	50.8	38.5	115.2
AR	29.1	39.9	2878	18.3	51.5	38.8	100.3
CA	28.1	62.6	4493	7.0	50.8	39.7	123.0
CO	26.2	63.9	3855	3.0	50.7	31.1	124.8
CT	29.1	56	4917	6.0	51.5	45.5	120.0
DE	26.8	54.6	4524	14.3	51.3	41.3	155.0
DC	28.4	55.2	5079	71.1	53.5	32.6	200.4
FL	32.3	52.6	3738	15.3	51.8	43.8	123.6
GA	25.9	40.6	3354	25.9	51.4	35.8	109.9
HI	25.0	61.9	4623	1.0	48.0	36.7	82.1
ID	26.4	59.5	3290	0.3	50.1	33.6	102.4
IL	28.6	52.6	4507	12.8	51.5	41.4	124.8
IN	27.2	52.9	3772	6.9	51.3	32.2	134.6
IO	28.8	59.0	3751	1.2	51.4	38.5	108.5
KA	28.7	59.9	3853	4.8	51.0	38.9	114.0
KY	27.5	38.8	3112	7.2	50.9	30.1	155.8
LA	24.8	42.2	3090	29.8	51.4	39.3	115.9
ME	28.6	54.7	3302	.3	51.3	38.8	128.5
MD	27.1	52.3	4309	17.8	51.1	34.2	123.5
MA	29.0	58.5	4340	3.1	52.2	41.0	124.3
MI	26.3	52.8	4180	11.2	51.0	39.2	128.6
MN	26.8	57.6	3859	.9	51.0	40.1	104.3
MS	25.1	41.0	2626	36.8	51.6	37.5	93.4
MO	29.4	48.8	3781	10.3	51.8	36.8	121.3
MT	27.1	59.2	3500	.3	50.0	43.7	111.2
NB	28.6	59.3	3789	2.7	51.2	34.7	108.1
NV	27.8	65.2	4563	5.7	49.3	44.0	189.5
NH	28.0	57.6	3737	.3	51.1	34.1	265.7*
NJ	30.1	52.5	4701	10.8	51.7	41.7	120.7
NM	23.9	55.2	3077	1.9	50.7	41.7	90.0
NY	30.3	52.7	4712	11.9	52.2	41.7	119.0
NC	26.5	38.5	3252	22.2	51.0	29.4	172.4
ND	26.4	50.3	3086	0.4	49.5	38.9	93.8
OH	27.7	53.2	4020	9.1	51.5	38.1	121.6
OK	29.4	51.6	3387	6.7	51.3	39.8	108.4
OR	29.0	60.0	3719	1.3	51.0	29.0	157.0
PA	30.7	50.2	3971	8.0	52.0	44.7	107.3
RI	29.2	46.4	3959	2.7	50.9	40.2	123.9

(Continued)

TABLE 4.11 *(Continued)*

Cigarette Consumption by State in 1970

State	Age	Education	Income	Percentage of Blacks	Percentage of Females	Price	Consumption
SC	24.8	37.8	2990	30.5	50.9	34.3	103.6
SD	27.4	53.3	3123	0.3	50.3	38.5	92.7
TN	28.1	41.8	3119	15.8	51.6	41.6	99.8
TX	26.4	47.4	3606	12.5	51.0	42.0	106.4
UT	23.1	67.3	3227	0.6	50.6	36.6	65.5
VT	26.8	57.1	3468	0.2	51.1	39.5	122.6
VA	26.8	47.8	3712	18.5	50.6	30.2	124.3
WA	27.5	63.5	4053	2.1	50.3	40.3	96.7
WV	30.0	41.6	3061	3.9	51.6	41.6	114.5
WI	27.2	54.5	3812	2.9	50.9	40.2	106.4
WY	27.2	62.9	3815	0.8	50.0	34.4	132.2

Source: Chatterjee, S., and B. Price, *Regression Analysis by Example,* John Wiley & Sons Inc., New York, 1991.

TABLE 4.12

Bayesian Analysis for Cigarette Consumption Data

Parameter	Mean	SD	Error	2½	Median	97½
α	120.2	28.81	1.325	64.68	120.7	175.9
β_1	.02168	.004872	.0001583	.01238	.0216	.0315
β_2	−2.171	.7161	.03237	−3.528	−2.188	−.7488
σ^2	449.5	97.37	.8215	298.2	436.6	677.1
τ	.002324	.000482	<.000001	.0001477	.00229	.003354

On the other hand, the effect of price per pack is estimated as .02168, implying that consumption increases by .02335 packs (packs sold per capita) for a 1 c. increase in price, for a given value of income (and for all income levels). MCMC errors are small and one had confidence 55,000 observations are enough for the simulation.

I am beginning to suspect that the model is only a fair fit to the data, but one should plot the predicted values of consumption versus the observed consumption values (Figure 4.10). This plot is shown in Table 4.10.

The lowess curve shows an increasing trend in that as the consumption increases so does the predicted cigarette consumption, and a reasonable conclusion is that the multiple linear regression model with price and income as independent variables has a reasonably good fit to the data. Incidentally, the Pearson correlation between observed and predicted consumption is .576.

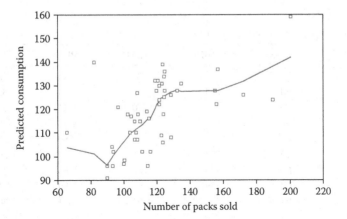

FIGURE 4.10
Predicted consumption versus observed consumption number.

4.4 Weighted Regression

When the assumption of equal variance of the dependent variable (for all values of the independent variables) is not reasonable, weighted regression models are implemented to estimate the regression coefficients of a model. Often, the larger the measurement, the larger the variance. Consider an example from the study by Chatterjee and Price,[1] which is analyzed with a weighted regression. A weighted regression is appropriate if the variance of the dependent variable is not constant for all observations. The information in Table 4.13 is state expenditure data for education for all 50 states and is of interest to the health administrators at the state medical and public health schools, including departments of epidemiology. The variables (columns) are defined as follows: URB70 is the number of residents per thousand living in urban areas in 1970, SE75 is the per capita expenditures for education in 1975, PI73 is the per capita income in 1973, and Y74 is the number of residents per thousand under 18 years of age in 1974.

Is an unweighted regression reasonable?

Using BUGS CODE 4.5, the information from Table 4.13, and the multiple linear regression model

$$y_i = \alpha + \beta_1 x_{1i} + \beta_2 x_{2i} + \beta_3 x_{3i} + e_i \tag{4.11}$$

for state $i = 1, 2, \ldots, 50$, the residuals are computed. In the aforementioned model, the dependent variable y_i is the projected expenditure for education for state i in 1975, x_{1i} is the number of residents per thousand living in an urban area for state i, x_{2i} is the per capita income for 1973, and x_{3i} is the number of residents per thousand under 18 years of age.

TABLE 4.13

State Expenditure for Education

Case	State	URB70	SE75	PI73	Y74
1	ME	508	235	3944	325
2	NH	564	231	4578	323
3	VT	322	270	4011	328
4	MA	846	261	5233	305
5	RI	871	300	4780	303
6	CT	774	317	5889	307
7	NY	856	387	5663	301
8	NJ	889	285	5759	310
9	PA	715	300	4894	300
10	OH	753	221	5012	324
11	IN	649	264	4908	329
12	IL	830	308	5753	320
13	MI	738	379	5439	337
14	WI	659	342	4634	328
15	MN	664	378	4921	330
16	IA	572	232	4869	318
17	MO	701	231	4672	309
18	ND	443	246	4782	333
19	SD	446	230	4296	330
20	NB	615	268	4827	318
21	KS	661	337	5057	304
22	DE	722	344	5540	328
23	MD	766	330	5331	323
24	VA	631	261	4715	317
25	WV	390	214	3828	310
26	NC	450	245	4120	321
27	SC	476	233	3817	342
28	GA	603	250	4243	339
29	FL	805	243	4647	287
30	KY	523	216	3967	325
31	TN	588	212	3946	315
32	AL	584	208	3724	332
33	MS	445	215	3448	358
34	AR	500	221	3680	320
35	LA	661	244	3825	355
36	OK	680	234	4189	306
37	TX	797	269	4336	335
38	MT	534	302	4418	335
39	ID	541	268	4323	344
40	WY	605	323	4813	331
41	CO	785	304	5046	324

TABLE 4.13 *(Continued)*

State Expenditure for Education

Case	State	URB70	SE75	PI73	Y74
42	NM	698	317	3764	366
43	AZ	796	332	4504	340
44	UT	804	315	4005	378
45	NV	809	291	5560	330
46	WA	726	312	4989	313
47	OR	671	316	4697	305
48	CA	609	332	5438	307
49	AK	484	546	5613	386
50	HI	831	311	5309	333

Source: Chatterjee, S., and B. Price, *Regression Analysis by Example*, John Wiley & Sons Inc., New York, 1991.

The Bayesian analysis is executed with BUGS CODE 4.5 and 65,000 observations for the simulation, a burn-in of 5000, and a refresh of 100. The dependent variable is the projected per capita state expenditure for education, and the independent variables are x_1 (the number of residents per thousand living in urban areas), x_2 (per capita income for 1973), and x_3 (the number of residents per thousand under 18 years of age).

The first list statement is the data from Table 4.13, the second is data for a weighted regression when the dependent variables have been transformed using the logarithmic scale, and the third list statement contains the data for a weighted regression when the dependent variable is the square root of *y*.

BUGS CODE 4.5

```
model;
{
for(i in 1:n){y[i]~dnorm(mu[i], tau)}
for(i in 1:n){z[i]~dnorm(mu[i], tau)}
for(i in 1:n){mu[i]<-alpha+beta[1]*x1[i]+beta[2]*x2[i]+beta
  [3]*x3[i]}
# alpha is intercept
# uninformative priors
# y is the projected per capita expendure for education
# x1 is the number of residents per thousand living in urban
  areas

# x2 is the 1973 per capita income
# x3 is the number of residents per thousand under 18
# regression coefficients
beta[1]~dnorm(0,.0001)
beta[2]~dnorm(0,.0001)
beta[3]~dnorm(0,.0001)
```

```
alpha~dnorm(0.0,.0001)
# prior for precision
tau~dgamma(.000001,.000001)
# sigsq is variance about regression line
sigsq<-1/tau
}
# data for state education expenditure data
list(n = 50,
# URB70
x1 = c(508,564,322,846,871,774,856,889,715,753,649,830,738,
   659,664,
572,701,443,446,615,661,722,766,631,390,450,476,603,805,523,
588,584,445,500,661,680,797,534,541,605,785,698,796,804,809,
726,671,609,484,831),
# state expenditure education
y = c(235,231,270,261,300,317,387,285,300,221,264,308,379,
   342,378,
232,231,246,230,268,337,344,330,261,214,245,233,250,243,216,
212,208,215,221,244,234,269,302,268,323,304,317,332,315,291,
312,316,332,NA,311),
# PI73
x2 = c(
3944,4578,4011,5233,4780,5889,5663,5759,4894,5012,4908,5753,
5439,4634,4921,4869,4672,4782,4296,4827,5057,5540,5331,4715,
3828,4120,3817,4243,4647,3967,3946,3724,3448,3680,3825,4189,
4336,4418,4323,4813,5046,3764,4504,4005,5560,4989,4697,5438,
5613,5309),
# Y74
x3 = c(325,323,328,305,303,307,301,310,300,324,329,320,337,
   328,330,
318,309,333,330,318,304,328,323,317,310,321,342,339,287,325,
315,332,358,320,355,306,335,335,344,331,324,366,340,378,330,
313,305,307,386,333))


# weighted observations
# log transformation
list(n = 50,
# URB70
x1 = c(508,564,322,846,871,774,856,889,715,753,649,830,738,
   659,664,
572,701,443,446,615,661,722,766,631,390,450,476,603,805,523,
588,584,445,500,661,680,797,534,541,605,785,698,796,804,809,
726,671,609,484,831),
# state expenditure education
y = c(5.46,5.44,5.60,5.56,5.70,5.76,5.96,5.65,5.70,5.40,5.58,
   5.73,5.94,5.83,5.93,5.45,5.44,5.51,5.44,5.59,5.82,5.84,5.80,
   5.56,5.37,5.50,5.45,5.52,5.49,5.38,5.36,5.34,5.37,5.40,5.50,
   5.46,5.59,5.71,5.59,5.78,5.72,5.76,5.81,5.75,5.67,5.74,5.76,
   5.81,NA,5.74),
```

```
# PI73
x2 = c(
3944,4578,4011,5233,4780,5889,5663,5759,4894,5012,4908,5753,
5439,4634,4921,4869,4672,4782,4296,4827,5057,5540,5331,4715,
3828,4120,3817,4243,4647,3967,3946,3724,3448,3680,3825,4189,
4336,4418,4323,4813,5046,3764,4504,4005,5560,4989,4697,5438,
5613,5309),
# Y74
x3 = c(325,323,328,305,303,307,301,310,300,324,329,320,337,
   328,330,
318,309,333,330,318,304,328,323,317,310,321,342,339,287,325,
315,332,358,320,355,306,335,335,344,331,324,366,340,378,330,
313,305,307,386,333))

# square root transformation
# state expenditure education
# weighted observations
list(n = 50,
# URB70
x1 = c(508,564,322,846,871,774,856,889,715,753,649,830,738,
   659,664,
572,701,443,446,615,661,722,766,631,390,450,476,603,805,523,
588,584,445,500,661,680,797,534,541,605,785,698,796,804,809,
726,671,609,484,831),

y = c(15.33,15.20,16.43,16.16,17.32,17.80,19.67,16.88,17.32,
14.87,
16.25,17.55,19.47,18.49,19.44,15.23,15.20,15.68,15.17,16.37,
18.36,18.55,18.17,16.16,14.63,15.65,15.26,15.81,15.59,14.70,
14.56,14.42,14.66,14.87,15.62,15.30,16.40,17.38,16.37,17.97,
   17.44,17.80,18.22,17.75,17.06,17.66,17.78,18.22,NA,17.64),
# PI73
x2 = c(
3944,4578,4011,5233,4780,5889,5663,5759,4894,5012,4908,5753,
5439,4634,4921,4869,4672,4782,4296,4827,5057,5540,5331,4715,
3828,4120,3817,4243,4647,3967,3946,3724,3448,3680,3825,4189,
4336,4418,4323,4813,5046,3764,4504,4005,5560,4989,4697,5438,
5613,5309),
# Y74
x3 = c(325,323,328,305,303,307,301,310,300,324,329,320,337,
   328,330,
318,309,333,330,318,304,328,323,317,310,321,342,339,287,325,
315,332,358,320,355,306,335,335,344,331,324,366,340,378,330,
313,305,307,386,333))

# inital values
list(alpha = 0,beta = c(0,0,0), tau = 1)
```

The Bayesian analysis for the unweighted regression is as shown in Table 4.14.

TABLE 4.14

Bayesian Analysis for Multiple Linear Regression

State Expenditures for Education

Parameter	Mean	SD	Error	2½	Median	97½
α	−93.12	86.05	4.672	−257.2	−95.89	76.72
β_1	.06355	.04903	.001588	−.03248	.06329	.1615
β_2	.041	.01012	.000414	.02089	.04135	.06055
β_3	.4318	.2328	.01238	−.04971	.4391	.8851
σ^2	1399	311.7	3.802	917.3	1357	2127
τ	.000748	.0001593	.00000191	.0004702	.000737	.00109

Thus, the effect of per capita income is .041, that is, as per capita income increases by \$1 the average per capita expenditure increases by .041 dollars. Note that the 95% credible intervals for β_1 and β_3 include zero, implying that perhaps variables x_1 (the number of residents per thousand living in urban areas) and x_3 (the number of residents per thousand under 18 years of age) do not impact the average per capita state expenditure for education.

Figure 4.10 is a plot of absolute residuals versus observed state education expenditures and the lowess curve reveals an increasing trend, that is, as state expenditures for education increase so do the absolute values of the residuals. The residual is the difference between the observed and predicted expenditures.

The absolute value of the residual is an estimator of the variance of the observed expenditures; thus, it appears that the variance is not constant and a weighted linear regression is called for. Referring to BUGS CODE 4.5, note that the first statement is the regression of y on the three independent variables; the second generates the predicted values of y; and the third is the regression function, which is linear in the three independent variables.

Recall that the value of predicted expenditure is the mean of the posterior distribution of the projected expenditure listed in Table 4.13. The absolute residual is the absolute value of the predicted expenditure minus the observed expenditure. What are plotted above are the absolute values of the residuals versus the state expenditures for education. The absolute residual is a surrogate of the variance of observations.

Notice that the code is for a multiple linear regression model (Equation 4.11) with three independent variables where the unknown parameters are given uninformative prior distributions, namely, normal distributions for the regression coefficients and the precision about the regression line. The analysis to follow is for an unweighted regression, which is executed to compute the predicted values, z, and the residuals (observed minus predicted state expenditures). The code is modified for weighted regression analysis. How should the observations be weighted?

Weighted regression is accomplished by transforming the dependent variable and performing a multiple regression on the independent variables. The type of transformation depends on the association between the variance and the mean of the dependent variable. Table 4.15 guides the user in how to transform the dependent variable so that approximately the variance of the dependent variable is constant for all combinations of values of the independent variables.

Figure 4.10 suggests using a square root transformation. Thus, BUGS CODE 4.5 is executed again, but using the second list statement with a square root transformation for the dependent variable y, the projected state expenditures for education, and the results appear in Table 4.16.

Comparison of Tables 4.14 and 4.16 reveals that for weighted regression the three independent variables have less effect than the corresponding effects for the unweighted regression. This is true because the dependent variable (square root of the original) is smaller than the dependent variable for the unweighted regression. Did this transformation have an effect on the variance of the dependent variable? Consider a plot of the absolute residuals versus the predicted values of the square root of the dependent variable.

The lowess curve of Figure 4.11 shows that the variance of the dependent variable (the square root of y) has somewhat stabilized, compared to the

TABLE 4.15

Variance-Stabilizing Transformations for y

Relationship	Transformation
Variance constant	No transformation
Variance proportional to mean	Square root of y
Variance proportional to the square of the mean	Log (log) of y
Variance proportional to the cube of the mean	1 divided by the square root of y
Variance proportional to the fourth power of the mean	Reciprocal of y

Source: Montgomery et al., *Introduction to Linear Regression Analysis*, John Wiley & Sons Inc., New York, 2001.

TABLE 4.16

Bayesian Weighted Regression for State Education Expenditures

Square Root Transformation						
Parameter	Mean	SD	Error	2½	Median	97½
α	−.2643	4.104	.2452	−7.693	.3235	8.551
β_1	.001698	.001427	.0000466	−.001099	.001697	.004576
β_2	.001505	.000336	.00004156	.0008312	.001513	.002122
β_3	.02548	.0104	.0006148	.004202	.02548	.04636
σ^2	1.175	.2608	.00305	.7721	1.14	1.784
τ	.8911	.1886	.002062	.5607	.8769	1.295

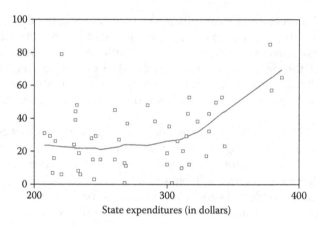

FIGURE 4.11
Absolute values of residuals versus state expenditures for education.

trend shown in Figure 4.10. The variance (as measured by the absolute value of the residual) still depends to some extent on the predicted values, but not to the extreme as the variance of the original observations; thus, the square root transformation has done its job.

4.5 Ordinal and Other Regression Models

Ordinal regression models are appropriate when the dependent variable assumes several nominal or ordinal values. For example, Broemeling (p. 117)[8] performs an ordinal regression analysis that estimates the accuracy of sentinel lymph biopsy for assessing the extent of metastasis in melanoma patients.

The dependent variable assumes five values: (1) indicates absolutely no evidence, (2) no evidence, (3) very little evidence, (4) some evidence, and (5) definite evidence of metastasis. The independent variables are four radiologists who are assessing the degree of metastasis on a five-point scale.

Of course, there are many examples of ordinal regression appropriate for epidemiology, and the topic is explained in more detail in a Chapter 8 devoted to advanced modeling techniques. Also in that chapter, nonlinear regression is introduced and many examples are used to illustrate the Bayesian methodology.

4.6 Comments and Conclusions

In this chapter, regression techniques applicable to studies in epidemiology are introduced. The reader is referred to the study by Rothman (p. 300)[9] for more examples involving the use of regression in assessing the association between disease and several confounders and other factors. Regression techniques, similar to other methods, are employed to estimate the association between disease (or morbidity) and various factors that impact morbidity. There are several types of regression models, and two are described in this chapter, namely, the logistic model and multiple regression models (including simple linear regression models).

Logistic regression models have dependent variables that are binary (yes or no, disease or no disease, etc.), although independent variables can be categorical or quantitative. On the other hand, the dependent variable of a normal multiple regression model is continuous or quantitative.

Bayesian techniques for the logistic model are illustrated with the Israeli Heart Disease Study, in which the dependent variable is the occurrence of myocardial infarction and the independent variables are age and SBP.

The model is defined and Bayesian inferences are briefly described and implemented, where the inferences are executed with WinBUGS, and consist of determining the posterior distribution of relevant parameters. The relevant parameters are the regression coefficients and the associated odds ratios of age and SBP. Age and SBP are dichotomized so that it is possible to calculate the odds ratios. For example, the odds of disease among those with a high SBP versus the odds among those without is the odds ratio of age and is expressed as an exponent of the appropriate regression coefficient. See Table 4.1 for the study data and Table 4.2 for the Bayesian analysis that includes the posterior distribution of the two odds ratios. Logistic regression Bayesian analysis is concluded with an example involving four independent variables and the occurrence of heart disease for the dependent variable. The goodness of fit of the model is assessed by plotting the predicted occurrence of heart disease with the actual occurrence of heart disease.

This chapter continues with the introduction of simple linear regression, where Bayesian analysis is illustrated by the Woolson[6] heart disease study. The dependent variable is the measured value of SBP, and the independent variable is age. In a simple linear regression model, there are two regression coefficients, the slope and the intercept, which are estimated by the posterior mean of their posterior distribution. The estimate of the slope assesses the change in the average value of SBP for a unit increase in age (1 year), and the estimate of the intercept assesses the average value of blood pressure when age is 0.

By plotting the predicted blood pressure versus the actual blood pressure values, the goodness of fit of the model can be determined. If the model does not fit the model well one should investigate certain alternatives, such as using a nonlinear model. Of course, the goodness of fit of a regression model can also be estimated by the R^2 value, where R is the correlation between the observed and predicted values of the dependent variable.

Simple linear regression is generalized into multiple linear regression models, which have more than one independent variable. The Woolson example is analyzed as a multiple linear regression with age and weight as the independent variables and SBP as the dependent variable, and a matrix plot reveals a linear association between blood pressure and age and blood pressure and weight. Execution of BUGS CODE 4.4 demonstrates that weight has an effect, where an increase of 1 lb in weight increases the average blood pressure by .81 mmHg. Multiple linear regression reveals a good fit to the data as portrayed in Figure 4.6, a plot of predicted blood pressure values versus observed values. Of interest to public health and epidemiology is the state cigarette consumption example, in which consumption is measured by packs sold per capita and the two most important independent variables are price per pack and per capita income (averaged over all people in the state). The Bayesian analysis showed that price per pack has an effect, that is, with an increase of 1 c. per pack the number of packs sold per capita decreases on the average of 2.17 on an annual basis.

Weighted regression is appropriate if the variance of the dependent variable is not the same for all settings of the independent variables. To detect unequal variances a multiple regression is performed the usual way, and if a plot of the absolute value of the residuals versus the predicted values of the dependent variable reveals a trend unequal variances are suspected and a weighted regression may be appropriate. The approach taken here is to transform the dependent variable, depending on the relationship between the variance of an observation and its mean. If this relationship can be determined by a plot, then Table 4.15 suggests the appropriate transformation. For example, if the variance (estimated by the absolute residual) is proportional to the mean (estimated by the predicted values of the dependent variable), the square root transformation is recommended. For example, on the basis of Figure 4.10, a square root transformation is recommended for the state education expenditure information of Table 4.13 and BUGS CODE 4.5 is executed to perform the Bayesian analysis reported in Table 4.16. Note that Figure 4.11 shows that square root transformation did indeed stabilize the variance. For additional information on advanced approaches to weighted regression, see the study by Carroll and Ruppert.[10]

This chapter concludes with a brief discussion of the ordinal and nonlinear regression models, topics that will be explored in a Chapter 8 on advanced modeling techniques.

Exercises

1. Using BUGS CODE 4.1 and the information in Table 4.1, perform a Bayesian analysis with 55,000 observations for the simulation with a burn-in of 5000 and a refresh of 100.

   a. Verify the posterior analysis of Table 4.2.

   b. Verify the plot of posterior density of the odds ratio for blood pressure.

    c. Is there an association between heart attack and SBP, adjusted for age?

    d. Is age a confounder? Does age affect the association between heart attack and SBP?

2. Refer to problem 1, the Bayesian analysis for the study of Table 4.1, where the odds ratios for the age and SBP factors are estimated with the logistic model. An alternative model is a logistic model with an interaction term, namely,

$$\log it(\theta_i) = \alpha + \beta_1 x_{1i} + \beta_2 x_{2i} + \beta_3 x_{1i} x_{2i} \qquad (4.12)$$

where θ_i is the probability of a heart attack for the ith subject, α is the intercept, and (x_{1i}, x_{2i}) are the values of the independent variables for the ith subject. The first variable is age, where $x_{1i} = 0$ if age <60 and $x_{1i} = 1$ otherwise, and the second one is SBP, where $x_{2i} = 0$ if SBP <140 and $x_{2i} = 1$ otherwise. The third regression coefficient β_3 is the effect of the interaction on the logit scale, where the interaction is the product of age and blood pressure. I used this model to reanalyze the study results of Table 4.1 using BUGS CODE 4.1 Alternative (Table 4.17). Does the inclusion of an interaction term change the posterior analysis of Table 4.2?

BUGS CODE 4.1

```
model;
{
# Bernoulli distribution for the observations

# theta is the probability of a heart attack

for (i in 1:400){y[i]~dbern(theta[i])}
# logistic regression of theta on age and systolic blood
  pressure
# with interaction
for (i in 1:400){logit(theta[i])<-alpha+beta[1]*x1[i]+beta[2]*
  x2[i]+beta[3]*x1[i]*x2[i]}
# prior distributions for the regression coefficients
# uninformative priors
alpha~ dnorm(0.0000,.00001)
for(i in 1:3){beta[i]~dnorm(.0000,.00001)}
ORage<-exp(beta[1])
ORsbp<-exp(beta[2])
}
# y is the occurence of a heart attack
# x1 is age
# x2 is systolic blood pressure
# age and blood pressure are coded with binary values
```

```
list(y = c(1,1,
            1,
            0,0,0,0,0,0,0,0,0,0,0,0,0,0,0,0,0,0,0,0,0,0,0,0,0,
              0,0,0,0,0,
            0,0,0,0,0,0,0,0,0,0,0,0,0,0,0,0,0,0,0,0,
            1,1,1,1,
            1,1,1,1,1,
            0,0,0,0,0,0,0,0,0,0,0,0,0,0,0,0,0,0,0,0,0,0,0,0,0,0,
            0,0,0,0,0,0,0,0,0,0,0,0,0,0,0,0,0,0,0,0,0,0,0,0,0,0,
            0,0,0,0,0,0,0,0,0,0,0,0,0,0,0,0,0,0,0,0,0,0,0,0,0,0,
            0,0,0,0,0,0,0,0,0,0,0,0,0,0,0,0,0,0,0,0,0,0,0,0,0,0,
            0,0,0,0,0,0,0,0,0,0,0,0,0,0,0,0,0,0,0,0,0,0,0,0,0,0,
            0,0,0,0,0,0,0,0,0,0,0,0,0,0,0,0,0,0,0,0,0,0,0,0,0,0,

            0,0,0,0,0,0,0,0,0,0,0,0,0,0,0,0,0,0,0,0,0,0,0,0,0,0,
            0,0,0,0,0,0,0,0,0,0,0,0,0,0,0,0,0,0,0,0,0,0,0,0,0,0,
            0,0,0,0,0,0,0,0,0,0,0,0,0,0,0,0,0,0,0,0,0,0,0,0,0,0,
            0,0,0,0,0,0,0,0,0,0,0,0,0,0,0,0,0,0,0,0,0,0,0,0,0,0,
            0,0,0,0,0,0,0,0,0,0,0,0,0,0,0,0,0,0,0,0,0,0,0,0,0,0,
            0,0,0,0,0,0,0,0,0,0,0,0,0,0,0,0,0,0,0,0,0,0,0,0,0,0,
            0,0,0,0,0,0,0,0,0,0,0,0,0,0,0,0,0,0,0,0,0,0,0,0,0,0,
            0,0,0,0,0,0,0,0,0,0,0,0,0,0,0,0,0,0,0,0,0,0,0,0,0,0,
            0,0,0,0,0,0,0,0,0,0,0,0,0,0,0,0,0,0,0,0,0,0,0,0,0,0,
            0,0,0,0,0,0,0,0,0,0,0,0,0,0,0,0,0),

x1 = c(1,1,
        1,
        1,1,1,1,1,1,1,1,1,1,1,1,1,1,1,1,1,1,1,1,1,1,1,1,1,
          1,1,1,1,
        1,1,1,1,1,1,1,1,1,1,1,1,1,1,1,1,1,1,1,
        0,0,0,0,
        0,0,0,0,0,
        0,0,0,0,0,0,0,0,0,0,0,0,0,0,0,0,0,0,0,0,0,0,0,0,0,0,
        0,0,0,0,0,0,0,0,0,0,0,0,0,0,0,0,0,0,0,0,0,0,0,0,0,0,
        0,0,0,0,0,0,0,0,0,0,0,0,0,0,0,0,0,0,0,0,0,0,0,0,0,0,
        0,0,0,0,0,0,0,0,0,0,0,0,0,0,0,0,0,0,0,0,0,0,0,0,0,0,
        0,0,0,0,0,0,0,0,0,0,0,0,0,0,0,0,0,0,0,0,0,0,0,0,0,0,
        0,0,0,0,0,0,0,0,0,0,0,0,0,0,0,0,0,0,0,0,0,0,0,0,0,0,

        0,0,0,0,0,0,0,0,0,0,0,0,0,0,0,0,0,0,0,0,0,0,0,0,0,0,
        0,0,0,0,0,0,0,0,0,0,0,0,0,0,0,0,0,0,0,0,0,0,0,0,0,0,
        0,0,0,0,0,0,0,0,0,0,0,0,0,0,0,0,0,0,0,0,0,0,0,0,0,0,
        0,0,0,0,0,0,0,0,0,0,0,0,0,0,0,0,0,0,0,0,0,0,0,0,0,0,
        0,0,0,0,0,0,0,0,0,0,0,0,0,0,0,0,0,0,0,0,0,0,0,0,0,0,
        0,0,0,0,0,0,0,0,0,0,0,0,0,0,0,0,0,0,0,0,0,0,0,0,0,0,
        0,0,0,0,0,0,0,0,0,0,0,0,0,0,0,0,0,0,0,0,0,0,0,0,0,0,
        0,0,0,0,0,0,0,0,0,0,0,0,0,0,0,0,0,0,0,0,0,0,0,0,0,0,
        0,0,0,0,0,0,0,0,0,0,0,0,0,0,0,0,0,0,0,0,0,0,0,0,0,0,
        0,0,0,0,0,0,0,0,0,0,0,0,0,0,0,0,0),
```

```
x2 = c(1,1,
          0,
          1,1,1,1,1,1,1,1,1,1,1,1,1,1,1,1,1,1,1,1,1,1,1,1,1,
            1,1,1,1,
          0,0,0,0,0,0,0,0,0,0,0,0,0,0,0,0,0,0,
          1,1,1,1,
          0,0,0,0,0,
          1,1,1,1,1,1,1,1,1,1,1,1,1,1,1,1,1,1,1,1,1,1,1,1,1,
          1,1,1,1,1,1,1,1,1,1,1,1,1,1,1,1,1,1,1,1,1,1,1,1,1,
          1,1,1,1,1,1,1,1,1,1,1,1,1,1,1,1,1,1,1,1,1,1,1,1,1,
          1,1,1,1,1,1,1,1,1,1,1,1,1,1,1,1,1,1,1,1,1,1,1,1,1,
          1,1,1,1,1,1,1,1,1,1,1,1,1,1,1,1,1,1,1,1,1,1,1,1,1,
          1,1,1,1,1,1,1,1,1,1,1,1,1,1,1,1,1,1,1,1,1,1,1,1,1,

          0,0,0,0,0,0,0,0,0,0,0,0,0,0,0,0,0,0,0,0,0,0,0,0,0,
          0,0,0,0,0,0,0,0,0,0,0,0,0,0,0,0,0,0,0,0,0,0,0,0,0,
          0,0,0,0,0,0,0,0,0,0,0,0,0,0,0,0,0,0,0,0,0,0,0,0,0,
          0,0,0,0,0,0,0,0,0,0,0,0,0,0,0,0,0,0,0,0,0,0,0,0,0,
          0,0,0,0,0,0,0,0,0,0,0,0,0,0,0,0,0,0,0,0,0,0,0,0,0,
          0,0,0,0,0,0,0,0,0,0,0,0,0,0,0,0,0,0,0,0,0,0,0,0,0,
          0,0,0,0,0,0,0,0,0,0,0,0,0,0,0,0,0,0,0,0,0,0,0,0,0,
          0,0,0,0,0,0,0,0,0,0,0,0,0,0,0,0,0,0,0,0,0,0,0,0,0,
          0,0,0,0,0,0,0,0,0,0,0,0,0,0,0,0,0,0,0,0,0,0,0,0,0,
          0,0,0,0,0,0,0,0,0,0,0,0,0,0,0))

# actual values of age and blood pressure

list( y = c(1,1,
               1,
               0,0,0,0,0,0,0,0,0,0,0,0,0,0,0,0,0,0,0,0,0,0,0,0,0,0,
                 0,0,0,
               0,0,0,0,0,0,0,0,0,0,0,0,0,0,0,0,0,0,0,
               1,1,1,1,
               1,1,1,1,1,
               0,0,0,0,0,0,0,0,0,0,0,0,0,0,0,0,0,0,0,0,0,0,0,0,0,0,
               0,0,0,0,0,0,0,0,0,0,0,0,0,0,0,0,0,0,0,0,0,0,0,0,0,0,
               0,0,0,0,0,0,0,0,0,0,0,0,0,0,0,0,0,0,0,0,0,0,0,0,0,0,
               0,0,0,0,0,0,0,0,0,0,0,0,0,0,0,0,0,0,0,0,0,0,0,0,0,0,
               0,0,0,0,0,0,0,0,0,0,0,0,0,0,0,0,0,0,0,0,0,0,0,0,0,0,
               0,0,0,0,0,0,0,0,0,0,0,0,0,0,0,0,0,0,0,0,0,0,0,0,0,

               0,0,0,0,0,0,0,0,0,0,0,0,0,0,0,0,0,0,0,0,0,0,0,0,0,0,
               0,0,0,0,0,0,0,0,0,0,0,0,0,0,0,0,0,0,0,0,0,0,0,0,0,0,
               0,0,0,0,0,0,0,0,0,0,0,0,0,0,0,0,0,0,0,0,0,0,0,0,0,0,
               0,0,0,0,0,0,0,0,0,0,0,0,0,0,0,0,0,0,0,0,0,0,0,0,0,0,
               0,0,0,0,0,0,0,0,0,0,0,0,0,0,0,0,0,0,0,0,0,0,0,0,0,0,
               0,0,0,0,0,0,0,0,0,0,0,0,0,0,0,0,0,0,0,0,0,0,0,0,0,0,
               0,0,0,0,0,0,0,0,0,0,0,0,0,0,0,0,0,0,0,0,0,0,0,0,0,0,
               0,0,0,0,0,0,0,0,0,0,0,0,0,0,0,0,0,0,0,0,0,0,0,0,0,0,
```

```
         0,0,0,0,0,0,0,0,0,0,0,0,0,0,0,0,0,0,0,0,0,0,0,0,0,
         0,0,0,0,0,0,0,0,0,0,0,0,0,0,0,0,0,0),

x1 = c(77,85,87,74,56,76,69,71,74,83,67,69,52,77,72,76,77,76,70,
   67,70,74,72,74,57,88,79,64,67,67,69,77,68,86,58,77,79,86,70,80,
   60,72,63,92,56,79,76,65,70,76,55,56,48,54,50,49,45,36,54,47,52,
   47,47,49,46,39,50,51,39,59,46,44,52,53,49,53,52,46,49,46,41,58,
   61,50,45,58,51,58,44,50,57,60,48,51,56,47,44,53,41,39,50,36,55,
   47,59,43,51,43,40,43,39,38,62,49,46,45,42,50,53,43,47,31,40,43,
   45,51,51,55,56,54,49,46,53,49,34,43,60,44,56,50,46,44,49,46,53,
   45,46,56,43,53,45,49,49,41,46,44,43,48,39,59,55,46,42,38,40,42,
   29,41,38,52,51,54,50,48,47,47,47,55,44,50,55,50,43,54,42,55,42,
   38,52,49,50,56,50,39,51,49,51,54,47,42,59,54,52,47,52,44,44,45,
   50,49,48,35,43,45,49,46,47,47,59,48,52,44,48,50,58,47,49,43,56,
   50,42,48,41,52,41,43,47,43,47,54,52,50,62,47,56,42,46,53,43,43,
   50,56,54,46,51,49,47,42,42,54,53,45,35,48,52,45,47,41,48,40,42,
   50,41,56,56,49,53,34,46,50,54,42,48,51,51,56,49,56,44,47,51,51,
   36,46,58,58,51,40,53,53,44,44,51,47,48,61,40,50,49,39,47,36,45,
   47,57,44,44,51,40,44,51,60,48,51,46,51,47,48,57,48,39,49,40,58,
   50,45,41,47,49,41,64,36,47,49,58,59,45,45,53,37,43,43,50,44,49,
   43,53,48,54,53,44,47,55,52,40,43,47,47,54,46,51,65,54,53,56,53,
   49,47,38,59,55,44,52,41,55,43,47,46,55,38,45,42,54,52,42,56,48,
   44,40,47,37,44,51,37,50,47,46,45,46,44,44,52,45,38,48,54,46,41,
   41,43,44,44,49,48,41,57,45,46,46,35,50,43,41,52,43,54,46,40,50,
   61,49,44,50,40,50,38,52,46,49,53),

x2 = c(158,164,150,157,161,165,166,147,169,159,150,166,161,161,
   167,164,167,163,166,159,162,166,145,163,161,162,156,160,178,164,
   154,115,114,128,117,123,122,114,116,114,119,112,123,119,122,130,
   119,128,120,115,163,166,162,154,167,163,155,160,157,154,149,159,
   167,162,159,160,162,157,158,168,168,164,157,171,150,158,150,163,
   163,159,171,165,148,156,164,168,158,167,153,167,149,158,159,144,
   167,151,156,150,154,164,158,147,151,167,151,151,160,161,160,160,
   156,168,162,171,155,150,159,156,147,162,163,170,158,163,168,153,
   151,150,156,162,164,164,159,152,164,159,158,163,162,152,162,157,
   158,162,149,166,142,151,160,154,168,165,161,152,151,159,155,169,
   169,168,166,163,158,150,151,150,152,167,155,150,157,152,158,168,
   158,152,162,165,154,168,173,158,164,157,155,161,160,160,159,155,
   162,164,160,158,164,157,168,159,169,159,155,161,150,119,114,120,
   114,128,126,116,119,109,121,114,121,120,122,112,113,116,126,117,
   109,120,111,123,113,119,110,112,120,127,116,124,111,134,137,113,
   124,127,125,123,120,119,121,115,119,110,123,123,124,126,128,119,
   122,118,115,121,114,125,117,125,124,133,115,121,120,127,120,118,
   136,125,119,120,111,119,121,111,125,130,117,131,127,109,124,114,
   120,112,130,127,118,119,121,119,117,117,117,123,122,121,119,125,
   113,115,131,126,116,126,120,119,125,113,124,127,116,119,128,119,
   129,125,122,121,113,118,113,130,119,117,125,130,120,115,118,131,
   121,117,130,103,136,127,118,122,125,121,123,117,110,126,118,125,
   123,128,129,116,127,109,118,127,134,123,124,115,113,126,120,116,
```

```
122,112,109,121,110,114,133,131,123,122,122,116,120,104,120,125,
117,121,125,121,121,116,115,121,126,121,113,126,128,112,112,126,
120,122,113,126,119,131,118,112,126,131,124,116,122,120,115,104,
115,127,122,123,134,120,124,104,113,110,118,117,118,128,121,120,
116,121,116,121,116,128,115,123,118,114,123,121,116,111,118,119,
127,129,114,122))

list(alpha = 0,beta = c(0,0,0))
```

   a. Execute a Bayesian analysis similar to that of Table 4.2 and base
      the analysis on BUGS CODE 4.1 Alternative. Use the study results
      of Table 4.1, and for the simulation generate 55,000 observations
      with a burn-in of 5000 and a refresh of 100.
   b. The main focus of this analysis is on the interaction coefficient β_3.
      What is the 95% credible interval for β_3?
   c. Does the inclusion of the interaction term in the model (Equation 4.12)
      alter the analysis reported in Table 4.12?
   d. What is the most appropriate model: the one with or the one
      without the interaction?
 3. Using the actual values of age and SBP, perform a Bayesian analysis
    with the logistic model containing an interaction term. The second
    list statement of BUGS CODE 4.1 Alternative contains the actual age
    and SBP values. Generate 55,000 observations for the simulation,
    with a burn-in of 5000 and a refresh of 100.
   a. Confirm the Bayesian analysis reported in Table 4.18.
   b. Compare Table 4.18 with Tables 4.10 and 4.3.
   c. After comparing Table 4.18 with Table 4.3, would you conclude
      that an interaction term is needed?
   d. After comparing Table 4.18 with Table 4.10, is it better to use the
      actual values of age and SBP? Explain.

TABLE 4.17

Posterior analysis of Heart Study (Based on the Information in Table 4.1)

Logistic Regression with Interaction						
Parameter	Mean	SD	Error	2½	Median	97½
Odds ratio age	2.808	3.671	.06011	.0541	1.605	12.85
Odds ratio blood pressure	1.301	1.063	.01729	.2384	1.013	4.124
α	-3.752	.477	.00704	-4.776	-3.718	-2.912
β_1	.2902	1.392	.03692	-2.917	.4734	2.554
β_2	.005863	.7201	.01135	-1.434	.01283	1.417
β_3	.5249	1.704	.04159	-2.613	.4451	4.14

TABLE 4.18

Bayesian Analysis for Heart Study Information

Logistic Regression with Interaction

Actual Values of Age and SBP

Parameter	Mean	SD	Error	2½	Median	97½
Odds ratio age	1.266	.3549	.008029	.6845	1.227	2.039
Odds ratio blood pressure	1.167	.1324	.003044	.9139	1.158	1.45
α	−28.7	18.02	.423	−64.97	−28.39	6.536
β_1	.1971	.2823	.006742	−.3791	.2046	.7142
β_2	.1484	.1126	.002613	−.07116	.1466	.3713
β_3	−.001001	.001771	.0000418	−.004821	−.001052	.002591

4. Refer to Table 4.3 and BUGS CODE 4.2 for an analysis of the association between race and coronary heart disease. The study results are provided by Hosmer and Lemeshow (p. 48).[3] Verify the results of Table 4.3 with a simulation using 55,000 observations, a burn-in of 5000, and a refresh of 100.

   a. The posterior mean and median of the odds ratio of coronary heart disease of blacks relative to those of the group consisting of other races are 2.477 and 2.084, respectively. Is the posterior distribution skewed? To estimate the odds ratio, what is your estimate? The usual estimate is $(20/10)/(10/10) = 2$.

   b. Suppose we make the first group (whites) the reference group for our analysis. Then one would have to code the independent variables as follows: $x_1 = (0, 1, 0, 0)$, $x_2 = (0, 0, 1, 0)$, and $x_3 = (0, 0, 0, 1)$. Using this representation, perform a Bayesian analysis based on BUGS CODE 4.2 with 55,000 observations for the simulation, a burn-in of 5000, and a refresh of 100. What is the odds ratio of coronary artery disease of Hispanics relative to that of whites? Find the posterior mean and median of the odds ratio.

5. Repeat the analysis for the Israeli heart study of Table 4.1 with BUGS CODE 4.2. Note that there are four groups of patients; thus, relative to the notation used in BUGS CODE 4.2 there will be four thetas. Therefore, code the groups with three vectors, $x_1, \ldots, x_3$, and make the last group, group 4, the reference group. The reference group contains subjects with age <60 and SBP <140. Note that each vector will have four components. Compare the results of your Bayesian analysis with Table 4.2, which is based on the results of Table 4.1, whereas the Bayesian analysis is based on 400 observations.

6. With BUGS CODE 4.2, verify Table 4.5, the Bayesian analysis for predicting the number with coronary artery disease for four racial groups.

TABLE 4.19

Posterior Mean for Predicted Values

Parameter	Mean
$z[1]$	3.036
$z[2]$	3.037
$z[3]$	3.030
$z[4]$	3.04
$z[5]$	3.037
$z[6]$	5.048
$z[7]$	5.046
$z[8]$	5.047
$z[9]$	5.047
$z[10]$	5.046
$z[11]$	7.055
$z[12]$	7.056
$z[13]$	7.055
$z[14]$	7.054
$z[15]$	7.055

7. Table 4.19 gives the posterior analysis for the 15 predicted y values of Table 4.6. Verify this table using BUGS CODE 4.3 with 55,000 observations for the simulation, a burn-in of 5000, and a refresh of 100. The first list statement of the code gives the data for Table 4.6, and the second statement of the code generates the predicted y values. This is only part of the output because I did not include the standard deviation, the MCMC error, the median, and the upper and lower 2½ percentiles of the posterior distribution.

   a. Plot the predicted values, y, versus the actual values, x.

   b. Is the simple linear regression model a good fit to the data?

   c. Plot the posterior density of $z[1]$.

   d. What is the 95% credible interval for $z[1]$?

8. Verify Table 4.7, the Bayesian analysis for the association between blood pressure and age, using BUGS CODE 4.2.

   a. What is your estimate of the intercept?

   b. What is your estimate of the slope?

   c. Is the model a good fit to the data?

9. Using BUGS CODE 4.4 and the information about cigarette consumption, verify Table 4.12, the Bayesian analysis for the multiple regression of consumption on income and price per pack. Use 55,000 observations with a burn-in of 5000 and a refresh of 100. Note that the second statement of the code generates 51 predicted values of cigarette consumption. You must use z as the node, and the predicted

values will be generated. Use the posterior mean of the predicted values as the predicted values. I deleted the value 265.7 for the consumption of NH, the state of New Hampshire, which is considered an outlier and is not used in the Bayesian analysis.

   a. Does the model provide a good fit to the observed cigarette consumption?

   b. What is the 95% credible interval for the variance about the regression line?

   c. What is the interpretation for the estimated intercept term of the model?

   d. Verify Figure 4.10, a plot of predicted consumption versus observed consumption.

10. Based on BUGS CODE 4.1, perform a Bayesian regression analysis with 55,000 observations for the simulation, a burn-in of 5000, and a refresh of 100. Recall that the first three statements of the code are relevant for the regression analysis, with y as the dependent variable and x_1 and x_2 as the independent variables. The z vector consists of the predicted SBP values.

   a. Compute the residuals, that is, the differences: SBP minus the corresponding predicted values z.

   b. Graph the absolute value of the residuals versus the predicted value. The graphs should look similar to Figure 4.12.

   c. Is there a trend in absolute residuals as the predicted values increase?

   d. If there is a trend, perform a weighted regression by transforming the dependent variable (SBP) by the appropriate transformation.

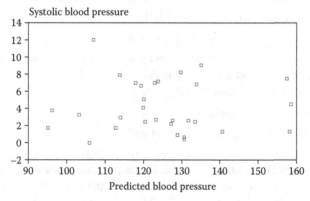

FIGURE 4.12
Absolute residuals versus predicted values.

To choose the appropriate transformation of the dependent variable, refer to Table 4.15.

e. Does the transformation stabilize the variance of the dependent variable?

f. Is the weighted regression model a good fit to the data?

11. a. Verify Table 4.14, the Bayesian analysis for the state education expenditure data for 50 states. Use BUGS CODE 4.5 to execute the regression of state expenditure on the three independent variables. Note that the first list statement is the information for the unweighted regression. Use 65,000 observations for the simulation, with a burn-in of 5000 and a refresh of 100.

b. Verify Table 4.15, the Bayesian weighted regression with the square root of the dependent variable (state education expenditure data) on the three independent variables. The second list statement gives the data for the weighted regression. Use 65,000 observations for the simulation, with a burn-in of 5000 and a refresh of 100.

c. Did the square root transformation stabilize the variance? Why?

d. Is the weighted regression model a good fit to the data? Plot the predicted values of the dependent variable versus the independent variables.

12. Table 4.20 is a previous study, which was carried out before the study reported in Table 4.1.

Using BUGS CODE 4.1, execute a Bayesian analysis with 55,000 observations using Table 4.20 as prior information and the information in Table 4.1. Use a burn-in of 5000 and a refresh of 100.

TABLE 4.20

Myocardial Infarction versus SBP by Age

A Previous Study			
	MI	No MI	
Age ≥60			
SBP ≥140	1	15	16
SBP <140	0	8	8
Total	2	23	24
Age <60			
SBP ≥140	2	80	82
SBP <140	3	118	121
Total	5	198	203

    a. What is the posterior distribution of the odds ratio for age?

    b. What is the posterior distribution of the odds ratio for SBP?

    c. Compare the results with those reported in Table 4.2.

    d. How much does the standard deviation of the posterior distribution for the odds ratio decrease?

References

1. Chatterjee, S., and Price, B. *Regression Analysis by Example*, John Wiley & Sons Inc., 1991, New York.
2. Ntzoufras, I. *Bayesian Modeling Using WinBUGS*, John Wiley & Sons Inc., 2009, New York.
3. Hosmer, D.W., and Lemeshow, S. *Applied Logistic Regression*, John Wiley & Sons Inc., 1989, New York.
4. Agresti, A. *Categorical Data Analysis*, John Wiley & Sons Inc.,1990, New York.
5. Congdon, P. *Bayesian Models for Categorical Data*, John Wiley & Sons Inc., 2005, New York.
6. Woolson, R.F. *Statistical Methods for the Analysis of Biomedical Data*, John Wiley & Sons Inc., 1987, New York.
7. Montgomery, D.C., Peck, E.A., Vining, G. G. *Introduction to Linear Regression Analysis*, John Wiley & Sons Inc., 2001, New York.
8. Broemeling, L.D. *Advanced Bayesian Methods for Medical Test Accuracy*, CRC Press, 2012, Boca Raton, FL.
9. Rothman, K.J. *Modern Epidemiology*, Little Brown and Company, 1986, Toronto, Canada.
10. Carroll, R.J., and Ruppert, D. *Transformation and Weighting in Regression*, Chapman and Hall, 1988, New York.

5

A Bayesian Approach to Life Tables

5.1 Introduction

A ubiquitous epidemiologic technique is the estimation of survival by the life table, and the method will be described with a Bayesian approach. Life tables play an important role in many areas, including estimating survival in clinical trials and screening tests, estimating survival for use in the insurance industry, and measuring survival for use of pension funds.

For example, many retired people take lifetime annuities in the form of a monthly or annual payment and it is of essential interest to the pension fund to have a good idea of the life expectancy of the recipient. Knowing the life expectancy of the pensioner is essential in setting the payout amount of the pension. Of course, the same problem is faced by the Social Security Administration and other insurance entities. When a person takes out a whole life policy, the insurance company needs to know the survival time to set the premiums of the policy. Needless to say, survival analysis is an industry by itself. I was employed by the University of Texas MD Anderson Cancer Center and was involved in designing Phase I and Phase II clinical trials. In a Phase II trial, the objective is to compare a therapy with a historical control in regard to the response to therapy, and a life table is essential to estimate the response rate.

The chapter begins with a definition of the basic life table, where a certain number of individuals are followed for a fixed period of time and the survival is measured at the end of each interval, where there are many intervals. For example, a cohort is followed for 10 years and the mortality measured at the end of each year. The basic table bases mortality on the number who die during a given interval, assuming the there is no loss to follow-up and that each mortality is from the same cause and not some other factor. The basic life table is generalized to include subjects lost to follow-up and patients that die from other causes (not the disease of interest).

The Bayesian approach is to assume the number of deaths over a given interval follows a binomial distribution with an unknown probability of mortality; thus, the joint distribution of the number of deaths for all intervals has a multinomial distribution and the joint distribution of the probability of death for all intervals is Dirichlet (assuming a uniform prior density for the probabilities of mortality).

Various generalizations allow for more realistic survival studies. The first generalization is to assume a random number of individuals are lost to follow-up in each interval and that the probability of survival is the ratio of the number who died divided by the number alive at the beginning of the interval minus the number who withdrew (for various reasons) during that interval of time. The next generalization is to allow for those who withdraw and those that die from other causes, other than the cause of interest (e.g., lung cancer).

Several examples will illustrate Bayesian inferences for estimating the survival experience of a cohort of subjects. For example, a cohort of melanoma patients is followed over the course of therapy and for a fixed period after the termination of the trial the response time is estimated by the life table technique. The estimation of mortality is based on the joint posterior of mortality for each interval, then the overall mortality is also easily estimated. As in earlier chapters, the foundation of the analysis is the WinBUGS code, which can be used by the reader to learn the fundamentals of an important Bayesian methodology.

Chapter 7 of Kahn and Sempos[1] is a good introduction to life tables; Chiang[2] presents a more advanced treatment of the subject; and Rothman[3] gives an account from the point of view of an epidemiologist. Of interest to the biostatistician and epidemiologist is the book by Newman.[4]

5.2 Basic Life Table

A group of subjects is followed over time and their survival is measured at the end of each period. Consider the following notation for a life table:

n: the length of each period

t: the time at the beginning of the period

O_t: the number under observation at exact time t

$_n m_t$: the mortality during the period t to t + n

$_n p_t$: the probability of surviving from time t to time t + n

$_n q_t$: the probability of dying from time to time t + n

$_n P_t$: the probability of surviving over the period t to t + n (for an interval larger than a single period)

Suppose a group of 1000 patients who had a heart attack is followed for 10 years and the length of each period is a year.

The information in Table 5.1 is based on a chart from the National Institutes of Health (NIH)[5] and can be accessed at the link cited in the reference. However, the information reported in Table 5.2 is somewhat hypothetical and is based on the NIH table. Using the notation defined in the preceding discussion,

TABLE 5.1

Summarization of Survival for Subjects with Coronary Heart Disease

Year	Under Observation at Start of Period	Mortality during Period
1996	1000	6
1997	994	5
1998	989	5
1999	984	5
2000	979	5
2001	974	4
2002	970	5
2003	965	5
2004	960	5
2005	955	5

Source: National Institutes of Health. National Heart, Lung, and Blood Institute. Chart 3–26, Death and Age-Adjusted Death Rates for Coronary Heart Disease, 1980–2006. http://www.nhlbi.nih.gov/resources/docs/2009_ChartBook.pdf.

TABLE 5.2

Life Table Calculations for Survival of Coronary Heart Disease Subjects

Time at Beginning of Period t	Under Observation at Time t O_t	Mortality during Period $_1m_t$	Probability of Dying in Interval $_1q_t$	Probability of Surviving through Period $_1p_t$
1 (1996)	1000	6	.006	.994
2 (1997)	994	5	.0050301810	.9949698
3 (1998)	989	5	.0050556117	.9949443
4 (1999)	984	5	.0050813008	.9949186
5 (2000)	979	5	.0051072522	.9948927
6 (2001)	974	4	.0041067761	.9958932
7 (2002)	970	5	.0051546391	.9948453
8 (2003)	965	5	.0051813471	.9948186
9 (2004)	960	5	.0052083333	.9947916
10 (2005)	955	5	.0052356020	.9947643

the information is displayed as follows. The reader should be aware that the information is related to the NIH information but is hypothetical.

Note that the probability of surviving from time 0 (beginning of the year 1996) through the end of period 9 (the end of the year 2005) is given by

$$_9P_0 = \prod_{i=1}^{i=10} {}_1p_i = .955$$

which equals the product of the 10 values in the last column of Table 5.2.

Thus, the overall 10-year survival is 95.5%. Why use the product of 10 numbers to calculate the 10-year survival when it is obvious that the answer is given by 950/1000? The life table analysis assumes that there were no withdrawals. This assumption will be relaxed in future analyses of life tables. Note that it is also assumed that for a given period, the probability of death is the same for each individual entering that period and that the event of death is independent among the n individuals entering the cohort.

What is the Bayesian approach to estimate the individual period survivals and the overall survival? One approach is to assume that the mortality $_nm_t$ (the number who die in a given period) has a binomial distribution with parameters$(_nq_t, O_t)$, where O_t is the number under observation at the beginning of time t (for an interval of length n) and $_nq_t$ is the probability of death of an individual person during the interval from time t to time t + n. Note that for coronary heart disease, n = 1 and t = 1, 2, ..., 10. The unknown parameters are the mortality probabilities $_nq_t$, and if one assumes an improper prior density, namely,

$$f(_1q_1, _1q_2, _1q_3, ..., _1q_{10}) \propto \frac{1}{_1q_1, _1q_2, _1q_3, ..., _1q_{10}} \tag{5.1}$$

for $0 < _1q_t < 1$, for t = 1, 2, ..., 10, the posterior density of the mortality probability $_1q_t$ is a beta distribution with parameter $(_1m_t, O_t - _1m_t)$, where O_t is the total number of subjects available at time t. That is to say, the mortality probabilities are jointly independent and

$$_1q_t \sim \text{beta}(_1m_t, O_t - _1m_t) \tag{5.2}$$

t = 1, 2, ..., 10.

Consider Table 5.2 that reports the survival experience of 1000 coronary artery disease patients from 1996 through 2005. The code given as follows will be used to perform a life table analysis from a Bayesian viewpoint.

BUGS CODE 5.1

```
model;
{
# posterior distribution of mortality probabilities
for(i in 1:a){q[i]~dbeta(m[i], b[i])}
for(i in 1:a){b[i]<- O[i]-m[i]}
# survival probabilities
for(i in 1:a){p[i]<-1-q[i]}
# overall survival
s <- p[1]*p[2]*p[3]*p[4]*p[5]*p[6]*p[7]*p[8]*p[9]*p[10]
}
# data from table 5.2
# the O vector is the number available at the beginning of the
  period
```

```
list(a = 10, m = c(6,5,5,5,5,4,5,5,5,5),
O = c(1000,994,989,984,979, 974,970,965,960,955,950))
# initial values
list(q = c(.5,.5,.5,.5,.5,.5,.5,.5,.5,.5))
```

The first statement of the code determines the posterior beta distribution of the individual mortality probabilities, and the first list statement gives the information from Table 5.2, where q is the vector of mortality probabilities and p the corresponding vector of survival probabilities. The vector m is the vector of deaths, and the vector O gives the number of individuals entering the 10 periods over a 10-year interval. For the simulation, 55,000 observations are generated with a burn-in of 5000 and a refresh of 100, and the results are portrayed in Table 5.3.

The Bayesian analysis provides an estimate of uncertainty for each posterior distribution, namely, the posterior standard deviation, and the uncertainty is also reflected with the corresponding 95% credible interval. For example, the probability of a death for the first period (1996) is estimated as .005985 with a 95% credible interval of (.002214, .01164). Of primary interest is the overall 10-year survival P, estimated as .9501 with a 95% credible interval of (.9355, .9626). When referring to Table 5.2, one notices the similarity in the probability of death given by the posterior mean and the "standard" calculation of the table. Also shown is the posterior density of the probability of death for the fourth period (2009) (Figure 5.1).

Note the slight right skewness of the distribution, which should be expected since the posterior mean is .005075 compared to a posterior median of .004743!

TABLE 5.3

Bayesian Life Table Analysis for 1000 Subjects with Coronary Artery Disease

Parameter	Mean	SD	Error	2½	Median	97½
$_1q_1$	.005985	.002438	<.00001	.002214	.005654	.01164
$_1q_2$	.005035	.002246	<.000001	.001639	.004708	.0103
$_1q_3$	.005045	.00225	<.000001	.001619	.004726	.01026
$_1q_4$	.005075	.002273	<.000001	.00164	.004743	.01042
$_1q_5$	.005117	.002273	<.00001	.001677	.004785	.01042
$_1q_6$	.004104	.00205	<.000001	.001119	.003768	.008996
$_1q_7$	.005141	.002294	<.00001	.001657	.004797	.0106
$_1q_8$	.00517	.0023	<.00001	.00169	.004827	.01056
$_1q_9$	.00522	.002311	<.00001	.001692	.004903	.01062
$_1q_{10}$	.00522	.002311	<.000001	.001707	.004877	.01063
P	.9501	.006929	<.00001	.9355	.9504	.9626

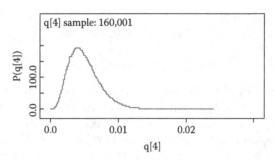

FIGURE 5.1
Posterior density of q[4].

5.2.1 Life Table Generalized

As was stated in Section 5.2, the basic life table assumes that for a given period, the probability of death is the same for all individuals and that there are no withdrawals. The second restriction will be relaxed to allow for withdrawals from the study. When the cohort is followed through the study time, an individual can drop out for various reasons: moving away, leaving the study for various reasons, the study terminates, and so on.

To accommodate this eventuality, consider the scenario with a cohort of 450 people who have recently had an operation for lung cancer, but in addition, suppose the number of subjects that withdraw is denoted by $_2w_t$, the number in the interval from time t to time t + 2 months. The divisor for the probability of death is defined by

$$O'_t = (O_t - {}_2w_t) \tag{5.3}$$

namely, the number under observation at time t minus the number who withdrew from time t to time t + 2. The first time period is from time 1 to time less than or equal to 3 months, and the second time period is from 3 months to less than 5 months. The probability of dying in the period from time t to time t + 2 has numerator $_2m_t$ and denominator O'_t. The period of observation is 2 months over six 2-month intervals.

Note that the death of an individual might be something other than complications of lung cancer. The subject might have died of suicide, murder, accident, or from some other disease. Life tables appropriate for disease-specific causes of death are discussed in Section 5.3. What is the overall survival over the 1-year interval? It is given by the product of the six survival probabilities $_2p_t$, t = 1, 2, 3, 4, 5, 6, or

$$P = \prod_{i=1}^{i=6} {}_2p_t = .3101 \tag{5.4}$$

What is the Bayesian approach to the analysis of a life table (Table 5.4)? BUGS CODE 5.2 is quite similar to BUGS CODE 5.1, where the latter has as a denominator of the mortality probability the revised number, namely, the

TABLE 5.4

Postoperative Survival Experience of 450 Subjects with Lung Cancer

Time at Beginning of Period t	Under Observation at Time t O_t	Number of Deaths in Interval $_2m_t$	Number Withdrawn in Interval $_2w_t$	Adjusted O_t $(O_t - _2w_t)$ O_t'	Mortality during Interval $_2m_t/O_t'$ $_2q_t$	Survival $_2p_t$
1	450	207	8	442	.4684	.5316
3	235	41	10	225	.1822	.8177
5	184	9	9	175	.0514	.9485
7	166	7	11	155	.0416	.9548
9	148	18	12	136	.1323	.8676
11	118	10	9	109	.0917	.9082
7	99					

number observed at the beginning of the period minus the number of withdrawals during the period. Also, the first list statement contains the data from Table 5.4, where the vector m is the number of deaths, the vector w the number who withdrew, and O the observed number at the beginning of each 2-month period.

BUGS CODE 5.2

```
Model;
{

# posterior distribution of mortality probabilities
for(i in 1:a){q[i]~dbeta(m[i], b[i])}
for(i in 1:a){b[i]<-Op[i]-m[i]}
# the below statement applies to Table 5.4
#for(i in 1:a){Op[i]<-O[i]-w[i]}
# below is applied when only half the withdrawals are used
# applies to Table 5.6
#for(i in 1:a){Op[i]<-O[i]-w[i]/2}
# cause specific net data
for(i in 1:a){Op[i]<-O[i]-w[i]/2-do[i]}
# survival probabilities
for(i in 1:a){p[i]<-1-q[i]}
# overall survival
s <- p[1]*p[2]*p[3]*p[4]*p[5]*p[6]
# overall probability of response problem 4
#P<- p[1]*p[2]*p[3]*p[4]*p[5]

}
# data from table 5.4
# the O vector is the number available at the beginning of the
  period
# w vector is the number of withdrawals
```

```
# m vector is the number of deaths
list(a = 6, m = c(207,41,9,7,18,10),
O = c(450,235,184,166,148,118),
w = c(8,10,9,11,12,9))
# data for crude cause specific information
list(a = 6, m = c(51,9,5,6,11,5),
O = c(450,235,184,166,148,118),
w = c(8,10,9,11,12,9))
# data for net cause specific information
list(a = 6, m = c(51,9,5,6,11,5),
O = c(450,235,184,166,148,118),
w = c(8,10,9,11,12,9),
do = c(156,32,4,1,7,5))

# data for clinical trial problem 4
list(a = 5, m = c(1,1,3,4,5),
O = c(34,30,27,20,12),
w = c(3,2,4,5,1))

# initial values
list(q = c(.5,.5,.5,.5,.5,.5))
# initial values for problem 4
list(q = c(.5,.5,.5,.5,.5))
```

Based on BUGS CODE 5.2, the Bayesian analysis is executed with 55,000 observations for the simulation, a burn-in of 5000, and a refresh of 100, and the analysis reported in Table 5.5.

One sees the posterior mean of the mortality probabilities is quite similar to the calculations of Table 5.5 and that the standard deviation of their posterior distributions is quite small, as are the Monte Carlo Markov chain (MCMC) errors. Overall, survival has a posterior mean of .3104, which agrees quite well with the estimate of .3101 calculated directly from Table 5.4. One should be very confident that the Bayesian estimates are accurate estimates of the relevant parameters of the life table. If one ignores the withdrawals, the overall survival is .3511, compared to .3104 when withdrawals are accounted for.

TABLE 5.5

Bayesian Life Table for Lung Cancer Patients

Parameter	Mean	SD	Error	2½	Median	97½
$_2q_1$	.4683	.02375	<.0001	.4221	.4683	.515
$_2q_3$	.1822	.02569	<.0001	.1347	.1811	.2352
$_2q_5$	.0514	.01664	<.00001	.024	.04971	.08852
$_2q_7$	.04508	.01664	<.00001	.0184	.04313	.08315
$_2q_9$	.1325	.02905	<.0001	.08074	.1309	.1939
$_2q_{11}$	.09159	.0273	<.0001	.04506	.08912	.1515
P	.3104	.0234	<.0001	.2658	.3101	.3567

5.2.2 Another Generalization of the Life Table

For the life table (Table 5.4), it is assumed that the withdrawals occur at exactly the beginning of each time period and that the probability of death is the same as for those remaining under observation. The first restriction will be relaxed, by considering three possible scenarios proposed by Kahn and Sempos:[1]

1. All withdrawals have died by the end of the interval.
2. None have died by the end of the interval.
3. They have the same probability of death after withdrawal as those that remain under observation.

The first two alternatives are unrealistic; thus, the third will be adopted under the proposition that the withdrawals are uniformly distributed over each time period and that half of them will die over a given time period. The revised denominator for the probability of death is

$$O_t'' = \left(O_t - \frac{{}_n w_t}{2} \right) \tag{5.5}$$

where O_t is the number under observation at the beginning of the period and ${}_n w_t$ is the number of withdrawals; thus the probability of death over the interval is

$$ {}_n q_t = \frac{{}_n m_t}{O_t''} \tag{5.6}$$

Suppose that the withdrawals occur uniformly from time t to time t + n, which is probably realistic in many applications, but there are situations where it can be questioned. For example, if one is following the general population over the life span of the cohort, and the first period is from birth to 5 years, one would expect more deaths immediately following birth compared to the end of the period. One could shorten the time period or adopt a model that allows for this expected pattern of withdrawals.

There is only a slight change in the mortality and survival probabilities and it can be shown that the overall survival is estimated as .3183. To do the Bayesian analysis, execute BUGS CODE 5.2 with 55,000 observations, a burn-in of 5000, and a refresh of 100. Note that the code includes two statements for computing the denominator of the mortality probabilities. One is

```
for(i in 1:a){Op[i]<-O[i]-w[i]}          (5.7)
```

and applies to Table 5.4, while the other is

```
for(i in 1:a){Op[i]<-O[i]-w[i]/2}        (5.8)
```

and is appropriate for Table 5.6.

TABLE 5.6

Postoperative Survival Experience of 450 Subjects with Lung Cancer

Time at Beginning of Period t	Under Observation at Time t O_t	Number of Deaths in Interval $_2m_t$	Number Withdrawn in Interval $_2w_t$	Adjusted O_t $(O_t - {_2w_t}/2)$ O''	Mortality During Interval $_2m_t / O_t'$ $_2q_t$	Survival $_2p_t$
1	450	207	8	446	.4641	.5358
3	235	41	10	230	.1782	.8217
5	184	9	9	179.5	.0501	.9498
7	166	7	11	160.5	.0436	.9563
9	148	18	12	142	.1267	.8732
11	118	10	9	113.5	.0881	.9118
	99					

TABLE 5.7

Bayesian Life Table for Lung Cancer Patients

Parameter	Mean	SD	Error	2½	Median	97½
$_2q_1$	.4641	.02365	<.0001	.418	.4641	.5107
$_2q_3$	.1782	.02518	<.0001	.1318	.1771	.2302
$_2q_5$	.05011	.01632	<.00001	.02333	.04847	.08637
$_2q_7$	.04352	.01607	<.00001	.01776	.04164	.08018
$_2q_9$	.1269	.02789	<.0001	.07729	.1253	.1857
$_2q_{11}$	.08797	.02627	<.0001	.04315	.08562	.1455
P	.3186	.0232	<.0001	.2741	.3183	.3648

One of these statements needs to be deactivated to execute the analysis. Deactivate the statement by putting the # symbol before the statement. This is the only revision required to perform the Bayesian analysis, and the posterior analysis is slightly changed. For example, the overall survival is changed from .3101 to .3183 as computed directly from Table 5.6. For more information about the Bayesian analysis, see Table 5.7.

There is excellent agreement between the direct calculation of the survival probabilities from Table 5.6 and the corresponding posterior means of Table 5.7, and furthermore, the distributions appear to be symmetric about the posterior means.

5.3 Disease-Specific Life Tables

For the preceding analyses, the cause of death was of no concern, but in many clinical trials, one is primarily interested in the number who die from a specific disease. For example, with a Phase II trial for melanoma, where the therapy is an immunotherapy, one wants to know the response rate of that therapy and also wants to know the time to recurrence of each patient. Also for the life table (Table 5.6), one would want to estimate the survival probabilities for death from lung cancer. Recall the study begins with a cohort of 450 patients, who underwent surgery to remove the primary tumor. Thus, returning to the lung cancer survival study, consider a column that records the number who die from complications of lung cancer.

On following the subjects, one must know the cause of death. Note that the denominator of the probability of dying from lung cancer and from dying from other causes is adjusted by one half the number of withdrawals. The computations are clear cut, and the estimated probabilities of dying from lung cancer are easily computed and referred to as crude probabilities of dying from lung cancer.

The analysis is easily executed with BUGS CODE 5.2, where the second list statement contains the information from Table 5.8. I used 55,000 observations for the simulation, with 5000 for the burn-in and 100 for the refresh. As is seen, the results are almost identical to those of Table 5.7.

It is interesting to see that the probability of surviving death from lung cancer for 1 year is estimated at .7025 with a 95% credible interval (.6408, .7588); however, it should be noted that a person who does not die from lung cancer could die from other causes (Table 5.9).

Consider a variation of the preceding scenario for estimating the probability of dying from complications due to lung cancer. If the denominator of the probability of dying from lung cancer is defined as

$$O_t''' = \left(O_t - \frac{{}_2 w_t}{2} - \frac{{}_2 od_t}{2} \right) \tag{5.9}$$

TABLE 5.8

Postoperative Survival Experience of 450 Subjects with Lung Cancer (Crude Mortality)

Period T	Under Observation at Time t O_t	Total Deaths	Lung Cancer	Other	Number Withdrawn in Interval ${}_2 w_t$	Adjusted O_t $(O_t - {}_2 w_t/2)$ O_t''	Lung Cancer Mortality during Interval	Other
1	450	207	51	156	8	446	.1143	.3497
2	235	41	9	32	10	230	.0391	.1391
3	184	9	5	4	9	179.5	.0278	.0222
4	166	7	6	1	11	160.5	.0373	.00623
5	148	18	11	7	12	142	.0774	.0492
6	118	10	5	5	9	113.5	.0440	.0440
7	99							

TABLE 5.9

Bayesian Analysis for Postoperative Survival of Lung Cancer Patients

Parameter	Mean	SD	Error	2½	Median	97½
$_2q_1$	.1144	.01506	<.00001	.08668	.1138	.1455
$_2q_3$	.03908	.01281	<.00001	.01806	.03771	.0679
$_2q_5$	.02777	.01218	<.00001	.00920	.02605	.05615
$_2q_7$	.03731	.01486	<.00001	.01391	.03543	.07144
$_2q_9$	.07737	.0224	<.00001	.03948	.0754	.1266
$_2q_{11}$	.04401	.0192	<.00001	.01439	.04138	.08861
P	.7025	.3022	<.0001	.6408	.7034	.7588

TABLE 5.10

Bayesian Analysis for Net Probability of Death from Lung Cancer

Parameter	Mean	SD	Error	2½	Median	97½
$_2q_1$	.1759	.02234	<.00001	.1345	.1752	.2217
$_2q_3$	.04536	.01478	<.00001	.02107	.04379	.07854
$_2q_5$	.02482	.01246	<.00001	.009384	.02669	.05746
$_2q_7$	.03756	.01496	<.00001	.01404	.03568	.07198
$_2q_9$	.08138	.02351	<.0001	.04154	.07924	.1332
$_2q_{11}$	.04609	.02009	<.00001	.01513	.04335	.09276
P	.6446	.03204	<.0001	.5797	.6453	.7053

where $_2w_t$ is the number of withdrawals over the period t to t + 2, and $_2od_t$ is the number of deaths from other causes, one in effect is treating the number who die from other causes the same as the number who withdraw.

One is also assuming the probability of death for those who die of other causes is the same as the probability of death for those observed at the beginning of the interval t to t + 2. The resulting probability of death from lung cancer is called the net probability. The analysis is executed using BUGS CODE 5.2 with 55,000 observations for the simulation, a burn-in of 5000, and a refresh of 100. Note that the third list statement of the code has the relevant data for performing the Bayesian estimation of the net probability of dying from lung cancer with the following results.

One sees that the net probability of death is larger for each time period compared to the crude probability of death from lung cancer (Table 5.10). This is true because the denominator (Equation 5.9) of the net probability is usually smaller than (Equation 5.5) for the crude probability.

5.4 Life Tables for Medical Studies

5.4.1 Introduction

The objective of this section is to present an adaptation of life tables to follow-up medical studies, where the program is carried out over a period of years beginning at time 0 with N_0 patients, namely, the number of patients alive at the beginning of the study. The notation and description of follow-up medical studies follows Chiang,[2] who explains the time axis refers to the time of observation since time 0 and x denotes the time in years that the patients have been followed (p. 270). A time period of 1 year will be used but, of course, the period length can be arbitrary; thus, the number of deaths, the number of withdrawals, and the number alive at time x will be measured over the period from x to x + 1. Other symbols important to remember are p_x and q_x, where the former is the probability that a subject alive at time x will survive over the period (x, x + 1) and the probability a subject will die over the same period is $q_x = 1 - p_x$. Other notation is defined as follows.

N_x the number alive at time x, which is the number of survivors of those that entered the study at least x years before the closing time of the study. Obviously N_x decreases through time because of deaths and withdrawals (due to the termination of the study or lost to follow-up)

Among the N_x patients, there are two groups. Let m_x be the number of patients who enter the program more than x + 1 years before the close and therefore will be observed for the entire period of one year. Of these patients, let d_x and s_x be the number who die and survive, respectively, during the period (x, x + 1).

Let n_x be the number who enter the program less than x + 1 years before the close of the study; therefore, these patients will not be observed in the period beginning at time x + 1, due to withdrawals at the close of the study.

What happens to the n_x individuals in the period (x, x + 1)? Denote by d'_x the number who die before the close of the study, and let w_x be the number who will survive to the closing date. Let the total number of deaths in the period be

$$D_x = d_x + d'_x$$

For the Bayesian approach, what variables in the study will be considered random variables? The basic variables are d_x, s_x, d'_x, and w_x. For many studies, the objective is to study the survival experience of subjects that die from a specific cause; however, people die of many causes. For example, thousands of patients will be followed, and it is the main concern to determine the survival experience of those that die of complications due to breast cancer, but, of course, they can die from other causes as well. For an overall view of such a study, see Table 5.11.

TABLE 5.11

Survival, Withdrawals, and Deaths in (x, x + 1)

	Total Number of Patients	Number to be Observed for the Entire (x, x + 1)	Withdrawals during (x, x + 1)
Total	N_x	m_x	n_x
Survivors	$s_x + w_x$	s_x	w_x
Deaths	D_x	d_x	d'_x
Deaths due to Competing Risks			
R_1	D_{x1}	d_{x1}	d'_{x1}
R_2	D_{x2}	d_{x2}	d'_{x2}
R_r	D_{xr}	d_{xr}	d'_{xr}

Source: Chiang, C.L., *Introduction to Stochastic Processes in Biostatistics*, Table 3, p. 281, John Wiley & Sons Inc., New York, 1968.

The California study provides numbers to the symbols in Table 5.11. Note that there are r causes of death. For the Bayesian analysis, the approach is to assign distributions to the number of deaths for each cause of death for the individuals of the two groups of patients m_x and n_x. Recall that m_x is the number of patients who will be observed for the entire period, because they entered the study more than x + 1 years before the study close, whereas n_x patients entered the study less than x + 1 years before the close.

It will be assumed that among the m_x, $(s_x, d_{x1}, d_{x2}, ..., d_{xr})$ follow a multinomial distribution with parameter $(p_x, q_{x1}, q_{x2}, ..., q_{xr})$, and that independently, among the n_x, $(w_x, d'_{x1}, d'_{x2}, ..., d'_{xr})$ follow a multinomial distribution with parameter $(P'_x, q'_{x1}, q'_{x2}, ..., q'_{xr})$. Now assume the unknown parameters follow an improper prior distribution; then the posterior distribution of

$$(P_x, q_{x1}, q_{x2}, ..., q_{xr}) \sim \text{Dirichlet}(s_x, d_{x1}, d_{x2}, ..., d_{xr}) \qquad (5.10)$$

and independently, the posterior distribution of

$$(P'_x, q'_{x1}, q'_{x2}, ..., q'_{xr}) \sim \text{Dirichlet}(w_x, d'_{x1}, d'_{x2}, ..., d'_{xr}) \qquad (5.11)$$

If a uniform distribution is assumed for the unknown probabilities

$$(P_x, q_{x1}, q_{x2}, ..., q_{xr}) \sim \text{Dirichlet}(s_x + 1, d_{x1} + 1, d_{x2} + 1, ..., d_{xr} + 1) \qquad (5.12)$$

and independently,

$$(P'_x, q'_{x1}, q'_{x2}, ..., q'_{xr}) \sim \text{Dirichlet}(w_x + 1, d'_{x1} + 1, d'_{x2} + 1, ..., d'_{xr} + 1) \qquad (5.13)$$

On the basis of these two Dirichlet posterior distributions, the survival experience of 20,858 patients, recorded by the California Tumor Registry, will be analyzed from a Bayesian approach.

5.4.2 California Tumor Registry 1942–1963

To illustrate the methodology of the Bayesian approach, $N_0 = 20,858$ females diagnosed with breast cancer entered California hospitals beginning on January 1, 1942, and patients could enter at anytime from that date to a year before the closing date of December 31, 1962.

See Tables 5.12 and 5.13 for detailed information about the California study. Subjects died of breast cancer, or other cancers, or other causes than

TABLE 5.12

Survival Experience Following Diagnosis of Breast Cancer Numbers Not due for Withdrawal in Period $(x, x + 1)$

Interval since Admission	N_x	m_x	s_x	Breast Cancer d_{x1}	Other Cancer d_{x2}	Other Causes d_{x3}	Lost Causes d_{x4}
0–1	20858	20858	17202	2381	56	649	570
1–2	17202	16240	14052	1689	35	352	112
2–3	14052	13134	11563	11161	28	282	100
3–4	11563	10769	9521	863	29	257	99
4–5	9521	8741	7881	533	33	187	107
5–6	7881	7223	6486	432	33	184	88
6–7	6486	5880	5392	270	17	131	70
7–8	5392	4798	4385	195	24	128	66
8–9	4385	3888	3527	177	13	100	71
9–10	3527	3120	2865	106	17	85	47
10–11	2865	2470	2293	84	10	50	33
11–12	2293	1942	1798	44	11	59	30
12–13	1798	1499	1376	47	5	46	25
13–14	1376	1168	1071	25	12	39	21
14–15	1071	894	817	21	2	36	18
15–16	817	652	599	13	1	25	14
16–17	599	460	419	16	—	13	12
17–18	419	312	284	9	1	9	9
18–19	284	202	192	1	—	6	3
19–20	192	130	117	—	3	6	4
20–21	117	55	49	2	1	2	1
21–22	49	0	0	—	—	—	—

Source: Chiang, C.L. *Introduction to Stochastic Processes in Biostatistics*, John Wiley & Sons Inc., 1968, New York.

TABLE 5.13

Survival Experience Following Diagnosis of Breast Cancer Numbers due for Withdrawal in Period (x, x + 1)

Interval since Admission	n_x	w_x	Breast Cancer d'_{x1}	Other Cancer d'_{x2}	Other Causes d'_{x3}	Lost Causes d'_{x4}	N_x
0–1	—	—	—	—	—	—	20858
1–2	962	745	59	1	5	152	17202
2–3	918	726	192	4	11	133	14052
3–4	794	610	31		8	145	11563
4–5	780	596	26	3	11	144	9521
5–6	658	504	17		9	128	7881
6–7	606	461	17	1	6	121	6486
7–8	594	438	20	1	5	130	5392
8–9	497	392	8		3	94	4385
9–10	407	321	7		6	73	3527
10–11	395	313	3		7	72	2865
11–12	351	256	15		3	77	2293
12–13	299	228	4		5	62	1798
13–14	208	157			1	50	1376
14–15	177	145	1		2	29	1071
15–16	165	118	3		1	43	817
16–17	139	111	2		3	23	599
17–18	107	82	1		1	23	419
18–19	82	69			3	10	284
19–20	62	43	1		3	15	192
20–21	62	45		1	1	15	117
21–22	49	40			8	49	

Source: Chiang, C.L. *Introduction to Stochastic Processes in Biostatistics,* John Wiley & Sons Inc., 1968, New York.

cancer, or those lost to follow-up. All patients were observed the first year of the study, none were withdrawn because of study close; 17,202 survived the first year, and 2381 died of breast cancer. During the first year, none were lost due to withdrawal, but during the second year, 962 withdrew from the study and among those 745 withdrew alive and 59 died of breast cancer, and 152 died of unknown causes. It is interesting to observe that 49 patients who registered for the study in 1942 were still alive 22 years later on December 31, 1963. It is important to realize that each patient was carefully monitored for their duration of study.

The Bayesian analysis is based on BUGS CODE 5.3.

BUGS CODE 5.3

```
model;
{

# Dirichlet distribution of those that do not withdraw in
interval
# i is the year number
for (i in 1:21) {for (j in 1:5){x[i,j]~dgamma(m[i,j],2)}
for (i in 1:21) {s[i]<-sum(x[i,])}
for (i in 1:21) {for (j in 1:5){q[i,j]<-x[i,j]/s[i]}
# q[i,j] have a Dirichlet distribution
# q[1,1] is the probability of survival for year 1 interval
  (0,1)
# q[i,j] the probability of category j of year i

# Dirichlet Distribution of those that do withdraw in interval
# i is the year number
for (i in 1:21) {for (j in 1:5){y[i,j]~dgamma(n[i,j],2)}
for (i in 1:21) {sw[i]<-sum(y[i,])}
for (i in 1:21) {for (j in 1:5){qw[i,j]<-y[i,j]/sw[i]}
# qw[i,j] have a dirichlet distribution
# qw[1,1] is the probability of survival for year 1 interval
  (0,1)
# qw[i,j] is the probability of category j of year i

}

list(n = structure(.Data = c(

746,60,2,6,153,
727,45,5,12,134,
611,32,1,9,146,
597,27,4,12,145,
505,18,1,10,129,
462,18,2,7,122,
439, 21,2,6,131,
393,9,1,4,95,
322,8,1,7,74,
314,4,1,8,73,
257,16,1,4,78,
229,5,1,6,63,
158, 1,1,2,51,
146,2,1,3,30,
119,4,1,2,44,
112,3,1,4,24,
83,2,1,2,24,
```

```
70,1,1,4,11,
44,2,1,4,16,
46,1,2,2,16,
41,1,1,2,9),.Dim = c(21,5)),

m = structure(.Data = c(

17203,2382,57,650,571,
14053,1690,36,351,113,
11564,1162,29,283,101,
9522,864,30,258,100,
7882,534,34,188,108,
6487,433,34,185,89,
5393,271,18,132,71,
4386,196,25,129,67,
3528,178,14,101,72,
2866,107,18,86,49,
2294,85,11,51,34,
1799,45,12,60,31,
1377, 48,6,47,26,
1072,26,13,40,22,
818,22,3,37,19,
600,14,2,26,15,
420,17,1,14,13,
285,10,2,10,10,
193,2,1,7,4,
118, 1,4,7,5,
50,3,2,3,1),.Dim = c(21,5)))

# activate initial values from specification tool
```

Execution of the Bayesian analysis utilizes BUGS CODE 5.3 and the information in Tables 5.12 and 5.13 with 55,000 observations for the simulation, a burn-in of 5000, and a refresh of 100. Reported are the probabilities of death from breast cancer for those who did not withdraw during the interval. A uniform prior is assumed for all unknown parameters, and for each annual period, the posterior mean, standard deviation, median, and 95% credible interval are computed. One may report the posterior analysis of the yearly probability of death from breast cancer for those who withdraw from the interval. This is left as an exercise.

Based on the posterior distribution, it is seen for the first year that the estimated probability of dying from breast cancer is .1142 with a standard deviation of .002198 and a 95% credible interval of (.1099, .1186) (Table 5.14). It is also seen that there is a trend in the posterior means, which become smaller with time from the start of the study. Employing maximum likelihood estimation, Chiang[2] reveals similar estimates of the probability of dying from breast cancer (p. 291, table 5).

TABLE 5.14

Posterior Distribution of the Probability of Death from Breast Cancer among Those Who Did Not Withdraw during the Period

Parameter	Mean	SD	Error	2½	Median	97½
q[1,2]	.1142	.002198	<.000001	.1099	.1141	.1185
q[2,2]	.1041	.002399	<.000001	.09935	.104	.1088
q[3,2]	.08845	.002484	<.000001	.08365	.08842	.0933
q[4,2]	.080190	.007611	<.000001	.07513	.08015	.08539
q[5,2]	.06107	.007561	<.000001	.05613	.06103	.06618
q[6,2]	.0599	.00278	<.000001	.05461	.05986	.06549
q[7,2]	.04606	.002725	<.000001	.04089	.04602	.05184
q[8,2]	.04081	.002853	<.000001	.03541	.04074	.04657
q[9,2]	.04573	.00335	<.00001	.03936	.04566	.05249
q[10,2]	.03424	.00326	<.00001	.02814	.03414	.04089
q[11,2]	.03436	.003653	<.00001	.02753	.03424	.04185
q[12,2]	.02311	.003393	<.00001	.01694	.02294	.03026
q[13,2]	.0319	.004531	<.00001	.02362	.03168	.04135
q[14,2]	.02216	.004299	<.00001	.01455	.02189	.03133
q[15,2]	.02448	.005154	<.00001	.01545	.02413	.03561
q[16,2]	.02132	.005623	<.00001	.01171	.02085	.0336
q[17,2]	.0366	.008716	<.00001	.02151	.03589	.0555
q[18,2]	.03155	.00977	<.00001	.01534	.03056	.05334
q[19,2]	.009652	.006773	<.00001	.001188	.008101	.02678
q[20,2]	.007409	.007337	<.00001	.000189	.005163	.007409
q[21,2]	.05082	.02831	<.00001	.01075	.04585	.119

5.5 Comparing Survival

5.5.1 Introduction

There are several ways to compare the survival experience of several life tables. Among them are the standard approaches including the log-rank test, the Mantel–Haenszel (MH) test, and the Kaplan–Meier method. In what is to follow, Bayesian adaptations of the standard approaches are taken, but the more direct Bayesian methods are in general much easier to interpret and implement.

This section begins with an explanation of the direct Bayesian approach, where the posterior distribution of the difference in the overall survival of two groups is determined. Bayesian interpretations of the standard tests will also be presented and all techniques illustrated with examples introduced earlier in the chapter.

5.5.2 Direct Bayesian Approach for Comparison of Survival

Recall Table 5.4 that reports the postoperative survival experience of 450 lung cancer patients and suppose that a similar study is done at another institution involving some 410 patients. Table 5.15 reports the survival experience of the study with 450 patients, while Table 5.16 details the results for the 410 patients in a similar study done at another medical center.

Remarks inserted between the statements of BUGS CODE 5.4 identify the various steps of the Bayesian analysis. For example, the probability of death is adjusted for withdrawals and this is noted by the remark #. A # symbol is placed before the statement for(i in1:a){Op1[i]<-O1[i]-w1[i]/2}, which deactivates the analysis and adjusts the denominator of the probability of death by half the number of withdrawals of each 2-month period.

TABLE 5.15

Postoperative Survival Experience of 450 Subjects with Lung Cancer (Medical Center 1)

Time at Beginning of Period t	Under Observation at Time t O_t	Number of Deaths in Interval $_2m_t$	Number Withdrawn in Interval $_2w_t$	Adjusted O_t $(O_t - _2w_t)$ O_t'	Mortality during Interval $_2m_t/O_t'$ $_2q_t$	Survival $_2p_t$
1	450	207	8	442	.4684	.5316
3	235	41	10	225	.1822	.8177
5	184	9	9	175	.0514	.9485
7	166	7	11	155	.0416	.9548
9	148	18	12	136	.1323	.8676
11	118	10	9	109	.0917	.9082
13	99					

TABLE 5.16

Postoperative Survival Experience of 410 Subjects with Lung Cancer (Medical Center 2)

Time at Beginning of Period t	Under Observation at Time t O_t	Number of Deaths in Interval $_2m_t$	Number Withdrawn in Interval $_2w_t$	Adjusted O_t $(O_t - _2w_t)$ O_t'	Mortality During Interval $_2m_t/O_t'$ $_2q_t$	Survival $_2p_t$
1	410	200	8	402	.4975	.5316
3	202	39	9	193	.2020	.7979
5	154	10	7	147	.0680	.9319
7	137	16	7	130	.1230	.8769
9	114	24	3	111	.2162	.7837
11	87	15	6	81	.1851	.8148
13	66					

BUGS CODE 5.4

```
Model;
{

# Study 1
# posterior distribution of mortality probabilities
for(i in 1:a){q1[i]~dbeta(m1[i], b1[i])}
for(i in 1:a){b1[i]<-Op1[i]-m1[i]}
for(i in 1:a){Op1[i]<-O1[i]-w1[i]}
# below is applied when only half the withdrawals are used
# for(i in 1:a){Op1[i]<-O1[i]-w1[i]/2}
# survival probabilities
for(i in 1:a){p1[i]<-1-q1[i]}
# overall survival
s1 <- p1[1]*p1[2]*p1[3]*p1[4]*p1[5]*p1[6]

# Study 2
# posterior distribution of mortality probabilities
for(i in 1:a){q2[i]~dbeta(m2[i], b2[i])}
for(i in 1:a){b2[i]<-Op2[i]-m2[i]}
for(i in 1:a){Op2[i]<-O2[i]-w2[i]}
# below is applied when only half the withdrawals are used
# for(i in 1:a){Op2[i]<-O2[i]-w2[i]/2}

# survival probabilities
for(i in 1:a){p2[i]<-1-q2[i]}

# overall survival
s2 <- p2[1]*p2[2]*p2[3]*p2[4]*p2[5]*p2[6]
d<-s1-s2

}

# data from Table 5.4
# the O vector is the number available at the beginning of the
  period
# w vector is the number of withdrawals
# m vector is the number of deaths
list(a = 6, m1 = c(207,41,9,7,18,10),
O1 = c(450,235,184,166,148,118),
w1 = c(8,10,9,11,12,9),
m2 = c(200,39,10,16,24,15),
O2 = c(410,202,154,137,114,87),
w2 = c(8,9,7,7,3,6))

# data for crude cause specific information Table 5.8
list(a = 6, m = c(51,9,5,6,11,5),
O = c(450,235,184,166,148,118),
w = c(8,10,9,11,12,9))

# data for net cause specific information Table 5.8
list(a = 6, m = c(51,9,5,6,11,5),
O = c(450,235,184,166,148,118),
```

```
w = c(8,10,9,11,12,9),
do = c(156,32,4,1,7,5))
# initial values
list(q = c(.5,.5,.5,.5,.5,.5))
```

Based on BUGS CODE 5.4 and the information from Tables 5.15 and 5.16, the analysis is executed with 55,000 observations with a burn-in of 5000 and a refresh of 100.

The posterior mean of the survival for the first medical center is .3103, compared to a posterior mean of .2093 for the second, and the 95% credible interval for the difference in survival between the two is (.03932,.1623), which implies that the survival rates are not the same (Table 5.17).

The estimated survival is based on the death of lung cancer patients, but death could be from causes other than complications from lung cancer. Death could be due to competing risks such as death from other cancers, death from heart attack, and so on (Figure 5.2).

5.5.3 Indirect Bayesian Comparison of Survival

5.5.3.1 Introduction

In contrast to the direct approach to comparing two life tables, there are other non-Bayesian methods, including the MH odds ratio, the log-rank test, and the Kaplan–Meier product-limit test. The approach for indirect methods is to present a Bayesian analog where the posterior distribution of the relevant parameter is determined. For example, for the MH approach, the odds ratio is defined

TABLE 5.17

Comparing Survival among Lung Cancer Patients of Two Medical Centers

Parameter	Mean	SD	Error	2½	Median	97½
d	.101	.03155	<.0001	.03932	.101	.1623
s_1	.3103	.02323	<.00001	.2659	.31	.3567
s_2	.2093	.02143	<.00001	.1689	.2088	.2528

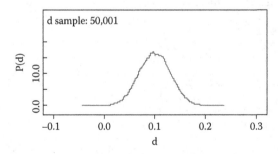

FIGURE 5.2
Posterior density of difference in survival.

as an unknown parameter depending on the posterior distribution of the number of patients who survive and the number who will die. The unknown probabilities of death in each stratum are considered unknown parameters with known posterior distributions that induce a posterior distribution of the odds ratio. Two life tables are compared with regard to death rates based on the posterior distribution of the difference in the two odds ratios. A similar approach is taken for the log-rank and Kaplan–Meier tests and illustrated with examples that were examined in previous sections of this chapter.

5.5.3.2 Mantel–Haenszel Odds Ratio

Referring to Kahn and Sempos,[1] there are several traditional ways to compare the survival of two groups of patients. To begin the section, a Bayesian analog of the MH procedure is developed. Recall that the MH method creates an odds ratio, where each time period is considered a stratum. Consider the example of the lung cancer survival experience of two medical centers reported in Tables 5.15 and 5.16, where the survivors and deaths are portrayed in Table 5.18.

Recall that each stratum is a 2-month time period, and for the Bayesian analysis, each cell entry is a random variable. It is assumed that the number of deaths for each period has a binomial distribution; thus, the number of deaths and the number of survivors have a multinomial distribution.

In general, for stratum t and the first study,

$$\left({}_2 m_t^1, O_t^1 - {}_2 m_t^1\right) \sim \text{multinomial}\left({}_2 q_t^1, {}_2 P_t^1\right) \tag{5.14}$$

while for the second study, and independently of the first,

$$\left({}_2 m_t^2, O_t^2 - {}_2 m_t^2\right) \sim \text{multinomial}\left({}_2 q_t^2, {}_2 P_t^2\right) \tag{5.15}$$

for t = 1, 3, 5, 7, 9, 11.

For the Bayesian approach, if one assumes an improper prior for the probabilities of death, for the first study

$$\left({}_2 q_t^1, {}_2 P_t^1\right) \sim \text{Dirichlet}\left({}_2 m_t^1, O_t^1 - {}_2 m_t^1\right) \tag{5.16}$$

TABLE 5.18

Deaths and Survivors of Lung Cancer Patients at Two Medical Centers

Number of Deaths in Interval Study 1 ${}_2 m_t^1$	Number of Deaths in Interval Study 2 ${}_2 m_t^2$	Survivors $(O_t^1 - {}_2 m_t^1)$ Study 1	Survivors $(O_t^2 - {}_2 m_t^2)$ Study 2	Total	Time at Beginning of Period t (Months)
207	200	243	200	850	1
41	39	194	163	437	3
9	10	175	149	343	5
7	16	159	121	303	7
18	24	130	99	271	9
10	15	108	72	205	11

Lastly, for the second study

$$\left(_2q_t^2{,_2}P_t^2\right) \sim \mathrm{Dirichlet}\left(_2m_t^2, O_t^2 -_2m_t^2\right) \tag{5.17}$$

The MH odds ratio parameter is

$$\theta_{\mathrm{OR}} = \frac{\left(\dfrac{\displaystyle\sum_i {_2}m_i^1\left(o_i^2 -_2m_i^2\right)}{t_i}\right)}{\left(\dfrac{\displaystyle\sum_i {_2}m_i^2\left(o_i^1 -_2m_i^1\right)}{t_i}\right)} \tag{5.18}$$

where t_i is the total of all observations of stratum i and the sum is over the six strata, $i = 1, 2, \ldots, 6$. θ_{OR} is considered an unknown parameter whose posterior distribution will be determined. For additional information about the MH odds ratio for stratified studies, see Rothman[3] (p. 206) and Mausner and Kramer (p. 173).[6]

BUGS CODE 5.5 is annotated with several remarks indicated with a #, which makes clear the posterior analysis for estimating the MH odds ratio. Assuming an improper prior for the unknown probabilities, the code begins with the posterior beta distributions of the probabilities of death for the six strata (the six 2-month time periods) of the two studies (the two medical centers). The posterior distribution of the probabilities of death and survival induces the posterior distribution of the number of deaths and the number of survivors.

BUGS CODE 5.5

```
model;
{

# probabilities of death first study
q1[1]~dbeta(207,243)
q1[2]~dbeta(41,194)
q1[3]~dbeta(9,175)
q1[4]~dbeta(7,159)
q1[5]~dbeta(18,130)
q1[6]~dbeta(10,108)

# probabilites of death second study
q2[1]~dbeta(200,200)
q2[2]~dbeta(39,163)
q2[3]~dbeta(10,149)
q2[4]~dbeta(16,121)
q2[5]~dbeta(24,99)
q2[6]~dbeta(15,72)
```

```
# survival probabilities first study
p1[1]~dbeta(243,207)
p1[2]~dbeta(194,41)
p1[3]~dbeta(175,9)
p1[4]~dbeta(159,7)
p1[5]~dbeta(130,18)
p1[6]~dbeta(108,10)

# survival probabilities second study
p2[1]~dbeta(200,200)
p2[2]~dbeta(163,39)
p2[3]~dbeta(149,10)
p2[4]~dbeta(121,16)
p2[5]~dbeta(99,24)
p2[6]~dbeta(72,15)

# posterior distribution of the number of deaths and survivals
for (i in 1:6){m1[i]<-q1[i]*T1[i]}
for (i in 1:6){m2[i]<-q2[i]*T2[i]}
for (i in 1:6){s1[i]<-p1[i]*T1[i]}
for (i in 1:6){s2[i]<-p2[i]*T2[i]}
# numerator of odds ratio
ORN<- m1[1]*s2[1]/850+m1[2]*s2[2]/437+m1[3]*s2[3]/343+m1[4]*s2
   [4]/303+m1[5]*s2[5]/271+m1[6]*s2[6]/205

# denominator of odds ratio

ORD<-
m2[1]*s1[1]/850+m2[2]*s1[2]/437+m2[3]*s1[3]/343+m2[4]*s1[4]/30
   3+m2[5]*s1[5]/271+m2[6]*s1[6]/205
# odds ratio

OR<-ORN/ORD

}

# see Table 5.18
list(T1 = c(450,235,184,166,148,118),
T2 = c(400,202,159,137,123,87))
# initial values
list(p1 = c(.5,.5,.5,.5,.5,.5),p2 = c(.5,.5,.5,.5,.5,.5),
q1 = c(.5,.5,.5,.5,.5,.5),q2 = c(.5,.5,.5,.5,.5,.5))
```

Bayesian computations are based on Table 5.18 and BUGS CODE 5.5 with 55,000 observations for the simulation, a burn-in of 5000, and a refresh of 100, and Table 5.19 reveals the results.

The overall approach of the Bayesian analysis is to induce a posterior distribution for the number of deaths and the number of survivors.

TABLE 5.19

Posterior Distribution of the MH Odds Ratio of Death for Lung Cancer Patients from Two Medical Centers

Parameter	Mean	SD	Error	2½	Median	97½
θ_{OR}	.7549	.06155	<.0001	.6408	.7521	.8821
θ_{ORD}	107.4	5.919	.02358	96.01	107.3	119.2
θ_{ORN}	80.79	4.83	.01906	71.54	80.7	90.49
m1[2]	41	5.819	.02118	30.2	40.77	53.0

θ_{OR} is the odds ratio with a numerator of θ_{ORN} and denominator θ_{ORD}, and the posterior mean of the odds ratio is .7549 with a 95% credible interval of (.6408, .8821). Thus, the odds of death for those from the first medical center is about 25% less than the odds of death for those from the second medical center. Since the credible interval for the odds ratio does not include 1, I am inclined to believe that there is indeed a difference in the two odds and that the survival experience of patients (over 1 year) from the first medical center is somewhat more favorable than that for the second. Also, this is the conclusion from Table 5.17, where the survival experience is estimated directly with the posterior mean of the difference in the overall survival between patients of the two medical centers. It is interesting to know that the posterior distribution of the number of deaths for the second period of the first medical center has a mean of 41 and a 95% credible interval of (30.2, 53.0) and that the actual observed value is indeed 41.

5.6 Kaplan–Meier Test

Kaplan and Meier[7] proposed estimating the survival of a group of subjects by estimating the survival of the group when each subject died or withdrew. In other words, one does not have to group the individuals into time periods. Suppose there is a group of 26 patients with the following survival experience.

Kaplan–Meier allows for withdrawals and loss due to follow-up. Note that when a withdrawal occurs, the probability of death is computed as 0 and of survival 1. The last column of Table 5.20 is the probability of survival at that time point (day) and is computed as a product-limit, hence, the name of the procedure. It is noted that the median survival appears to occur at approximately day 63.

Figure 5.3 depicts the survival experience of 26 subjects until the last patient dies at day 104 since the study began. The ordinate is the cumulative probability of survival computed as a product-limit appearing as the last column of Table 5.20. Note that the survival curve appears as a step function, where each step (vertical line) corresponds to a death, and where a cross corresponds to a censored observation (loss to follow-up).

TABLE 5.20

Kaplan–Meier Estimation of Survival (Study 1)

Time in Days t	Event Number	Number at Risk O_t	Event at Time t	Probability of Death q_t	Probability of Survival $1 - q_t$	Product-Limit Survival Function
2	1	26	1	.03846	.961538	.961538
5	2	25	1	.04	.96	.923076
8	3	24[a]	0	0	1	.9230764
9	4	23	1	.043478	.95652	.882942
14	5	22	1	.045454	.954545	.8428070
17	6	21	1	.047619	.952380	.8026725
26	7	20	1	.05	.95	.762538
28	8	19[b]	0	0	1	.762538
35	9	18	1	.055555	.944444	.7201752
39	10	17	1	.058823	.941176	.6778116
41	11	16	1	.0625	.9375	.635448
45	12	15	1	.066666	.933333	.593085
49	13	14[b]	0	0	1	.593085
58	14	13[a]	0	0	1	.593085
62	15	12	1	.083333	.916666	.543657
63	16	11	1	.090909	.909090	.494233
74	16	10	1	.100000	.900000	.444810
78	18	9	1	.111111	.888888	.3953827
82	19	8[a]	0	0	1	.3953827
88	20	7	1	.142857	.857142	.338800
91	21	6	1	.166666	.833333	.282415
94	22	5[b]	0	0	1	.282415
95	23	4	1	.25	.75	.211811
99	24	3	1	.33333	.66666	.141206
100	25	2	1	.5	.5	.070603
104	26	1	1	1	0	0

[a] Lost to follow-up.
[b] Withdrew.

How would the Bayesian estimate the survival curve? I would assume the probability of death at day t has a beta posterior distribution with parameter $(d_t, O_t - 1)$, where O_t is the number at risk at time t and d_t is the number of deaths at time t, where $d_t = 0, 1$. Of course, if there are no deaths, the probability of death is 0. Another approach is to assume the number of deaths at time t has a binomial distribution with parameters $(q[t], O_t)$, that is,

$$d[t] \sim \text{binomial}(q[t], O_t) \tag{5.19}$$

for day t and that the prior distribution is improper with density

$$f(q[t]) \propto \frac{1}{(q[t](1 - q[t]))} \tag{5.20}$$

for $0 < q[t] < 1$.

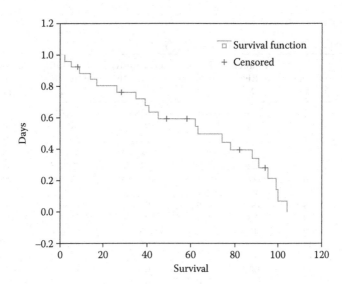

FIGURE 5.3
Kaplan–Meier: survival experience for 26 patients.

BUGS CODE 5.6 will determine the posterior distribution of the probability of death at each event (a death or a withdrawal). I put a beta (.01, .01) prior distribution on the probabilities of death. One cannot use a beta (0,0) distribution because the alpha and beta parameters must be positive, thus (.01, .01) are used. With this prior, the posterior means should be quite similar to the usual estimates.

BUGS CODE 5.6

```
model;
{

# binomial distributixon of the number of deaths
for (i in 1:a){d[i]~dbin(q[i],O[i])}
# prior distribution for probability of death
for (i in 1:a){q[i]~dbeta(.01,.01)}
# survival probabilities for each event
for (i in 1:a){p[i]<-1-q[i]}
# product limit survival probabilities
s[1]<-p[1]
for (i in 2:a){s[i]<-s[i-1]*p[i]}
for(i in 1:a){m[i]<-q[i]*O[i]}
# avg number of deaths
mm<-mean(m[])
# total number of deaths
tm<-a*mm
# average survival time
```

```
mt<-mean(t[])
# total survival time for a patient
tt<-a*mt
# average hazard
ah<-tm/tt

}

# data from table 5.20
list(a = 26, O = c(26,25,24,23,22,21,20,19,18,17,16,15,14,13,1
  2,11,10,9,8,7,6,5,4,3,2,1), d = c(1,1,0,1,1,1,1,0,1,1,1,1,0,
  0,1,1,1,1,0,1,1,0,1,1,1,1),
t = c(2,5,8,9,14,17,26,28,35,39,41,45,49,58,62,63,74,78,82,88,
91,94,95,99,100,104))
# data from table 5.22
list(a = 23,
O = c(23,22,21,20,19,18,17,16,15,14,13,12,11,10,9,8,7,6,5,4,3,
  2,1),
d = c(1,1,0,0,1,1,1,1,1,1,1,1,0,0,1,1,1,1,1,1,1,0,1),
t = c(2,5,7,11,12,25,31,32,37,40,51,61,63,68,75,78,92,
97,99,105,110,110,133))

# initial values table 5.20
list(q = c(.5,.5,.5,.5,.5,.5,.5,.5,.5,.5,.5,.5,.5,.5,.5,.5,.5,
  .5,.5,.5,.5,.5,.5,.5,.5,.5))
# initial values table 5.22
list(q = c(.5,.5,.5,.5,.5,.5,.5,.5,.5,.5,.5,.5,.5,.5,.5,.5,.5,
.5,.5,.5,.5,.5,.5))
```

The posterior analysis for the data of Table 5.20 is given in Table 5.21, which reports the posterior distribution of the product-limit survival probabilities. The analysis is executed with 55,000 observations for the simulation, a burn-in of 5000, and a refresh of 100.

All MCMC simulation errors are less than .0001, and it is seen that the posterior distributions of the survival probabilities are slightly skewed to the right. Thus, I recommend using the posterior median as a point estimate. As before, it appears that the median survival time occurs at day 63. The objective of this section is to compare the survival experience of two groups of patients. To that end, suppose the results of Table 5.20 are for a group of 26 patients who recently had coronary bypass surgery. The survival experience of a comparable group of 23 patients is reported in Table 5.22. It is supposed that the two groups differ in regard to the type of bypass surgery.

Using BUGS CODE 5.6 and the information in Table 5.22, 55,000 observations are generated for the simulation with a burn-in of 5000 and a refresh of 100, and the analysis is reported in Table 5.23. Note that the second list statement contains the data for study 2 portrayed in Table 5.22.

The analysis reveals very good agreement with Table 5.22, because the posterior mean is almost the same as the traditional estimate given in the last column of Table 5.22.

TABLE 5.21

Posterior Distribution of the Product-Limit Survival Probabilities (Study 1)

Parameter	Time	Mean	SD	2½	Median	97½
s[1]	2	.9611	.0372	.862	.9721	.9989
s[2]	5	.9222	.05146	.7958	.933	.9899
s[3]	8	.9218	.05156	.7953	.9326	.9897
s[4]	9	.8816	.06236	.736	.8915	.9731
s[5]	14	.8411	.07104	.6814	.8498	.9526
s[6]	17	.8011	.07747	.63	.8094	.9293
s[7]	26	.7604	.08306	.5806	.7667	.902
s[8]	28	.7599	.08315	.5798	.7662	.9016
s[9]	35	.7173	.0884	.5295	.7233	.8728
s[10]	39	.6747	.09218	.4825	.6792	.8417
s[11]	41	.6319	.09532	.4374	.6349	.8078
s[12]	45	.5895	.09762	.3933	.5921	.7723
s[13]	49	.5891	.09762	.3931	.5916	.7719
s[14]	58	.5886	.09764	.3925	.591	.7714
s[15]	62	.5394	.1006	.3407	.5408	.73
s[16]	63	.4903	.1021	.2911	.4907	.6883
s[17]	74	.4413	.1021	.2467	.44	.6435
s[18]	78	.392	.1014	.2032	.3891	.5974
s[19]	82	.3915	.1013	.2028	.3884	.597
s[20]	88	.3349	.1004	.1537	.3304	.5422
s[21]	91	.2789	.0968	.1099	.2728	.4841
s[22]	94	.2784	.09676	.1094	.2722	.4838
s[23]	95	.2087	.09212	.0573	.1998	.4128
s[24]	99	.139	.08167	.02091	.1256	.3289
s[25]	100	.06904	.06171	.002096	.0518	.2292
s[26]	104	.000622	.006077	0	0	.002871

TABLE 5.22

Product-Limit Kaplan–Meier Estimation of Survival

Time in Days t	Event Number	Number at Risk O_t	Event at Time t	Probability of Death q_t	Probability of Survival $1 - q_t$	Product-Limit Survival Function
2	1	23	1	.0434	.9565	.9565
5	2	22	1	.04545	.954545	.913022
7	3	21[a]	0	0	1	.913022
11	4	20[b]	0	0	1	.913022
12	5	19	1	.05263	.94736	.86496
25	6	18	1	.047619	.952380	.82377
31	7	17	1	.05882	.94117	.77530

TABLE 5.22 (*Continued*)

Product-Limit Kaplan–Meier Estimation of Survival

Time in Days t	Event Number	Number at Risk O_t	Event at Time t	Probability of Death q_t	Probability of Survival $1 - q_t$	Product-Limit Survival Function
32	8	16	1	.0625	.9375	.72685
37	9	15	1	.06666	.93333	.67839
40	10	14	1	.07142	.92875	.63005
51	11	13	1	.07692	.92307	.58158
61	12	12	1	.08333	.91666	.53311
63	13	11[b]	0	0	1	.53311
68	14	10[a]	0	0	1	.53311
75	15	9	1	.11111	.88888	.47387
78	16	8	1	.125	.875	.41464
92	16	7	1	.14285	.85714	.35540
97	18	6	1	.16666	.83333	.29617
99	19	5	1	.2	.8	.23693
105	20	4	1	.25	.75	.17770
110	21	3	1	.33333	.66666	.118466
110	22	2[b]	0	0	1	.118466
133	23	1	1	1	0	0

[a] Lost to follow-up.
[b] Withdrew.

TABLE 5.23

Posterior Analysis for Product-Limit Survival Probabilities (Study 2)

Parameter	Day of Event	Mean	SD	2½	Median	97½
s[1]	2	.9562	.04155	.8451	.9686	.9989
s[2]	5	.9122	.05778	.7711	.9241	.9885
s[3]	7	.9117	.05793	.7703	.9236	.9883
s[4]	11	.9113	.05805	.7697	.9232	.9882
s[5]	12	.8632	.0714	.6955	.8742	.969
s[6]	25	.8148	.08122	.6333	.8247	.9439
s[7]	31	.7667	.08892	.5721	.7752	.9143
s[8]	32	.718	.09477	.5152	.7251	.8817
s[9]	37	.6698	.09944	.4629	.6746	.8467
s[10]	40	.6212	.1029	.4116	.625	.8098
s[11]	51	.5729	.1053	.3628	.5754	.7705
s[12]	61	.525	.1061	.3175	.5254	.7283
s[13]	63	.5245	.1061	.3171	.525	.7278
s[14]	68	.5239	.1061	.3165	.5244	.7275
s[15]	75	.4653	.1083	.2561	.4645	.6771

(*Continued*)

TABLE 5.23 (*Continued*)

Posterior Analysis for Product-Limit Survival Probabilities (Study 2)

Parameter	Day of Event	Mean	SD	2½	Median	97½
s[16]	78	.4069	.1087	.204	.404	.6248
s[17]	92	.3485	.1067	.1554	.3438	.5684
s[18]	97	.2898	.1028	.1111	.2824	.5077
s[19]	99	.2318	.09641	.07346	.2224	.443
s[20]	105	.1738	.08707	.04064	.1626	.3716
s[21]	110	.116	.07408	.01563	.1015	.2943
s[22]	110	.1154	.07391	.01545	.1009	.2937
s[23]	133	.001112	.009412	0	0	.007471

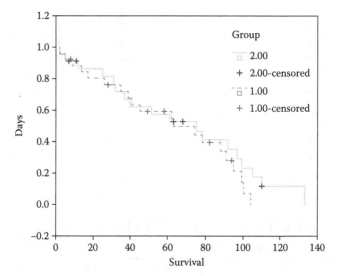

FIGURE 5.4

Survival plots: group 1 versus group 2.

Attention at this point is on comparing the survival experience between the two studies. Recall that the patients have undergone bypass surgery and one is interested in comparing the postoperative survival of the two. To this end, consider Figure 5.4, a plot of the two survival curves. Group 1 is the one with 26 patients and group 2 is the one with 23. It is apparent that there is little difference in survival between the two groups; however, it is noticed that patients of group 2 survive longer than those of group 1 (133 vs. 104 days). Namely, the last death for group 2 is at day 133 compared to day 104 for group 1, but over the period from day 0 to day 104, the survival probabilities are quite similar.

Chapter 5 introduces regression methods, including the Cox proportional hazards model, for estimating the survival experience of a group of patients, which allows a direct comparison between two groups.

5.7 Comments and Conclusions

A life table is an essential tool of the epidemiologist in estimating the survival of a group of patients, and this chapter presented the Bayesian approach. The chapter begins with the development of the Bayesian posterior distribution of the probability of death and survival for each period of the life table. The concept is then illustrated with a groups of 1000 coronary artery patients for 10 years.

Later, the life table is generalized to allow for withdrawals or loss to follow-up subjects. To calculate the probability of death or survival for a particular period, the denominator is the number at risk at the beginning of the interval minus the number of withdrawals. A cohort of 450 postoperative patients with lung cancer is followed for 7 years and a Bayesian analysis is conducted to estimate the period probability of survival and the overall 7-year survival.

An additional generalization of the life table takes into account the assumption that the withdrawals occur uniformly throughout each interval, which adjusts the probability of survival with a denominator that is expressed as the number at risk minus ½ the number of withdrawals.

When attention is focused on estimating disease-specific survival, the denominator of the probability of survival is the number at risk (at the beginning of the period) minus half the number of withdrawals minus half the number of patients that die of other causes. For example, the 450 postoperative patients with lung cancer are followed for 7 years, and the overall probability of survival due to complications of lung cancer is estimated using BUGS CODE 5.2. Bayesian estimation of the period survival and the overall survival is based on the posterior mean and posterior median as well as the 95% credible interval.

Next to be considered is a different adaptation of the life table due to Chiang,[2] where subjects enter the study at any time during the life of the study. For each interval beginning at time x, there are two types of patients: (1) those that will not withdraw during the 1-year interval from year x to year x + 1 (these are the patients that entered the study at least x + 1 years before the close of the study) and (2) those patients that enter the study less than x + 1 years before the close of the study (these patients will withdraw from the study during (x, x + 1)).

For a given interval (x, x + 1), r competing risks are taken into account, that is, a patient can die of r causes, and the Bayesian methodology is demonstrated with subjects from the California Tumor Registry over the period beginning on January 1, 1942 and closing on December 31, 1962.

Last to be presented is the comparison of survival between two groups of patients, and two approaches are considered, the direct Bayesian and the indirect Bayesian methods. With the direct method, the posterior distribution of the difference in the probability of overall survival is determined, and on the basis of the 95% credible interval, a decision is rendered as to the difference in survival of the two groups. Recall the lung cancer cohort with 450 lung cancer postoperative subjects. It is supposed that this study is conducted at medical center 1 and that a similar group of 410 is conducted at medical center 2. The interest is in comparing the 7-year overall survival of the two studies with the Bayesian approach.

Indirect methods are adaptations of classical epidemiologic methods. The first adaptation is the MH odds ratio, which is considered an unknown parameter. Recall that the MH odds ratio is applicable to studies with different strata, and in the case of a life table, the various time periods (or intervals) are considered strata. Thus, the comparison of two life tables is based on the odds ratio, where the odds of death from life table 1 is compared to the odds of death of life table 2. The two lung cancer groups of 450 and 410 patients of two medical centers are used to illustrate the Bayesian technique.

The chapter concludes with a Bayesian analog of the Kaplan–Meier product-limit method of estimating the survival curve. A group of patients is followed until the last patient dies or is lost to follow-up, and the method records the time of each event (death or withdrawal); thus, the entire survival experience of each group is known. Bayesian determinations of the survival curve of each group allow one to plot the probabilities of survival for each group and to make a decision about the similarity of the two.

Chapter 5 concludes with many exercises that provide the student with additional opportunities to reinforce what is being learned.

Exercises

1. a. Verify Table 5.3, the Bayesian analysis for the 10-year survival experience of 1000 coronary artery disease patients. Use BUGS CODE 5.1 with 55,000 observations for the simulation, a burn-in of 5000, and a refresh of 100.

   b. Determine the posterior distribution of the survival of 10 1-year periods. Your results should look similar to Table 5.24.

   c. Plot the posterior density of $_1p_1$.

   d. Is the posterior distribution of $_1p_1$ skewed?

   e. Compare the estimates through the posterior mean with the "usual" estimates of the survival probability P of Table 5.2.

2. Refer to Table 5.4 and BUGS CODE 5.2 and validate the Bayesian analysis reported in Table 5.5. Use 55,000 observations for the simulation with a burn-in of 5000 and a refresh of 100.

   a. Plot the posterior density of the probability of death for the third 2-month period. Is the distribution symmetric? If so, why?

   b. What is the posterior median of the overall survival probability?

TABLE 5.24

Posterior Distribution of the Probability of Survival

Parameter	Mean	SD	Error	2½	Median	97½
$_1p_1$	.994	.002439	<.00001	.9884	.9943	.9978
$_1p_2$	.995	.002247	<.000001	.9897	.9953	.9984
$_1p_3$	.9949	.002246	<.000001	.9897	.9953	.9984
$_1p_4$	.9949	.002265	<.000001	.9896	.9952	.9983
$_1p_5$	.9949	.002297	<.000001	.9895	.9952	.9984
$_1p_6$	.9959	.002052	<.000001	.991	.9962	.9989
$_1p_7$	.9949	.002298	<.000001	.9895	.9952	.9983
$_1p_8$	.9948	.002309	<.000001	.9895	.9951	.9983
$_1p_9$	.9948	.002331	<.000001	.9894	.9851	.9983
$_1p_{10}$	.9948	.00234	<.00001	.9893	.9951	.9983

c. Show the Bayesian estimate of the overall survival is the same as that computed directly from Table 5.4.

3. a. Based on Table 5.6 and BUGS CODE 5.2, execute a Bayesian analysis and verify Table 5.7. Use 55,000 observations for the simulation, a burn-in of 5000, and a refresh of 100.

b. Table 5.7 reports the posterior analysis for the mortality probabilities of each 2-month period and the overall survival. Include in your analysis the posterior analysis for the survival probabilities.

c. Compare the posterior means of the mortality probabilities computed in Table 5.7 with those reported in Table 5.10. Why are the differences so small? Carefully explain your answer.

d. Plot the posterior density of the $_2q_3$, the mortality for the period from month 3 to month 5. Does the posterior distribution appear symmetric about the posterior mean?

4. Based on the actual survival statistics for the year 2000 California population, Table 5.25 presents an abridged life table based on a group of 100,000 individuals.

Revise BUGS CODE 5.1 and execute a Bayesian analysis based on the survival of 100,000 individuals. Use 55,000 observations for the simulation, with a burn-in of 5000 and a refresh of 100. You will have to develop a new list statement that uses the information in Table 5.25 and revise the code to compute the overall survival. Note that there are 10 periods that vary in length.

a. What is the posterior mean of the proportion that die for the period 35–40 years? Note that the proportion surviving is calculated in the second column.

b. How do the posterior means of the 19 probabilities of dying compare to the corresponding values in column 2?

TABLE 5.25

Abridged Life Table for the 2000 California Population

Age Period	Proportion Dying during Period q_x	Number Alive at Beginning of Age Period al_x	Number Dying during Period d_x	Proportion Surviving during Period
0–1	.00543	100,000	543	?
1–5	.00103	99,457	102	
5–10	.00061	99,355	61	
10–15	.00090	99,294	89	
15–20	.00269	99,205	266	
20–25	.00392	98,939	388	
25–30	.00380	98,551	374	
30–35	.00441	98,177	433	
35–40	.00622	97,744	608	
40–45	.00972	97,135	944	
45–50	.01487	96,192	1,431	
50–55	.02181	94,761	2,067	
55–60	.03293	92,694	3,052	
60–65	.05004	89,641	4,485	
65–70	.07677	85,156	6,538	
70–75	.11983	78,618	9,421	
75–80	.18479	69,197	12,787	
80–85	.28878	56,410	16,290	
>85	1.000	40,120	40,120	

Source: Oreglia, A., *Methodology for Constructing Abridged Life Tables for California 1977*. Data Matters. Center for Health Statistics, California Department of Health Services, February 26, 1981.

    c. Determine the posterior distribution of the 19 survival probabilities.

    d. What is the posterior distribution of the overall survival of the group of 100,000 individuals?

    e. Plot the posterior distribution of the overall survival P.

  5. Consider a Phase II clinical trial involving 34 patients with stage III breast cancer, where the patients have been treated and are currently in remission. They are eligible to enroll in a study that will examine the efficacy of a new therapy to prolong their remission. Therapy consists of five cycles where treatment is administered on day 1 of the beginning of a cycle and a cycle is a period of 1 week (7 days). Efficacy of the therapy is to be measured by the tumor response, which is based on the growth of the tumors of the subjects. If the various target tumors decrease by at least 30% in size, the patient is said to respond. Table 5.26 details the outcomes to therapy. Stage III disease implies the disease has metastasized to various organs, and the patients have

TABLE 5.26

Response to Therapy for Breast Cancer, a Phase II Trial

Day at Beginning of Cycle	Under Observation at Time t O_t	Number of Responders in Interval $_2m_t$	Number Withdrawn in Interval $_2w_t$	Adjusted $O_t =$ $(O_t - {_2w_t})$ O_t'	Fraction Responding $_2m_t/O_t'$ $_2q_t$	Not Responding $_2p_t$
1	34	1	3	31	.0322	.9677
8	30	1	2	28	.0357	.9642
15	27	3	4	23	.1304	.8695
22	20	4	5	15	.2666	.7333
29	11	5	1	10	.5	.5
36	5					

several tumors that are referred to as target tumors. Using various imaging modalities, the size of the tumor is measured at baseline and then on day 7 of each cycle. Patients may withdraw for a number of reasons, including bad side effects to therapy, death, moving away, and refusing to continue therapy for various other reasons. See Broemeling[9] for additional information about Phase II clinical trials.

Using BUGS CODE 5.2, execute a Bayesian analysis with 65,000 for the simulation, a burn-in of 5000, and a refresh of 100. Note that the fourth list statement contains the information from Table 5.26.

a. For each cycle, determine the posterior distribution of the probability of responding to therapy.

b. What is the posterior distribution of the overall probability of responding?

c. Plot the posterior distribution of the overall probability of responding.

d. What is the uncertainty of the probability of overall response to therapy?

e. Is the therapy effective? Explain your answer.

6. A prior study of the postoperative survival experience of lung cancer patients revealed the following results. The 120 subjects are observed over six 2-month periods (Table 5.27).

a. Using BUGS CODE 5.2 and the information from Table 5.8, perform a Bayesian analysis with 65,000 observations, a burn-in of 5000, and a refresh of 100.

b. For each of the six 2-month periods, find the posterior distribution of the crude probability of dying from lung cancer. What is the posterior mean?

c. What is the overall crude probability of dying from complications of lung cancer?

TABLE 5.27

A Prior Study of Postoperative Survival Experience of 120 Subjects with Lung Cancer

Period t	Under Observation at Time t O_t	Total Deaths	Lung Cancer	Other	Number Withdrawn in Interval $_2 w_t$	Adjusted O_t $(O_t - _2 w_t / 2)$ O_t''	Lung Cancer Mortality during Interval
1	120	73	14	59	2	119	?
2	45	18	3	15	3	25.5	?
3	24	8	2	6	2	23	?
4	14	4	1	3	2	13	?
5	8	3	1	2	1	7.5	?
6	4	2	1	1	1	3.5	?
7	1						

d. What is the MCMC error for estimating the overall crude probability of death? Is 65,000 observations sufficient for the simulation?

e. Plot the posterior density of the crude probability of dying from lung cancer complications for the first 2-month period.

f. Compare your results with those of Table 5.9.

7. Based on BUGS CODE 5.3 and the information in Table 5.12 (this applies to those patients who did not withdraw from the interval), execute a Bayesian analysis with 65,000 observations for the simulation, with a burn-in of 5000 and a refresh of 100.

a. Verify the posterior analysis reported in Table 5.17.

b. What is the probability of dying from breast cancer for the period (3, 4)?

c. Plot the posterior distribution of the probability of dying from breast cancer for the period (3, 4).

d. What is the 95% credible interval of the probability of dying from breast cancer for the period (3, 4)?

e. What is the MCMC error of the posterior mean for estimating the probability of dying from breast cancer for the period (3, 4)?

8. Based on BUGS CODE 5.3 and the information in Table 5.13 (this applies to those patients who did withdrew from the interval), execute a Bayesian analysis with 65,000 observations for the simulation, with a burn-in of 5000 and a refresh of 100.

a. Plot the posterior density of the probability of dying from breast cancer during the interval (3, 4).

b. What is the mean of the posterior distribution of the probability of dying from breast cancer during the interval (3, 4)?

c. What is the 95% credible interval of the posterior distribution of the probability of dying from breast cancer during the interval (3, 4)?

d. Is there a trend (as the study progresses) in the posterior mean of the probability of dying from breast cancer during the interval (3, 4)?

e. What is the MCMC error of the posterior mean for estimating the probability of dying from breast cancer for the period (3, 4)?

9. Based on BUGS CODE 5.4 and Tables 5.15 and 5.16, execute an analysis with 55,000 observations with a burn-in of 5000 and a refresh of 100.

a. Verify the posterior analysis of Table 5.17.

b. Plot the posterior density of the difference d.

c. Is there a difference in the two survival rates of the two medical centers?

d. Repeat the Bayesian analysis by adjusting the denominator of the probability of death for one half the number of withdrawals. Activate the statement for(i in 1:a){Op2[i]<-O2[i]-w2[i]/2} in BUGS CODE 5.4 by eliminating the # sign in front of the statement. Also deactivate the statement for(i in 1:a){Op1[i]<-O1[i]-w1[i]} by putting a # sign in front of it.

10. Kahn and Sempos[1] present the comparison of two life tables corresponding to two similar studies (pp. 186–187). They employed the MH odds ratio to compare the two death rates (Table 5.28).

To perform the Bayesian analysis, use BUGS CODE 5.5 and put the information from Table 5.28 into another list statement. Use 55,000 observations for the simulation with a burn-in of 1000 and a refresh of 100.

TABLE 5.28

Two Life Tables: Deaths and Survivors for Two Studies

Number of Deaths in Interval Study 1 $_2m_t^1$	Number of Deaths in Interval Study 2 $_2m_t^2$	Survivors $(O_t^1 -_2m_t^1)$ Study 1	Survivors $(O_t^2 -_2m_t^2)$ Study 2	Total	Time at Beginning of Period t (Months)
193	165	104	132	594	1
25	30	72	102	229	3
12	15	51	87	165	5
8	15	33	72	128	7
4	11	19	61	95	9
3	17	6	44	70	11

Source: Kahn, H.A., Sempos, C.T., *Statistical Methods in Epidemiology*, Oxford University Press, New York, 1989.

For the Bayesian analysis, answer the following:

a. Verify the posterior analysis reported in Table 5.29.

b. Are 55,000 observations for the simulation sufficient?

c. What is the posterior distribution of the probability of death for the third period of study 1?

d. The posterior mean of the number of deaths in the first period of study 1 is 193. Is this estimate reasonable?

e. Note that the posterior median of the odds ratio is 1.398. This compares to the value of 1.37 reported by Kahn and Sempos (pp. 186–187).[1] I assert that the Bayesian approach agrees with the standard approach of Kahn and Sempos. Do you agree?

f. Plot the posterior density of the odds ratio.

g. Are the death rates of the two studies the same? Look at the 95% credible interval of the odds ratio.

11. Refer to problem 10. The two studies of that problem were preceded by preliminary studies reported in Table 5.30. Consider the

TABLE 5.29

Posterior Distribution of the MH Odds Ratio of Death for Lung Cancer Patients from Two Medical Centers

Parameter	Mean	SD	Error	2½	Median	97½
θ_{OR}	1.41	.1873	<.0001	1.077	1.398	1.81
θ_{ORD}	30.26	2.217	.008305	26.08	30.22	34.97
θ_{ORN}	42.29	2.786	.01149	36.97	42.25	47.87
q1[1]	.6497	.0277	<.0001	.5946	.6502	.703
m1[1]	193	8.207	.03367	176.7	193.1	208.8

TABLE 5.30

Two Life Tables: Deaths and Survivors for Two Studies, a Prior Study

Number of Deaths in Interval Study 1 $_2m_t^1$	Number of Deaths in Interval Study 2 $_2m_t^2$	Survivors $(O_t^1 - {_2}m_t^1)$ Study 1	Survivors $(O_t^2 - {_2}m_t^2)$ Study 2	Total	Time at Beginning of Period t (Months)
18	14	11	11	54	1
3	2	8	9	22	3
1	2	5	9	17	5
1	2	3	5	11	7
1	1	2	3	7	9
1	1	1	2	5	11

information in the table as prior information (prior to the information in problem 10) and perform a Bayesian analysis using 55,000 observations with a burn-in of 1000 and a refresh of 100.

a. What principle is used to combine the prior information of problem 11 with the information in problem 10?

b. Compare the posterior distribution of the odds ratio of problem 10 with the odds ratio based on the information from the table in problem 10 combined with the information in problem 11.

c. Plot the posterior density of the odds ratio.

d. Is the posterior distribution of the odds ratio skewed?

e. Are the odds ratios of the two studies the same? Explain your answer.

12. After an operation to repair the heart, 27 patients with coronary artery disease are followed until the last is dead. Table 5.31 records their survival experience. The event is either death designated by a 1 or a withdrawal indicated by a 0.

a. Complete the last three columns of the table.

b. Use BUGS CODE 5.6 and the information in the table and perform a Bayesian analysis with 55,000 observations for the simulation, a burn-in of 5000, and a refresh of 100. You will need a list statement that contains the number at risk and the number of deaths for the 27 events.

TABLE 5.31

Product-Limit Kaplan–Meier Estimation of Survival

Time in Days t	Event Number	Number at Risk O_t	Event at Time t	Probability of Death q_t	Probability of Survival $1 - q_t$	Product-Limit Survival Function
2	1	27	1			
5	2	26	1			
7	3	25[a]	0			
11	4	24[b]	0			
12	5	23	1			
25	6	22	1			
31	7	21	1			
32	8	20	1			
37	9	19	1			
40	10	18[a]	0			
51	11	17	1			
61	12	16	1			

(Continued)

TABLE 5.31 (*Continued*)

Product-Limit Kaplan–Meier Estimation of Survival

Time in Days t	Event Number	Number at Risk O_t	Event at Time t	Probability of Death q_t	Probability of Survival $1 - q_t$	Product-Limit Survival Function
63	13	15[b]	0			
68	14	14[a]	0			
75	15	13	1			
78	16	12	1			
92	16	11	1			
97	18	10	1			
99	19	9	1			
105	20	8	1			
110	21	7	1			
110	22	6[b]	0			
133	23	5	1			
135	24	4	1			
141	25	3	1			
145	26	2	1			
155	27	1[b]	0			

[a] Lost to follow-up.
[b] Withdrew.

    c. Determine the posterior distribution of the 27 product-limit survival probabilities.

    d. Are the posterior means the same as the last column of the Table 5.30?

    e. Plot the posterior means of the survival probabilities versus the time of the event (the first column).

    f. Estimate the median survival time (the time before which 50% of the deaths occur).

References

1. Kahn, H.A. and Sempos, C.T. *Statistical Methods in Epidemiology*, Oxford University Press, 1989, New York.
2. Chiang, C.L. *Introduction to Stochastic Processes in Biostatistics*, John Wiley & Sons Inc., 1968, New York.
3. Rothman, K.J. *Modern Epidemiology*, Little Brown & Co., 1986, Boston, MA.
4. Newman, S.C. *Biostatistical Methods in Epidemiology*, John Wiley & Sons Inc., 2001, New York.

5. National Institutes of Health. National Heart, Lung, and Blood Institute. Chart 3–26, Death and Age-Adjusted Death Rates for Coronary Heart Disease, 1980–2006. http://www.nhlbi.nih.gov/resources/docs/2009_ChartBook.pdf.

6. Mausner, J.S. and Kramer, S. *Epidemiology—An Introductory Text*, W.B. Saunders Company, 1985, Philadelphia, PA.

7. Kaplan, E.L. and Meier, P. Nonparametric estimation from incomplete observations, *J. Am. Stat. Assoc.*, 53,457–481, 1958.

8. Oreglia, A. *Methodology for Constructing Abridged Life Tables for California 1977.* Data Matters. Center for Health Statistics, California Department of Health Services, February 26, 1981, Sacramento, CA.

9. Broemeling, L.D. *Advanced Bayesian Methods for Medical Test Accuracy*, Taylor & Francis, 2012, Boca Raton, FL.

6

A Bayesian Approach to Survival Analysis

6.1 Introduction

Chapter 5 is now extended with another approach to estimating survival, where the main focus is on parametric models such as the Weibull and a nonparametric approach that includes the Cox proportional hazards model. These models use the survival time of each patient that experience an event (death, time to recurrence, etc.) and also, the time of survival for those patients that are censored, namely, those that are lost to follow-up. The survival time up to the point when the patient withdraws (lost to follow-up) is known and utilized to estimate the survival experience of a group of interest.

The chapter begins with the Kaplan–Meier curve of the survival experience, which allows a formal definition of three basic ideas that are fundamental to understanding survival models: (1) the hazard function, (2) the density of the survival times, and (3) the survival function. The Kaplan–Meier curve was introduced in Chapter 5, but the presentation in Chapter 6 will give the student a deeper understanding of this important concept. In Chapter 5, the main focus is on the life table method of estimating survival, where the times of death are grouped into various periods over the range of a person's life. On the other hand, survival models, including the Cox proportional hazards model, utilize the actual time of the event and the time of censoring.

Chapter 6 continues with the Kaplan–Meier curve and the log-rank tests for testing for a difference in the survival experience of two or more groups. The Bayesian version of the log-rank test is based on the posterior distribution of the observed minus the expected number of events in a particular group. Next, a parametric model for survival based on the Weibull distribution is introduced, followed by the Cox proportional hazards model. This is followed by a detailed outline of how to estimate the survival and hazard functions. Several interesting examples are explained, including one involving the survival experience of two groups of leukemia patients, where one group receives a treatment to extend the remission time, and where the other group receives a placebo.

Testing the proportional hazards assumption is an essential part of executing the Bayesian analysis and is based on the estimated survival curves.

The latter parts of the chapter are focused on more specialized topics such as the stratified Cox procedure that allows one to control by stratification when the proportional hazards assumption is not true. An interesting aspect of stratification is the test of the no interaction assumption. Every theme of this chapter is accompanied by examples that illustrate the important features of that theme. Exercises at the end of the chapter will develop a further understanding of the various topics.

The WinBUGS® package is utilized throughout for the Bayesian analyses, and the code is "borrowed" from several examples provided by the package. The following link will give the reader access to several examples involving survival analysis: www.mrc-bsu.cam.ac.uk/bugs/winbugs/contents.shtml. Examples from the package are used for survival analyses based on the Weibull parametric model as well as the Cox proportional hazards model.

A good introduction to survival analysis is given by Kleinbaum,[1] and more specialized Bayesian references are presented by Congdon.[2,3] I used the Congdon material to some extent because of the Bayesian nature of the approach. Newman[4] is especially appropriate for epidemiologists and develops a comprehensive technique to survival analysis that includes many realistic examples.

6.2 Notation and Basic Table for Survival

Suppose the survival experience of a group of n patients is recorded as follows:

The jth subject experiences an event at time t_j: either a failure when $\delta_j = 1$ or censoring when $\delta_j = 0$ (See Table 6.1). Also measured on each subject are the values of p covariates $X_1, X_2, ..., X_p$. If T is the random variable that represents the survival time of a patient, the probability that a patient survives more than t years is given by

$$S(t) = P(T > t) \qquad (6.1)$$

TABLE 6.1

Survival Experience for a Cohort of Individuals

No. of patient	t	Δ	X_1	X_2	...	X_p
1	t_1	δ_1	X_{11}	X_{12}	...	X_{1p}
2	t_2	δ_2	X_{21}	X_{22}	...	X_{2p}
.	.	.	.	.	...	.
j	t_j	δ_j	X_{j1}	X_{j2}	...	X_{jp}
.	.	.	.	.	...	.
n	t_n	δ_n	X_{n1}	X_{n2}	...	X_{np}

A plot of the $S(t_j)$ for $j = 1, 2, \ldots, n$ is the survival curve for this group of n individuals. However, note that the times t_j are not necessarily ordered from smallest to largest, that is, t_n is not necessarily the last recorded time, either for a failure or for a censored observation. Thus, $S(t)$ would be estimated by the proportion of n individuals who survive past time t. Another important function that is used in survival studies is the hazard function

$$h(t) = \lim_{\Delta t \to \infty} \frac{P(t \leq T < t + \Delta t \mid T \geq t)}{\Delta t} \qquad (6.2)$$

which is interpreted as the hazard that a person will fail in the next instant, given they have survived up to time t. Another interpretation is that the hazard at time t is the instantaneous rate of failure.

Thus, it appears that the hazard function and the survival function are inversely related, that is, as survival decreases over time, the hazard function tends to increase with time.

The survival function is related to the distribution function of survival, namely

$$F(t) = 1 - S(t) \qquad (6.3)$$

$$= P[T \leq t] \, for \, t > 0$$

Thus, the probability that a patient will not survive after time t is given by $F(t)$.

Three functions involving survival are related, and in particular,

$$h(t) = \frac{\left[dS(t)/dt \right]}{S(t)} \qquad (6.4)$$

that is, the hazard function at time t is the derivative of the survival function with respect to t, divided by the survival function at time t. The Kaplan–Meier curve allows us to illustrate the use of the hazard and survival functions; thus, consider the following study of Freireich et al.[5] Comparing the recurrence time of two groups of leukemia patients, where with one, the patients receive a treatment to prolong remission, while with the second, the patients receive a placebo. The study results are given in the first list statement of BUGS CODE 6.1, where there are 21 patients in each group (See Table 6.2).

TABLE 6.2

Recurrence of Leukemia Patients

For Group 1	6,6,6,7,10, 13, 16,22,23,6+,9+,10+,11+,17+,19+,20+,25+,32+,32+,34+,35+
For Group 2	1,1,2,2,3,4,4,5,5,8,8,8,8,11,11,12,12,15,17,23,23

For Group 1, the censored observations are superscripted by a+, where the first censored time is at 6 weeks. Censored means the subject is lost to follow up or withdrew, and recurrence time means the time from treatment to the time the patient goes out of remission. Note that there are no censored observations in Group 2. If a censored observation occurs at the end of the study, the accumulated remission time of the patient is measured.

The study results of the Freireich et al.[5] study are now presented in the format of Table 6.3, where the first column is the individual number, the second lists the failure or censored time in weeks, the third indicates whether or not the patient experiences a failure or is censored, and the fourth reports which group the patient belongs to: either the treatment group indicated with a 1 or the placebo group denoted by a 0. The next to last column is a covariate that gives the log of the white blood cell count of each patient and the last lists gender as a 1 for male and 0 for female.

TABLE 6.3

Comparing the Recurrence of Two Groups of Leukemia Patients

Individual #	t Weeks	Indicator:Failure Censored δ	Group	Log WBC	Gender
1	6	1	1	2.31	0
2	6	1	1	4.06	1
3	6	1	1	3.28	0
4	7	1	1	4.43	0
5	10	1	1	2.96	0
6	13	1	1	2.88	0
7	16	1	1	3.60	1
8	22	1	1	2.32	1
9	23	1	1	2.57	1
10	6	0	1	3.20	0
11	9	0	1	2.80	0
12	10	0	1	2.70	0
13	11	0	1	2.60	0
14	17	0	1	2.16	0
15	19	0	1	2.05	0
16	20	0	1	2.01	1
17	25	0	1	1.78	1
18	32	0	1	2.20	1
19	32	0	1	2.53	1
20	34	0	1	1.47	1
21	35	0	1	1.45	1
22	1	1	0	2.80	1
23	1	1	0	5	1
24	2	1	0	4.91	1
25	2	1	0	4.48	1

TABLE 6.3 (*Continued*)

Comparing the Recurrence of Two Groups of Leukemia Patients

Individual #	*t* Weeks	Indicator:Failure Censored δ	Group	Log WBC	Gender
26	3	1	0	4.01	1
27	4	1	0	4.36	1
28	4	1	0	2.42	1
29	5	1	0	3.49	1
30	5	1	0	3.97	0
31	8	1	0	3.52	0
32	8	1	0	3.05	0
33	8	1	0	2.32	0
34	8	1	0	3.26	1
35	11	1	0	3.49	1
36	11	1	0	2.12	0
37	12	1	0	1.5	0
38	12	1	0	3.06	0
39	15	1	0	2.30	0
40	17	1	0	2.95	0
41	22	1	0	2.73	0
42	23	1	0	1.97	1

Source: Freireich, E.J., Gehan, E., Frei, E., and Schroeder, L.R., *Blood*, 21, 699–716, 1963.

6.3 Kaplan–Meier Survival Curves

6.3.1 Introduction

In order to describe the Kaplan–Meier survival curve, the times indicating an event and those indicating censored times are ordered and appear as follows:

Note that the first column reports the ordered times (of recurrence or censored), the second gives the number of failures at that time, the third gives the number censored at that time, and the last column gives the number of patients that survive past that time (See Table 6.4). For example, $R(t_{(2)}) = 17$ is the number of patients that remain in remission at least 7 weeks, among which, one has a recurrence and one is censored, that is, 17 remain in remission at least 7 weeks. For the second group of patients, Table 6.5 reports the recurrence and censored times.

Each group consists of 21 leukemia patients and the two tables report their recurrence experience, thus, for Group 1, week 23 is the last recorded time of the study, where 1 subject went out of remission and 5 were censored, while for Group 2, the last recorded time of the study is also week 23, with 1 recurrence and no censored patients at that time. Of course, the main question is: Is there a difference in the recurrence experience between the treatment

TABLE 6.4

Times to Recurrence for Group 1 of Leukemia
Patients

t_0 in Weeks	d_i is the # Recurrences	c_i is # Censored	$R(t_{(j)})$
$t_{(0)} = 00$	0	0	21
$t_{(1)} = 6$	3	1	21
$t_{(2)} = 7$	1	1	17
$t_{(3)} = 10$	1	2	15
$t_{(4)} = 13$	1	0	12
$t_{(5)} = 16$	1	3	11
$t_{(6)} = 22$	1	0	7
$t_{(7)} = 23$	1	5	6

group and the placebo? The Freireich et al.[5] study is very important, because it is one of the first to report an efficacious treatment for acute leukemia. Note that at time 0, there were no reported recurrences or censored observations; however, it is possible to have a censored observation before time $t_{(1)}$. Overall estimates of the recurrence are the average survival time of Group 1 and Group 2, which are $\bar{T}_1 = 17.09$ and $\bar{T}_2 = 8.66$ weeks respectively, where the average is taken over all the 21 recurrence times, including the censored ones. Another estimate of overall recurrence is the average hazard of the two groups, namely $\bar{h}_1 = .025$ and $\bar{h}_2 = .115$ respectively, where the average hazard estimator is the number of recurrences divided by the total recurrence time including censored times. Later, the estimated hazard function will be plotted over the time range of each study. Note that the larger the average survival, the smaller the average hazard.

Recall Table 6.3. where the last column gives the log of the white blood cell count of each patient, and the average (SD) of the log blood cell counts for each group are 2.63 (.773) for the treatment group and 3.22 (.972) for the placebo. One could conclude that the difference in recurrence is due to the difference in the white blood cell count, because for the placebo group the average is higher than that for the treatment group. Such a scenario is called confounding, that is, it is difficult to say whether the difference in survival is due to the difference in the white blood cell count or if it is the effect of treatment.

Confounding is related to the idea of interaction, the interaction between the covariate and the treatment, that is, the effect of treatment may be different for different levels of the covariate. For the leukemia example, the effect of treatment on survival of Group 1 (which has a lower average white blood cell count) may be different than the effect of placebo (the group with the higher average blood cell count) on survival. Fortunately, the Cox proportional hazards model allows one to investigate the presence of interaction.

In order to calculate the survival rates through Kaplan–Meier, refer to Tables 6.4 and 6.5 for the two groups of leukemia patients. Two methods are presented for calculating the estimated survival probabilities: the direct approach and the product-limit method. Table 6.6 is Table 6.4 but with additional information, namely, the recurrence probabilities given by the $S(t_{(j)})$ and the conditional probabilities given by $P[T > t_{(j)} | T \geq t_{(j)}]$ reported in the last two columns respectively.

The last column of Table 6.4 gives the estimated probabilities of survival using the product-limit method.

TABLE 6.5

Times to Recurrence for Group 2 of Leukemia Patients

t_0	d_i is # Recurrences	c_i is # Censored	$R(t_{(j)})$
$t_{(0)} = 0$	0	0	21
$t_{(1)} = 1$	2	0	21
$t_{(2)} = 2$	2	0	19
$t_{(3)} = 3$	1	0	17
$t_{(4)} = 4$	2	0	16
$t_{(5)} = 5$	2	0	14
$t_{(6)} = 8$	4	0	12
$t_{(7)} = 11$	2	0	8
$t_{(8)} = 12$	2	0	6
$t_{(9)} = 15$	1	0	4
$t_{(10)} = 17$	1	0	3
$t_{(11)} = 22$	1	0	2
$t_{(12)} = 23$	1	0	1

TABLE 6.6

Times to Recurrence for Group 1

t_0	d_i is the # Recurrences	c_i is the # Censored	$R(t_{(j)})$	$S(t_{(j)})$	$P[T > t_{(j)} | T \geq t_{(j)}]$
$t_{(0)} = 0$	0	0	21	1	
$t_{(1)} = 6$	3	1	21	.8571	.8571
$t_{(2)} = 7$	1	1	17	.8067	.941
$t_{(3)} = 10$	1	2	15	.7529	.933
$t_{(4)} = 13$	1	0	12	.6901	.9166
$t_{(5)} = 16$	1	3	11	.6275	.909
$t_{(6)} = 22$	1	0	7	.5378	.857
$t_{(7)} = 23$	1	5	6	.4482	.833

Note that the number of censored observations at each time plays a role in the number at risk. Another way to express the estimated recurrence probabilities is with the Kaplan–Meier product limit, which is based on the formula

$$S(t_{(j)}) \ S(t_{(j)}) = \prod_{i=1}^{i=j} P\left[T > t_{(i)} \mid T \ge t_{(i)}\right] \ge t_{(i)}] \qquad (6.5)$$

for $j = 1,2,\ldots,n$.

Thus, the estimated survival probability at time $t_{(j)}$ is the product of j estimated conditional probabilities.

Consider Table 6.6 and the time $t_{(4)} = 13$ weeks. Then the proportion of recurrences occurring after 13 weeks is estimated as

$$S(4) = \left(\frac{21}{21}\right)\left(\frac{18}{21}\right)\left(\frac{16}{17}\right)\left(\frac{14}{15}\right)\left(\frac{11}{12}\right) = .690$$

Kaplan–Meier takes into account the censored observations.

Consider Table 6.7 and the time $t_{(4)} = 4$. Then the proportion of recurrences occurring after 4 weeks is estimated as

$$S(4) = \left(\frac{21}{21}\right)\left(\frac{19}{21}\right)\left(\frac{17}{19}\right)\left(\frac{16}{17}\right)\left(\frac{14}{16}\right) = \frac{14}{21} = .67$$

TABLE 6.7

Times to Recurrence for Group 2 of Leukemia Patients

t_0	d_i is # Recurrences	c_i is # Censored	$R(t_{(j)})$	$S(t_{(j)})$	$P\left[T > t_{(i)} \mid T \ge t_{(i)}\right]$
$t_{(0)} = 0$	0	0	21	1	
$t_{(1)} = 1$	2	0	21	.90	.904
$t_{(2)} = 2$	2	0	19	.8052	.894
$t_{(3)} = 3$	1	0	17	.7577	.941
$t_{(4)} = 4$	2	0	16	.6630	.875
$t_{(5)} = 5$	2	0	14	.5682	.857
$t_{(6)} = 8$	4	0	12	.3784	.666
$t_{(7)} = 11$	2	0	8	.2838	.75
$t_{(8)} = 12$	2	0	6	.1890	.666
$t_{(9)} = 15$	1	0	4	.1417	.75
$t_{(10)} = 17$	1	0	3	.0944	.666
$t_{(11)} = 22$	1	0	2	.0472	.50
$t_{(12)} = 23$	1	0	1	.00	.00

Note, that 19/21 estimates $P[T > 1 | T \geq 1]$, and that 14/16 estimates $P[T > 4 | T \geq 4]$, etc.

Note when there are no censored observations, such as for the placebo group of patients, the product-limit method agrees with the direct estimation of the recurrence probabilities. The direct method is given by

$$S(t_{(j)}) = \frac{[R(t_{(j)}) - d_i]}{n} \tag{6.6}$$

for $j = 1,2,\ldots,n$, where n is the total number of patients in the group.

Formula 6.5 can be expressed as

$$S(t_{(j)}) = S(t_{(j-1)})P[T > t_{(j)} | T \geq t_{(j)}] \tag{6.7}$$

for $j = 1,2,\ldots,m$, where m is the number of ordered times. This iteration formula is initiated with the known value $S(t_{(0)})$.

6.3.2 Bayesian Kaplan–Meier Method

What is the Bayesian approach to estimating the survival problems with the product-limit method? Referring to Tables 6.6 and 6.7, one must specify a posterior distribution for the unknown parameters, which are the probabilities of recurrence for the various ordered recurrence and censoring times. To that end, it assumed the number of recurrences at each of the ordered times has a binomial distribution, that is, for the ith ordered time

$$d[i] \sim \text{binomial } (q[i], R[i]) \tag{6.8}$$

where $q[i]$ is the probability of recurrence and $R[i]$ is the number at risk at the beginning of the ith ordered time. Assuming a beta prior for the $q[i]$, namely

$$q[i] \sim \text{beta } (.01, .01) \tag{6.9}$$

induces a posterior distribution for the probabilities of recurrence. By using a beta (.01,.01) prior distribution, the posterior mean will be quite similar to the usual estimators of the recurrence probabilities $S(t_{(i)})$ and the conditional probabilities $P[T > t_{(i)} | T \geq t_{(i)}]$ for $i = 1,2, \ldots, m$, where m is the number of order times (recurrence and censored). The number of censored observations are also given a binomial distribution, namely

$$c[i] \sim \text{binomial } (qc[i], R[i]) \tag{6.10}$$

and

$$qc[i] \sim \text{beta } (.01, .01) \tag{6.11}$$

If $r[i]$ is the number at risk at time i, then

$$r[i] = r[i-1] - c[i-1] - d[i-1] \qquad (6.12)$$

for $i = 2,3,\ldots,m$, where m is the number of distinct ordered times.

Thus, the distribution of the number of deaths and number of censored observations induces a distribution on the number at risk. Note also that the probability of an event and the probability of a censored observation are not the same at each time period, that is, it is not assumed that $q[i] = qc[i]$!

The following code will allow one to perform a Bayesian analysis that estimates the recurrence probabilities $S(t_{(i)})$ (computed by the product-limit method) and the conditional probabilities $P[T > t_{(i)} \,|\, T \geq t_{(i)}]$ both of which are part of the Kaplan–Meier method. The code gives the analysis for both groups of leukemia patients, where Group 1 is the treatment group and Group 2 the placebo.

BUGS CODE 6.1

```
model;
# leukemia patients two groups
{
# for group 1
# distribution of the number of recurrences
for (i in 1:m1){d1[i]~dbin(q1[i],r1[i])}
# distribution of the number of censored
for (i in 1:m1){ce1[i]~dbin(qc1[i],r1[i])}
# prior distribution of q1[i]
for (i in 1:m1){q1[i]~dbeta (.01,.01)}
for (i in 1:m1){qc1[i]~dbeta (.01,.01)}
# conditional probabilities
#for(i in 1:m1){p[i]<-(r[i]-d[i])/r[i]}
for (i in 1:m1){p1[i]<-1-q1[i]}
# number at risk
r1[1]<- 21
for(i in 2:m1){r1[i]<-r1[i-1]-d1[i-1]-ce1[i-1]}
# product-limit survival probabilities
for (i in 2:m1){s1[i]<-s1[i-1]*p1[i]}
s1[1]<-p1[1]
# for group 2
# distribution of the number of recurrences
for (i in 1:m2){d2[i]~dbin(q2[i],r2[i])}
# distribution of the number of censored
for (i in 1:m2){ce2[i]~dbin(qc2[i],r2[i])}
# prior distribution of q2[i]
for (i in 1:m2){q2[i]~dbeta (.01,.01)}
for (i in 1:m2){qc2[i]~dbeta (.01,.01)}
# conditional probabilities
#for(i in 1:m2){p2[i]<-(r2[i]-d2[i])/r2[i]}
for (i in 1:m2){p2[i]<-1-q2[i]}
```

```
# number at risk
r2[1]<- 21
for(i in 2:m2){r2[i]<-r2[i-1]-d2[i-1]-ce2[i-1]}
# product-limit survival probabilities
for (i in 2:m2){s2[i]<-s2[i-1]*p2[i]}
s2[1]<-p2[1]
}
# data for Freirich et al.
list(m1 = 7, d1 = c(3,1,1,1,1,1,1), ce1 = c(1,1,2,0,3,0,5),
m2 = 12, d2 = c(2,2,1,2,2,4,2,2,1,1,1,1),
ce2 = c(0,0,0,0,0,0,0,0,0,0,0,0))
# initial values
list(q1 = c(.5,.5,.5,.5,.5,.5,.5),
q2 = c(.5,.5,.5,.5,.5,.5,.5,.5,.5,.5,.5,.5))
```

The analysis is executed with 55,000 observations for the simulation, a burn-in of 5000, and a refresh of 100. What is reported in Tables 6.8 and 6.9 is the posterior distribution for the conditional and recurrence probabilities of Groups 1 and 2, respectively.

Note that the conditional probabilities are

$$P[i] = P\left[T > t_{(i)} \mid T \geq t_{(i)}\right] \qquad (6.13)$$

and the recurrence probabilities are

$$S[i] = S(t_{(i)}) \qquad (6.14)$$

TABLE 6.8

Posterior Distribution for Treatment Group

Parameter	Mean	SD	2½	Median	97½	$t_{(i)}$
$P[1]$	.8565	.07453	.6826	.8681	.9674	6
$P[2]$	.9407	.05539	.7949	.9568	.9983	7
$P[3]$	.9322	.06334	.7646	.9508	.9981	10
$P[4]$	.9162	.07685	.7158	.939	.9976	13
$P[5]$	.9088	.08304	.6907	.9328	.9974	16
$P[6]$	.8563	.1243	.5361	.8907	.9956	22
$P[7]$	.8323	.1405	.4774	.8685	.9948	23
$S[1]$	.8566	.07453	.6826	.8681	.9674	6
$S[2]$	.8057	.08478	.6154	.8157	.9406	7
$S[3]$	.7577	.09418	.5463	.7594	.909	10
$S[4]$	.6882	.104	.4677	.6954	.8697	13
$S[5]$	.6255	.1109	.3946	.6314	.8244	16
$S[6]$	.5355	.1233	.2879	.5394	.7619	22
$S[7]$	.4457	.1284	.2006	.4448	.6956	23

TABLE 6.9

Bayesian Analysis for the Placebo Group

Parameter	Mean	SD	2½	Median	97½	$t_{(i)}$
P[1]	.9045	.06289	.7507	.9174	.9878	1
P[2]	.8951	.06814	.7284	.9081	.9866	2
P[3]	.9404	.05544	.7941	.9566	.9983	3
P[4]	.8742	.08052	.6788	.8898	.9831	4
P[5]	.8571	.09016	.6414	.874	.9813	5
P[6]	.6667	.1309	.3904	.676	.8913	8
P[7]	.7501	.144	.4184	.771	.9634	11
P[8]	.6655	.1776	.2885	.6851	.9463	12
P[9]	.747	.1943	.2905	.79	.9914	15
P[10]	.6657	.2352	.1588	.7069	.9875	17
P[11]	.5021	.2871	.02652	.5035	.9746	22
P[12]	.01048	.07233	0	0	.08907	23
S[1]	.9045	.06289	.7507	.9174	.9878	1
S[2]	.8097	.08389	.6214	.8198	.9432	2
S[3]	.7614	.09095	.563	.77	.9139	3
S[4]	.6657	.101	.4535	.6714	.8453	4
S[5]	.5706	.1058	.3579	.5729	.768	5
S[6]	.3805	.104	.1908	.3769	.5933	8
S[7]	.2855	.09659	.1168	.2786	.4918	11
S[8]	.1901	.08381	.05677	.1802	.3789	12
S[9]	.1421	.07474	.03143	.1305	.3167	15
S[10]	.09476	.06284	.01209	.0817	.2505	17
S[11]	.04571	.04522	.001354	.03422	.1681	22
S[12]	<.0001	.004768	0	0	.002517	23

where

$$S[i] = S[i-1]P[i] \tag{6.15}$$

$i = 1,2,\dots,m$.

BUGS CODE 6.1 contains many remarks indicated by a # that identify the actual computation for the Bayesian analysis. One should compare Table 6.8 with 6.6 and note that the last column of Table 6.6 reports the conditional probabilities (that are estimated the usual way). When comparing this column with the corresponding posterior means of the $P[i]$ of Table 6.8, it is obvious that they are quite similar. However, it is important to recognize that the Bayesian analysis provides an estimate of uncertainty not provided by Table 6.6. For example, at time $t_{(4)} = 13$ weeks, the recurrence probability from Table 6.6 is .6901, compared to a posterior mean of .6882 with a posterior standard deviation of .1042. Note that the posterior median is .6954, which gives excellent agreement between the usual and Bayesian approaches to estimating the recurrence probabilities. Because the posterior distributions are skewed, I recommend the posterior median as the estimator of the recurrence probabilities. The simulation errors were <.001 for all parameters. The Bayesian analysis continues with the placebo group with Table 6.9.

Upon comparing Table 6.7 with 6.9, one observes excellent agreement with the Bayesian analysis of the latter compared to the conventional analysis of the former. For example, the posterior mean of the conditional probability corresponding to $t_{(8)} = 12$ is .6657 compared to the conventional value of .666 of Table 6.5. Simulation errors are not reported, but all are less than .0001. The reader is referred to Kleinbaum (pp. 51–58)[1] for the conventional analysis reported in Tables 6.6 and 6.7.

6.3.3 Kaplan–Meier Plots for Recurrence of Leukemia Patients

Using the information from Table 6.3, Figure 6.1 portrays the times to recurrence of the two groups of leukemia patients, where the abscissa is the time to recurrence in weeks and the ordinate is the probability of recurrence. Group 1 is the treatment group and Group 2 is the placebo, and the graph implies that the recurrence experience (time in remission) of the former is longer than that of the patients receiving placebo. Upon inspection of the plot, one would estimate the median time of remission to be approximately 9 weeks for the placebo group, whereas for the treatment group, the median was yet to be reached over the study period of about 23 weeks. One would conclude that the treatment is efficacious in that it appears to prolong remission in acute leukemia patients.

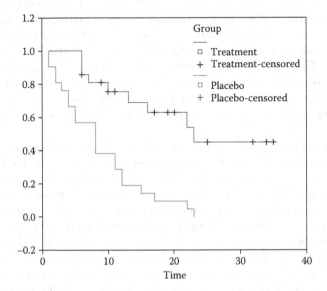

FIGURE 6.1
Times to recurrence, treatment versus placebo.

Note the vertical lines indicate the time at which a recurrence(s) occurs, for example, for the placebo group at time 8 weeks, four patients have a recurrence (came out of remission), thus, the relative "long" drop in the curve at 8 weeks. I used SPSS® (version 18) for the Kaplan–Meier curve, but there are many software packages that will execute survival plots.

Our next area of interest is to develop a Bayesian analysis that tests for a difference in the survival experience of two groups. One obvious test for this example is to compare the survival probability at week 23 between the two groups, which need not be performed because the survival probability at week 23 for the treatment group is estimated as .4464 with a 95% credible interval of (.1983, .6954), compared to a probability estimated as 0 for week 23 of the placebo group. The next section will present a Bayesian version of the log-rank test, which compares the overall survival of one group versus that of another. Although the idea of a Bayesian approach to the Kaplan–Meier curve is presented using two groups of leukemia patients, BUGS CODE 6.1 is easily revised to analyze any two (or more) groups of patients, where the time of the event is either survival, recurrence, or any time to event where the time of censored observations is taken into account.

6.3.4 Log-Rank Test for Difference in Recurrence Times

The log-rank test is one of the conventional procedures by which the overall survival experience of two groups is compared. The emphasis is on the word *overall*, that is, the two groups are compared from the first-ordered observation to the last-ordered observation (of time to event). A Bayesian version of the test is presented in that the Bayesian mimics to some extent the conventional; however, the interpretation of the test is strictly Bayesian. As in the previous section, the general idea is presented using the example of two groups of leukemia patients from the study of Freirich et al.[5]

Basically, the log-rank test consists of comparing the two groups by comparing the observed minus the expected number of events (recurrence, survival, etc.), normalized by the variance of the difference.

Recall the study of two groups of leukemia patients, but in the following format:

The ordered times of the two groups are combined and appear in the second column, the third column is the number of recurrences for the treatment group, and the recurrence times of the second group are reported in the fourth column. The fifth and sixth columns are the number at risk for the treatment and placebo group respectively, whereas the seventh and eighth are the average number of events (recurrences) for Group 1 and Group 2, respectively, assuming there is no difference in the overall recurrence between the two groups. The last two columns are the difference in the number of events minus the average number of events for the two groups. The study period for this version extends through week 23, but not beyond, thus the censored times 25, 32, 34, 35 are not included in the example.

The average or expected number of events for Group 1 is computed as follows:

$$a_{1i} = \left[\frac{R1(i)}{(R1(i) + R2(i))} \right](d_{1i} + d_{2i}) \tag{6.16}$$

where it is assumed that the probability of an event (and the probability of a censored observation) is the same for the two groups at each time period. Recall that the number of events and the number of censored observations for each time period is assumed to be a binomial random variable; thus, the number at risk is also a random variable. The number at risk will decrease by the number of events and the number of censored observations.

BUGS CODE 6.2 is for calculating the difference in the observed minus expected events for Group 1, assuming there is no difference in the probability of an event between the two. Note that the observed minus expected differences for Group 2 (the placebo) differ only in sign to those of the first group (treatment group). The number of deaths and number of censored observations are assumed to have a binomial distribution, whose corresponding probabilities are given beta (.01, .01) prior distributions. This induces a posterior distribution to the number of expected events given by (6.16), and thus induces a posterior distribution to the observed minus expected differences of Group 1. Of interest is testing for a difference in the overall recurrence between the two groups, which is expressed by the sum of the differences

$$\text{some 1} = \sum_{i=1}^{i=m} (d_{1i} - a_{1i}) \tag{6.17}$$

where m = 17 is the number of distinct observed times.

In a similar way, the sum of the observed minus expected differences for Group 2 is

$$\text{some 2} = \sum_{i=1}^{i=m} (d_{2i} - a_{2i}) \tag{6.18}$$

where

$$a_{2i} = \left[\frac{R2(i)}{(R1(i) + R2(i))} \right](d_{1i} + d_{2i}) \tag{6.19}$$

is the expected number of recurrences for Group 2 for the ith time.

If there is no difference in the event rate of Group 1 versus Group 2, one would expect half the differences to be positive and half to be negative, thus on the average, one would expect the sum to average 0. If the 95% credible interval for the sum of the differences (Equation 6.17) does not contain 0, one is inclined to believe that the two groups do indeed differ with respect to the rate of a recurrence. BUGS CODE 6.2 is well documented with statements indicated by a # sign and the code closely follows Equation 6.17

BUGS CODE 6.2

```
model;

# log-rank Bayesian approach
{
# for group 1
# distribution of the number of recurrences
for (i in 1:m1){d1[i]~dbin(q1[i],r1[i])}
# distribution of the number of censored
for (i in 1:m1){c1[i]~dbin(qc1[i],r1[i])}
# prior distribution of q1[i]
for (i in 1:m1){q1[i]~dbeta (.01,.01)}
for (i in 1:m1){qc1[i]~dbeta (.01,.01)}
# number at risk
r1[1]<- n1
for(i in 2:m1){r1[i]<-r1[i-1]-d1[i-1]-c1[i-1]}
# for group 2

# distribution of the number of recurrences
for (i in 1:m2){d2[i]~dbin(q2[i],r2[i])}
# distribution of the number of censored
for (i in 1:m2){c2[i]~dbin(qc2[i],r2[i])}
# prior distribution of q2[i]
for (i in 1:m2){q2[i]~dbeta (.01,.01)}
for (i in 1:m2){qc2[i]~dbeta (.01,.01)}
# number at risk
r2[1]<- n2
for(i in 2:m2){r2[i]<-r2[i-1]-d2[i-1]-c2[i-1]}
# mortality group1
for(i in 1:m1){s1[i]<- q1[i]*r1[i]}
# mortality group 2
for(i in 1:m2){s2[i]<- q2[i]*r2[i]}
# Expected number of events
for(i in 1:m1){e1[i]<- e11[i]*e12[i]}
for(i in 1:m2){e11[i]<- r1[i]/(r1[i]+r2[i])}
for(i in 1:m2){e12[i]<- s1[i]+s2[i]}
for(i in 1:m1){e2[i]<- e21[i]*e22[i]}
for(i in 1:m2){e21[i]<- r2[i]/(r1[i]+r2[i])}
for(i in 1:m2){e22[i]<- s1[i]+s2[i]}
# variance of difference
for(i in 1:m2){t1[i]<- r1[i]*r2[i]*(s1[i]+s2[i])*(r1[i]+
  r2[i]- s1[i]-s2[i])}
for(i in 1:m2){t2[i]<-(r1[i]+r2[i])*(r1[i]+r2[i])*(r1[i]+
  r2[i]-1)}
# difference observed minus expected
for(i in 1:m1){ome1[i]<- d1[i]-e1[i]}
```

```
for(i in 1:m1){vlr1[i]<-t1[i]/t2[i]}
for(i in 1:m2){ome2[i]<-d2[i]-e2[i]}
# sum of observed minus expected group 1
some1<- sum(ome1[])
# sum of expected group 1
se1<-sum(e1[])
# sum of expected group 2
se2<-sum(e2[])
# sum of observed minus expected group 2
some2<- sum(ome2[])
# likelihood ratio<- some1*some1/se1+some2*some2/se2
}
# data for Freirich et al. Table 6.6
list(n1 = 21, n2 = 21,m1 = 17, d1 = c(0,0,0,0,0,3,1,0,1,0,0,1,
    0,1,0,1,1),
# the number of censored observations group 1
c1 = c(0,0,0,0,0,1,0,1,1,1,0,0,0,0,3,0,0),
m2 = 17, d2 = c(2,2,1,2,2,0,0,4,0,2,2,0,1,0,1,1,1),
# the number of censored observations group 2
c2 = c(0,0,0,0,0,0,0,0,0,0,0,0,0,0,0,0,0))
# initial values Table 6.6
list(q1 = c(.5,.5,.5,.5,.5,.5,.5,.5,.5,.5,.5,.5,.5,.5,.5,.5,.5),
q2 = c(.5,.5,.5,.5,.5,.5,.5,.5,.5,.5,.5,.5,.5,.5,.5,.5,.5))


# data Exercise 5

list(n1 = 25,n2 = 25, m1 = 44,d1 = c(0,0,1,1,0,1,1,0,0,1,0,1,
    0,1,0,0,0,0,0,0,1,1,1,0,0,0,1,0,1,0,0,1,0,0,1,1,0,1,1,1,1,1,0,1),
    c1 = c(0,0,0,0,0,0,0,0,0,0,0,0,0,0,0,0,0,0,0,0,0,0,0,0,0,0,0,0,
    0,0,0,0,0,0,0,0,0,0,0,0,0,0,0,0),
m2 = 44, d2 = c(1,1,1,0,1,0,0,1,1,0,1,1,1,0,1,1,1,1,1,1,0,0,0,
    1,1,1,0,1,0,1,1,0,1,1,0,0,1,0,0,0,0,0,1,0),
c2 = c(0,0,0,0,0,0,0,0,0,0,0,0,0,0,0,0,0,0,0,0,0,0,0,0,0,0,0,
    0,0,0,0,0,0,0,0,0,0,0,0,0,0,0,0,0))
# initial values Exercise 5
list(q1 = c(.5,.5,.5,.5,.5,.5,.5,.5,.5,.5,.5,.5,.5,.5,.5,.5,
    .5,.5,.5,.5,.5,.5,.5,.5,.5,.5,.5,.5,.5,.5,.5,.5,.5,.5,.5,.5,
    .5,.5,.5,.5,.5,.5,.5,.5),
q2 = c(.5,.5,.5,.5,.5,.5,.5,.5,.5,.5,.5,.5,.5,.5,.5,.5,.5,
    .5,.5,.5,.5,.5,.5,.5,.5,.5,.5,.5,.5,.5,.5,.5,.5,.5,.5,.5,.5,
    .5,.5,.5,.5,.5,.5,.5))
```

The Bayesian analysis is based on Table 6.10 and executed through BUGS CODE 6.2 with 55,000 observations for the simulation, a burn-in of 5000, and a refresh of 100, where the main focus is on the posterior distribution of the sum (Equation 6.17) of the differences observed minus average number of events of the treatment group. See Table 6.11 for the Bayesian analysis.

Upon comparing the next to last column of Table 6.10 with the posterior mean and median of Table 6.11, one will see the similarity between the

TABLE 6.10

Combined Recurrence Experience of Two Groups of Leukemia Patients

i	$t_{(i)}$	d_{1i}	d_{2i}	$R1(i)$	$R2(i)$	a_{1i}	a_{2i}	$d_{1i} - a_{1i}$	$d_{2i} - a_{2i}$
1	1	0	2	21	21	1	1	−1	1
2	2	0	2	21	19	1.05	.95	−1.05	1.05
3	3	0	1	21	17	.5526	.4473	−.5526	.5526
4	4	0	2	21	16	1.135	.864	−1.135	1.135
5	5	0	2	21	14	1.2	.777	−1.2	1.2
6	6	3	0	21	12	1.909	1.099	1.09	−1.09
7	7	1	0	17	12	.586	.413	.414	−.414
8	8	0	4	16	12	2.285	1.714	−2.285	2.285
9	10	1	0	15	8	.652	.347	.348	−.348
10	11	0	2	13	8	1.238	.761	−1.238	1.238
11	12	0	2	12	6	1.333	.666	−1.333	1.333
12	13	1	0	12	4	.75	.25	.25	−.25
13	15	0	1	11	4	.733	.266	−.733	.733
14	16	1	0	11	3	.785	.214	.214	−.214
15	17	0	1	10	3	.769	.230	−.769	.769
16	22	1	1	7	2	1.555	.444	−.555	.555
17	23	1	1	6	1	1.714	.285	.714	−.714

TABLE 6.11

A Bayesian Analysis of Observed Minus Expected (Group 1, Treatment Group)

Parameter	Mean	SD	2½	Median	97½	Time $t_{(i)}$
diff(1)	−1.005	.6604	−2.629	−.8669	−.1297	1
diff(2)	−1.059	.6858	−2.727	−.9241	−.1376	2
diff(3)	−.5606	.5228	−1.953	−.4052	−.01628	3
diff(4)	−1.138	.7321	−2.909	−.9927	−.1482	4
diff(5)	−1.214	.7652	−3.048	−1.07	−.1634	5
diff(6)	1.084	.9926	−1.221	1.229	2.565	6
diff(7)	.4045	.5572	−1.081	.5692	.983	7
diff(8)	−2.303	.8953	−4.177	−2.239	−.7609	8
diff(9)	.3379	.6162	−1.281	.5153	.9809	10
diff(10)	−1.245	.7157	−2.879	−1.137	−.1879	11
diff(11)	−1.341	.7152	−2.881	−1.266	.2122	13
diff(12)	.2386	.6938	−1.575	.4363	.9776	13
diff(13)	−.7398	.569	−2.087	−.6119	−.02666	15
diff(14)	.2001	.7207	−1.707	.405	.9762	16
diff(15)	−.7746	.5476	−1.953	−.6815	−.03092	17
diff(16)	−.5606	.8083	−2.427	−.4753	.7197	22
diff(17)	−.7122	.732	−2.551	−.5193	.1228	23
Lr	16.32	3.591	11.99	15.41	25.56	
some1	−10.43	2.93	−16.43	−10.31	−4.98	
some2	10.15	1.776	6.461	10.23	13.41	
e1	19.43	2.93	13.98	19.31	25.43	
e2	10.85	1.776	7.59	10.77	14.54	

Bayesian analysis and the conventional. Note that the degree of uncertainty that accompanies the Bayesian analysis is expressed as the posterior standard deviation. Also, one notices the skewness in all the posterior distributions of the difference. At 11 weeks, the posterior mean of the difference in observed minus expected is −1.24 with a 95% credible interval of (−2.879, −.1879), but the parameter that measures the difference in the overall recurrence is the likelihood ratio with a posterior median of 16.32 and 95% credible interval (11.99, 25.56), which is similar to the usual value of the likelihood ratio test, indicating there is a difference in the recurrence rate between the treatment group and the placebo group.

The log-rank test is the conventional way of testing for a difference in the recurrence of two groups, but what is presented is a Bayesian analog. A conventional procedure, based on the chi-square distribution under the null hypothesis of no difference in the recurrence of the two groups, is presented by Kleinbaum (pp. 58–62)[1] and the value of the likelihood ratio is the same as the posterior mean of the likelihood ratio of Table 6.11. The conventional interpretation of the likelihood ratio is not the same as the interpretation the Bayesian analysis would present.

A more conventional approach to the log-rank test can be found with Kahn and Sempos[6] and Peto et al.[7] where the latter reference gives a discussion of the early development of the log-rank test and its applications to cancer clinical trials.

For the Bayesian, the likelihood ratio parameter is

$$Lr = \frac{some1^* \, some1}{e1} + \frac{some2^* \, some2}{e2} \tag{6.20}$$

where some1 is given by Equation 6.17, some2 by Equation 6.18,

$$e1 = \sum_{i=1}^{i=m} a_{1i} \tag{6.21}$$

is the total number of expected recurrences for Group 1, and

$$e2 = \sum_{i=1}^{i=m} a_{2i} \tag{6.22}$$

is the total expected number of recurrences for Group 2.

For the conventional form of the likelihood ratio, see Klenbaum (p. 71).[1]

The Bayesian analysis is based on the some1 parameter (Equation 6.17), which is the sum of the differences of the observed number of recurrences minus the expected (under the assumption the overall recurrence pattern is the same for the two groups) number of recurrences for all distinct ordered values of recurrence and censored times of the treatment group. Its posterior distribution is reported in Table 6.11 with a posterior mean (SD) of −10.43 (3.591), a 95% credible interval (−16.43, −4.98), and a posterior median of −10.31. Because the credible

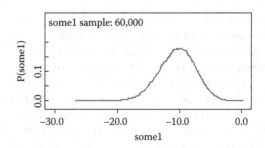

FIGURE 6.2
Posterior density.

interval does not contain 0, the implication is that there is indeed a difference in the recurrence rate of the two groups. Note, one could have used some2, the sum of the observed minus expected recurrences for the placebo group, to investigate the difference in the recurrence of the two Group. Also, observe from Table 6.10 that the observed minus expected differences for Group 2 are the same those for Group 1, except for sign! The posterior density of some1 appears in Figure 6.2. If there is no difference in the overall recurrence pattern between the two groups, one would expect some1 to tend to 0 on the average.

As with the likelihood ratio, the posterior distribution of some1 (Equation 6.17) implies a difference in the recurrence between the two groups.

6.4 Survival Analysis

6.4.1 Introduction

Survival models are an extension of the life table, and at this point, several models will be introduced for assessing the survival experience of several groups of patients. First to be introduced are parametric models where the survival times follow a known parametric distribution. Recall that survival times record the actual survival time of a subject or the time of censoring for a patient. Note that for an actual time of survival (e.g., death or recurrence), one knows the time at which the event occurs, but for a censored time, one knows the subject has survived at least as long as the recorded censored time. Both scenarios are used to estimate the unknown parameters of the survival distribution. Special survival distributions are the exponential, gamma, and Weibull distributions. Assuming the Weibull distribution, several examples will illustrate the possible Bayesian inferences with an emphasis on comparing the median survival of several groups of patients.

The final section of the chapter introduces Bayesian inference with the Cox proportional hazards model. This model is one of the most useful for survival analysis and is ubiquitous in the medical literature, including those publications devoted to the results of clinical trials. For example, the journal *Cancer* presents the results of many Type I and II clinical trials for the efficacy of new and experimental therapies, mostly chemotherapies. In such studies, the main end point is the survival time of a group of patients receiving the new treatment and this treatment is to be compared to the survival time of a placebo or an historical control.

Of course, sometimes survival time is not the main end point for such trials, but instead it is the recurrence time (or the time spent in remission). Such a trial was encountered previously with the Freireich et al.[5] study where a new experimental therapy was compared to a group of patients receiving a placebo. The study will be used for illustrating the Bayesian approach to inference with a Weibull model and the Cox proportional hazards model. As will be seen, the advantage of the Cox model is that it is appropriate for a large number of survival distributions and no particular survival distribution is assumed; thus, it is referred to as a nonparametric approach to survival analysis. Thus, for example, if the Weibull distribution is indeed the appropriate survival distribution, then the Cox model will provide results quite similar to that using the Weibull distribution. In order to provide a suitable Bayesian analysis, WinBUGS code is adopted or developed.

6.4.2 Parametric Models for Survival Analysis

The Weibull distribution is often used as the distribution of the survival times of a group of patients. It has density

$$f(t) = \lambda k t^{k-1} \exp - \lambda t^k \qquad (6.23)$$

where t is time that an event occurs, and k and λ are unknown positive parameters. The corresponding hazard function is

$$h(t) = \lambda k t^{k-1} \qquad (6.24)$$

and the survival function is

$$\begin{aligned} S(t) &= P[T > t] \\ &= \exp(-\lambda t^k) \end{aligned} \qquad (6.25)$$

One may refer to Ibrahim, Chen, and Sinha (p. 35)[8] and Congdon (p. 363)[3] for additional information about the Weibull distribution.

Recall the relation between the survival function and the hazard function

$$h(t) = \frac{[dS(t)/dt]}{S(t)} \qquad (6.4)$$

Thus, Equation 6.24 can be derived from Equation 6.25 using Equation 6.4.

If $k > 1$, the hazard is a monotone increasing function of t, while if $k < 1$, it a monotone decreasing function of the survival time.

The mean and median of the Weibull distribution are

$$E(T) = \left(\frac{1}{\lambda}\right)^{1/k} \Gamma\left(1 + \frac{1}{k}\right) \qquad (6.26)$$

and

$$\text{Med}(T) = \left(\frac{1}{\lambda}\right)^{\frac{1}{k}} [\text{In}(2)]^{1/k} \qquad (6.27)$$

respectively, where Γ is the gamma function. Often the median is of special importance in survival studies, where the median survival time for one group of subjects is compared to the median survival of another group. In order to implement a Bayesian analysis, a Weibull survival distribution is assumed for one group and a different Weibull is assumed for the other, then the posterior distribution of the difference in the median survival of the two groups is determined. Of course, one much specify the prior distribution of the parameters (k and λ) of the two Weibull distributions. Then by employing Bayes' theorem, the relevant posterior distributions will be determined.

The behavior of the Weibull distribution is portrayed in Figure 6.3 for various combinations of the two parameters with $\lambda = 1$ and $k = .5, 1, 1.5, 5$. Note the variation as k varies from .5 to 5, where for the first setting, the density

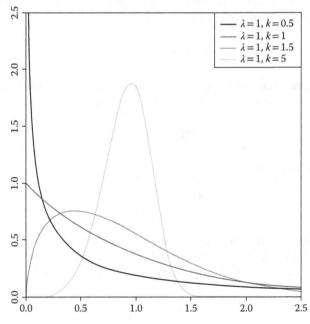

FIGURE 6.3
Plots of the Weibull density. http://en.wikipedia.org/wiki/Weibull_dist

is quite similar to the exponential density, but for the case $k = 5$, the density appears almost symmetric about its mean. For the intermediate scenario when $k = 1.5$, the density is similar to that of a gamma or chi-square distribution, thus it seems that the Weibull distribution is very versatile and can serve as a distribution for a large number of survival studies.

Recall the Freireich et al.[5] comparison of two groups of leukemia patients, where previously the analysis consisted of a Bayesian version of the log-rank test and the Kaplan–Meier plots displayed by Figure 6.1. The survival experience of the two groups of patients will be compared assuming the survival times follow a Weibull distribution.

The code below is the Bayesian analysis for the Freireich et al.[5] study with two groups of leukemia patients, namely a treatment group and a placebo. There are two sets of data in the list statements of the code and the leukemia study is the second set of list statements. For the list statements containing data, there are two sets of data: (1) the recurrence time of the patients is given by the 2 by 21 matrix t and (2) the corresponding censored times are given by the 2 by 21 matrix t.cen. Note when a patient is censored, the NA symbol appears in the t matrix and the actual censored time appears in the corresponding part of the t.cen matrix. If the patient is not censored, a zero appears in the t.cen matrix. The data is the time to recurrence of the disease and appears in Table 6.3. Note the code below is a slight revision of the mice example of Volume 1 of WinBUGS. The two parameters of the Weibull are k and lambda, defined by the Weibull density (Equation 6.23).

BUGS CODE 6.3

```
model
  {
  # Weibull distribution for survival times
  # right censored
          for(i in 1 : M) {
                  for(j in 1 : N) {
                  t[i, j] ~ dweib(k[i], lamda[i])I(t.cen[i, j],)
                  culmative.t[i, j] <- culmative(t[i, j], t[i,j])
                  }
          lamda[i] <- exp(beta[i])
          beta[i] ~ dnorm(0.0, 0.001)
          # median survival times
          median[i] <- pow(log(2) * exp(-beta[i]), 1/k[i])
      }
      for(i in 1:M){
      k[i] ~ dexp(0.01)}
      # difference in median survival
      # note the number of betas is the number of groups
      diff<- median[1] - median[2]
   }
      # leukemia data of Freireich et al.
```

```
# note there are two betas corresponding to two groups
list(t = structure(.Data =
 c(6,6,6,7,10,13,16,22,23,NA,NA,NA,NA,NA,NA,NA,NA,
   NA,NA,NA,NA,
               1,1,2,2,3,4,4,5,5,8,8,8,8,11,11,12,12,15,
               17,22,23),
             Dim = c(2, 21))),
 t.cen = structure(.Data =
 c(0,0,0,0,0,0,0,0,0,6,9,10,11,17,19,20,
   25,32,32,34,35,
               0,0,0,0,0,0,0,0,0,0,0,0,0,0,0,0,0,0,0,0,0),
             Dim = c(2, 21))),
 M = 2, N = 21)
 # initial values
 list(beta = c(-1, -1))
```

The analysis is executed with 65,000 observations for the simulation, with a burn-in of 5000 and a refresh of 100, and the results appear in Table 6.12. The parameters appearing in the table are the two parameters of the Weibull, the two posterior medians of the recurrence times of the two groups, and the difference in the two medians. It is assumed that the two groups of patients have different Weibull recurrence time distributions, namely for the first group, the parameters are $k[1]$ and lamda[1], while for the second they are $k[2]$ and lamda[2]. The median survival for the first treatment group is median[1] and median[2] for the second, respectively. The prior distributions for the betas are vague normal distributions, while the $k[i]$ parameters are given an exponential distribution with parameter .01.

Parameters of the Weibull distribution for the treatment groups are $k[1]$ and lamda[1] with posterior means 1.429 and .01209, respectively, compared to $k[2]$ and lamda[2] for the placebo group, with posterior means 1.403 and .0498, respectively. It appears that the two shape parameters $k[1]$ and $k[2]$ have a common posterior mean. Note that some of the posterior distributions are skewed, such as that for lamda[1] with a posterior mean of .01209 versus a median of .006914.

TABLE 6.12

Posterior Analysis of the Recurrence of Leukemia Patients

Parameter	Mean	SD	Error	2½	Median	97½
$k[1]$	1.429	.3751	.02036	.7601	1.406	2.235
$k[2]$	1.403	.2377	.00744	.9682	1.394	1.894
lamda[1]	.01209	.01546	<.0001	.000465	.006914	.0571
lamda[2]	.0498	.03245	<.0001	.0104	.04187	.1316
median[1]	28.26	9.276	.09622	19.96	26.42	50.86
median[2]	7.541	1.413	.03372	4.977	7.477	10.51
diff	20.72	9.379	.1017	9.001	18.94	43.38

The main emphasis for the comparison of the recurrence between the treatment and placebo groups is the difference in the two median recurrence times diff = median[1]–median[2]. Note the posterior mean of the median recurrence for the treatment group is 28.26 weeks compared to 7.541 weeks for placebo, which implies a large difference in the median recurrence between the two groups. A 95% credible interval for diff is (9.001, 43.38) implying that the two median recurrence times are indeed not the same and that the recurrence time of the treatment group is larger than that for placebo. Of course, this is also evident from Figure 6.1.

Is it reasonable to assume that the recurrence times follow a Weibull distribution? Figure 6.4 is a P–P plot of the 21 observations from the placebo group of the Freireich et al.[5] study, and it appears that it is not unreasonable to believe that the recurrence times do indeed have a Weibull distribution. Of course, one should always check to see whether the distribution is appropriate for the analysis!

The Cox model will also be used to compare the recurrence times between the two groups, but for the time being, I feel that the conclusion that the median recurrence for treatment is much larger than that for placebo is valid.

As another example for the Weibull distribution, consider the study of McGilchrist and Aisbett[9] who display the times (days) to infection of 38 patients undergoing dialysis for kidney disease. Table 6.13 displays the time to the first infection and the status of the observations (censored and not censored). Recurrence is indicated by a 1 and censored by a 0, while for gender, 1 denotes male and 2 denotes female.

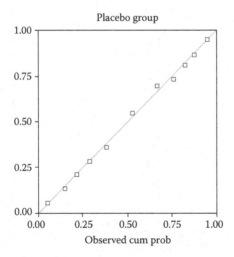

FIGURE 6.4
P–P plot recurrence times.

TABLE 6.13

Recurrence Times for Kidney Patients

Patient	Recurrence Time	Event	Gender
1	8	1	1
2	15	1	2
3	447	1	2
4	13	1	2
5	7	1	1
6	113	0	2
7	15	1	1
8	2	1	1
9	96	1	2
10	5	0	2
11	17	1	1
12	119	1	2
13	6	0	2
14	33	1	2
15	30	1	1
16	511	1	2
17	7	1	2
18	149	0	2
19	185	1	2
20	22	1	2
21	22	0	1
22	402	1	2
23	24	1	1
24	12	1	1
25	53	1	2
26	34	1	2
27	141	1	2
28	27	1	2
29	536	1	2
30	190	1	2
31	292	1	2
32	63	1	1
33	152	1	1
34	39	1	2
35	132	1	2
36	130	1	2
37	152	1	2
38	54	0	2

Source: McGilchrist, C.A., and Aisbett, C.W., *Biometrics*, 47, 461–466, 1991.

In order to estimate the median survival of the time to infection, BUGS CODE 6.3 is revised as follows:

BUGS CODE 6.4

```
model
    {

    # Weibull distribution for time to events
    # right censored
        for(i in 1 : M) {
            for(j in 1 : N) {
                t[i, j] ~ dweib(k[i], lamda[i])I(t.cen[i, j],)
                culmative.t[i, j] <- culmative(t[i, j], t[i, j])
            }
            lamda[i] <- exp(beta[i])
            beta[i] ~ dnorm(0.0, 0.001)
            # median survival times
            median[i] <- pow(log(2) * exp(-beta[i]), 1/k[i])
        }

        for(i in 1:M){
        k[i] ~ dexp(0.01)}
    }
# McGilchrist and Aisbett[9]
# time to infection
list(t = structure(.Data =
                    c(8,15,447,13,7,NA,15,2,96,NA,17,119,
                       NA,33,30,511,
                       7,NA,185,
      NA,22,402,24,12,53,34,141,27,536,190,292,63,152,39,132,130,
         152,NA),
                        .Dim = c(1, 38)),
                    t.cen = structure(.Data =
                c(0,0,0,0,0,113,0,0,0,5,0,0,6,0,0,0,0,149,0,
                   22,0,0,0,0,0,0,0,0,0,0,0,0,0,0,0,0,0,54),
                    .Dim = c(1,38)),
                    M = 1, N = 38)

                # initial values
                list(beta = c(-1))
```

The first list statement contains the 1 by 38 matrix *t* of times to infection, while the 1 by 38 *t*.cen matrix gives the censored times. The second list statement gives the initial values for beta. I executed the simulation with 110,000 observations, a burn-in of 5000, and a refresh of 100, and the results of the posterior analysis are reported in Table 6.14, which estimates the two parameters *k*[1] and lamda[1] of the Weibull, and the main parameter of interest, namely the median of the time to infection of the kidney patients.

TABLE 6.14

Bayesian Analysis for Estimating Time to Infection

Parameter	Mean	SD	Error	2½	Median	97½
beta[1]	−3.971	.5933	.02	−5.207	−3.946	−2.891
k[1]	.8221	.1091	.00353	.6215	.8178	1.045
lamda[1]	.0223	.01332	<.0001	.00547	.01934	.05552
median[1]	81.46	20.04	.4371	47.88	79.61	126

Inspection of Table 6.14 reveals an estimate of 81.46 days for the median time to infection, while the two parameters of the Weibull are estimated as .8211 for the shape parameter, and .0223 for the scale parameter lamda[1]. Also, the MCMC error for estimating the posterior mean of the median is .4371, implying that the estimate of 81.46 days is within .4371 days of the true posterior mean (with a level of 95% confidence). Note the uncertainty in estimating the median recurrence, because the 95% credible interval varies from 47.88 days to 126 days. Also notice the skewness of the posterior distribution of lamda[1] with a posterior mean of .0223 and a posterior median of .01934.

It is interesting that the Weibull assumption is somewhat confirmed by the P–P plot of Figure 6.5. Note that some of the points do not lie on the line; nevertheless, I suspect that there is not enough evidence to suggest the Weibull is not an appropriate distribution for the times to first infection. In a later section, alternative parametric models will be examined for modeling the times to infection of the McGilchrist and Aisbett[9] study.

Remember that the Bayesian approach allows one to utilize prior information for the analysis. When there is little prior information, one may employ a vague type prior, but when prior information is in the form of previous related studies, one may combine the two data sets for the analysis. Vague prior information is used in the above analysis where a normal distribution with large variance was used for the beta parameter and a vague exponential prior for the k parameter of the Weibull. Exercise 13 illustrates the incorporation of prior information in the form of a previous related study.

As with all regression models, the Weibull will accommodate several independent variables. Up to this point, only one independent variable has been used, and this is illustrated by the McGilchrist and Aisbett[9] example with kidney patients experiencing dialysis.

This study is quite complex because many measurements are made on the 38 subjects, including the age of each, the time to the first infection, the time to the second infection, the event status (time to infection or a censored time), the gender of each, and the disease status of each. The patients are classified into one of three disease states: acute nephritis, polycystic kidney disease, and other. In this study, the time to first infection is measured from the time the catheter is inserted, after which the subject is treated for infection, and after the infection is cured, the catheter is inserted and the time to the second infection is measured.

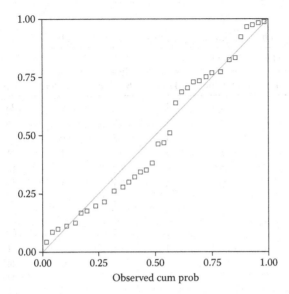

FIGURE 6.5
Times to infection observed.

Recall that the previous analysis focused on estimating the median time to the first and second infection, but for now, the emphasis will be on estimating the effect of age, gender, and disease status on the time to infection. The complete dataset is portrayed in Table 6.15.

Table 6.15 lists the patient number, the time to the first infection, the time to the second, the event status of the first infection (0 censored, 1 otherwise), the event status of the second, the gender (0 male, 1 female), the age in years, and the disease status.

Recall that all patients have kidney disease; however, they do differ with regard to the type. The analysis will present the effect of age, gender, and disease status on the time to infection. No differentiation is made between time to first infection and time to second.

The code below is taken from the kidney example of Volume 1 from WinBUGS, and the code is slightly revised and is executed with 55,000 for the simulation, a burn-in of 5000, and a refresh of 100. This is an example of Weibull regression, and the important aspect of the regression is portrayed by the following code:

```
t[i,j] ~ dweib(r, lamda[i,j]) I(t.cen[i, j],);
            log(mu[i,j]) <- alpha + beta.age * age[i, j]
                         + beta.sex *sex[i]
                         + beta.dis[disease[i]],
```

where the $t[i,j]$ are the time to infection and I(t.cen[i,j]) is the censoring mechanism.

TABLE 6.15

Gender, Age, and Disease Status of Kidney Dialysis Patients

Patient	Time to First Infection	Time to Second Infection	Event First	Event Second	Gender	Age	Disease Status
1	8	16	1	1	0	28	1
2	23	13	1	0	1	48	2
3	22	28	1	1	0	32	1
4	447	318	1	1	1	31	1
5	30	12	1	1	0	10	1
6	24	225	1	1	1	16	1
7	7	9	1	1	0	51	2
8	511	30	1	1	1	55	2
9	53	196	1	1	1	69	3
10	15	154	1	1	0	51	2
11	7	333	1	1	1	44	3
12	1141	8	1	0	1	34	1
13	96	38	1	1	1	35	3
14	149	70	0	0	1	42	3
15	536	25	1	0	1	17	1
16	17	4	1	0	0	60	3
17	185	177	1	1	1	60	1
18	292	114	1	1	1	43	1
19	22	159	0	1	1	53	2
20	15	108	1	0	1	44	1
21	152	562	1	1	0	46	4
22	402	24	1	0	1	30	1
23	13	66	1	1	1	62	3
24	39	46	1	0	1	42	3
25	12	40	1	1	0	43	3
26	113	201	0	1	1	57	3
27	132	156	1	1	1	10	2
28	34	30	1	1	1	52	3
29	2	25	1	1	0	53	2
30	130	26	1	1	1	54	2
31	27	58	1	1	1	56	3
32	5	43	0	1	1	50	3
33	152	30	1	1	1	57	4
34	190	5	1	0	1	44	2
35	119	8	1	1	1	22	1
36	54	16	0	0	1	42	1
37	6	78	0	1	1	52	4
38	63	8	1	0	0	60	4

Source: McGilchrist, C.A., and Aisbett, C.W., *Biometrics, 47*, 461–466, 1991.

The regression function is specified by the log of the second Weibull parameter lamda, and the regression coefficients are alpha, beta.age, beta. sex, and beta.dis[*i*], *i* = 2,3,4.

```
log(lamda[i,j]) <- alpha + beta.age * age[i, j]
                            + beta.sex *sex[i]
                            + beta.dis[disease[i]].
```

Prior information for the regression coefficients is expressed as a normal (0, .0001) distribution, and the first parameter *k* of the Weibull is given an uninformative gamma.

BUGS CODE 6.5

```
model
{
    for (i in 1 : N) {
        for (j in 1 : M) {
# Survival times bounded below by censoring times:
            t[i,j] ~ dweib(r, mu[i,j]) I(t.cen[i, j],);
            log(mu[i,j]) <- alpha + beta.age * age[i, j]
                            + beta.sex *sex[i]
                            + beta.dis[disease[i]] + b[i];
            culmative.t[i,j] <- culmative(t[i,j], t[i,j])
        }
# Random effects:
        b[i] ~ dnorm(0.0, tau)
    }
# Priors:
    alpha ~ dnorm(0.0, 0.0001);
    beta.age ~ dnorm(0.0, 0.0001);
    beta.sex ~ dnorm(0.0, 0.0001);
#   beta.dis[1] <- 0; # corner-point constraint
    for(k in 2 : 4) {
        beta.dis[k] ~ dnorm(0.0, 0.0001);
    }
    tau ~ dgamma(1.0E-3, 1.0E-3);
    r ~ dgamma(1.0, 1.0E-3);
    sigma <- 1/sqrt(tau); # s.d. of random effects
}
# kidney data

list(N = 38, M = 2,
        t = structure(
        .Data = c( 8, 16,
        23, NA,
        22, 28,
        447, 318,
        30, 12,
```

```
                24, 245,
                7, 9,
                511, 30,
                53, 196,
                15, 154,
                7, 333,
                141, NA,
                96, 38,
                NA, NA,
                536, NA,
                17, NA,
                185, 177,
                292, 114,
                NA, NA,
                15, NA,
                152, 562,
                402, NA,
                13, 66,
                39, NA,
                12, 40,
                NA, 201,
                132, 156,
                34, 30,
                2, 25,
                130, 26,
                27, 58,
                NA, 43,
                152, 30,
                190, NA,
                119, 8,
                NA, NA,
                NA, 78,
                63, NA),.Dim = c(38, 2)),
t.cen = structure(
                .Data = c( 0, 0,
                0, 13,
                0, 0,
                0, 0,
                0, 0,
                0, 0,
                0, 0,
                0, 0,
                0, 0,
                0, 0,
                0, 0,
                0, 8,
                0, 0,
                149, 70,
                0, 25,
                0, 4,
```

```
        0,  0,
        0,  0,
        22, 159,
        0,  108,
        0,  0,
        0,  24,
        0,  0,
        0,  46,
        0,  0,
        113, 0,
        0,  0,
        0,  0,
        0,  0,
        0,  0,
        0,  0,
        5,  0,
        0,  0,
        0,  5,
        0,  0,
        54, 16,
        6,  0,
        0,  8),.Dim = c(38, 2)),
age = structure(
        .Data = c(28, 28,
        48, 48,
        32, 32,
        31, 32,
        10, 10,
        16, 17,
        51, 51,
        55, 56,
        69, 69,
        51, 52,
        44, 44,
        34, 34,
        35, 35,
        42, 42,
        17, 17,
        60, 60,
        60, 60,
        43, 44,
        53, 53,
        44, 44,
        46, 47,
        30, 30,
        62, 63,
        42, 43,
        43, 43,
        57, 58,
        10, 10,
```

```
        52,  52,
        53,  53,
        54,  54,
        56,  56,
        50,  51,
        57,  57,
        44,  45,
        22,  22,
        42,  42,
        52,  52,
        60,  60),.Dim = c(38, 2)),
               beta.dis = c(0, NA, NA, NA),
sex = c(0, 1, 0, 1, 0, 1, 0, 1, 1, 0, 1, 1, 1, 1, 1,
        0, 1, 1, 1, 1, 0, 1, 1, 1, 0, 1, 1, 1, 0, 1, 1, 1, 1,
        1, 1, 1, 1, 0),
disease = c(1, 2, 1, 1, 1, 1, 2, 2, 3, 2, 3, 1, 3, 3, 1, 3, 1,
   1,2, 1, 4, 1, 3, 3,
               3, 3, 2, 3, 2, 2, 3, 3, 4, 2, 1, 1, 4, 4))

# initial values
list(beta.age = 0, beta.sex = 0, beta.dis = c(NA,0,0,0),
alpha = 0, r = 1, tau = 0.3)
```

Based on Table 6.16, how should one interpret the Bayesian analysis for the Weibull regression of time to infection on age, sex, and disease status? Based on the 95% credible intervals, it appears that the intercept term alpha is not zero, but on the other hand, the age coefficient beta.age does not have an effect on the time to infection. Note the 95% credible intervals are (−5.243, −2.748) for alpha and (−.0190, .02473) for the age coefficient, and that the latter contains zero. It does seem that gender does have an impact on time to infection, because its 95% credible interval (−2.388, −.941) does not include zero! In a similar fashion, beta.dis[4] corresponding to the fourth disease classification does impact the time to infection with a 95% credible interval of (−2.702, −.2991).

TABLE 6.16

Weibull Regression for the Kidney Dialysis Patients

Parameter	Mean	SD	Error	2½	Median	97½
alpha	−3.952	.6364	.0158	−5.243	−3.933	−2.748
beta.age	.00246	.01127	<.0001	−.01905	.002341	.02473
beta.dis[2]	.01639	.4063	.00429	−.8285	.02457	.799
beta.dis[3]	.5234	.4014	.00469	−.2643	.5347	1.31
beta.dis[4]	−1.463	.6115	.009304	−2.702	−1.454	−.2911
beta.sex	−1.67	.3679	.0068	−2.388	−1.675	−.941
k	1.046	.0989	.00363	.8578	1.044	1.246

As for gender, the mean time to first infection for males is 32.8 days with a standard deviation of 45.4 and a median time to first infection of 16 days. With regard to females, the mean (SD) time to first infection is 175.6 (245) days with a median of 104 days. Using time to second infection shows the mean (SD) to be 85 (173) with a median of 20.5 days for males, but for females the mean (SD) time to second infection is 92.8 (93) days with a median of 52 days. When calculated separately for times to first and second infections, and females, it appears that males experience a shorter time to infection.

When combining the first and second times to infections, mean and standard deviation of the time to infection is mean (SD) = 59.3 (126) with a median of 16.5 days for males, while for females, the mean and standard deviation are mean (SD) = 134.2 (188), with a median of 62 days. Thus, it appears that males experience a longer time to infection than females.

The mean, standard deviation, and median for times to infection are calculated using censored and uncensored times to infection, and it should be understood that it is more appropriate to use a method that properly accounts for the censored observations. Such is method is based on the hazard function

$$h(t) = k\lambda t^{k-1} \qquad (6.24)$$

Consider the ratio of the hazard function for females divided by the hazard function for males, namely the hazard ratio

$$\text{hr(females vs. males)} = \frac{h(\text{females}, t)}{h(\text{males}, t)} \qquad (6.28)$$
$$= \exp(\text{beta sex})$$

Note that the hazard ratio does not depend on time t or the parameter k.

From the BUGS CODE 6.5, which includes the statement hr<-exp(beta.sex), one may verify that the posterior mean (SD) of the hazard ratio of females versus males is .2016 (.0778) with a median of .1873 and a 95% credible interval of (.09184, .3902). Consider a particular time t, and a female who has not experienced an infection and a male at the same time t who also has not experienced an infection; then compared to the male, the female is 80% less likely to experience an infection in the next instant. This estimate is based on the posterior mean of the hazard ratio. Of course, this confirms the increased hazard of the males who experience a shorter time to infection, compared to females. Based on the 95% credible interval, the decreased hazard for a female compared to a male can vary from 91% to 61%. For additional examples of estimating the effect of covariates on the time to an event, see Exercise 14.

6.4.3 Cox Proportional Hazards Model

The Cox proportional hazards model is one of the most useful in biostatistics and appears in many of the major medical journals.

It is defined as

$$h(t, X) = h_0(t)e^{\sum_{i=1}^{i=p} \beta_i X_i}$$

where $h_0(t)$ is the baseline hazard function, the β_i are unknown regression parameters, and the X_i are known covariates or independent variables. Note that the baseline hazard function is a function of time only, but that the covariates are not functions of t. The time t is the time to the event of interest, which is usually the survival time of a group of patients or the time to recurrence, or some other event measured by time. Recall that the regression function is defined in the terms of the hazard function

$$h(t) = \frac{\lim\limits_{\Delta t \to \infty} P(t \le T < t + \Delta t \,|\, T \ge t)}{\Delta t} \tag{6.2}$$

which in turn is related to the survival function $S(t)$ by the relation in order to provide a suitable Bayesian analysis

$$h(t) = \frac{[dS(t)/dt]}{S(t)} \tag{6.4}$$

In addition, the survival function is

$$S(t) = P(T > t) \tag{6.1}$$

where T denotes the survival time of a subject. In survival studies, the Cox regression model is expressed as a hazard, whereas the usual way to express a regression is more directly using T as a function of unknown regression coefficients. One reason the Cox model is so popular is its versatility; for example, if the actual survival time has an exponential distribution or a Weibull distribution, the Cox model will provide similar results.

With the Cox model, the time variable T is not assumed to have a specific distribution; thus, the model is quite general in that it can be applied in a large variety of time to event studies. Also note that the p covariates $X_1, X_2, ..., X_p$ are not functions of t; however, there are cases where one would have time-dependent covariates, in which case, a more general Cox model is appropriate. The most important assumption of the Cox model is that if one is comparing the survival of two groups, the corresponding hazard functions must be proportional. Such a case will also be studied in a later section of the chapter.

The most important parameter in survival studies is the hazard ratio

$$\text{HR} = \frac{h(t, X^*)}{h(t, X)} \tag{6.29}$$

between two individuals, one with the covariate measurements X^*, and the other with the measurement X, on the p covariates. Note that it is easy to show that the hazard ratio (Equation 6.24) is equivalent to

$$\text{HR} = e^{\sum_{i=1}^{i=p} \beta_i (X_i^* - X_i)} \tag{6.30}$$

where both X^* and X are known. It is important to remember that to the Bayesian, the HR hazard ratio (Equation 6.30) is an unknown parameter because it depends on p unknown parameters

$$\beta = (\beta_1, \beta_2, ..., \beta_p) \tag{6.31}$$

Thus, the Bayesian must specify a prior distribution for β, then, through Bayes' theorem, determine the posterior distribution of β and any function of β such as the hazard ratio.

As an example of the hazard ratio, consider the two groups of leukemia patients, where Group 1 is the treatment group and Group 2 the placebo, with one covariate X, where $X^* = 0$ denotes the treatment group and $X = 1$ the placebo. The hazard ratio (Equation 6.25) then reduces to

$$\text{HR}(\beta) = e^{\beta} \tag{6.32}$$

where one individual is a patient from the treatment group and the other a patient from the placebo. Thus, Equation 6.32 expresses the effect of the treatment as a hazard ratio. Once one estimates β, one has an estimate of the hazard ratio, or for our interest, once one has the posterior distribution of β, one has the posterior distribution of HR.

As the first example of using the Cox model, consider the comparison of the treatment with the placebo group in the Freireich et al.[5] study, where the main focus will be on estimating the hazard ratio (Equation 6.32) between the two. The following BUGS CODE 6.6 follows the leukemia example of volume I of WinBUGS.

BUGS CODE 6.6

```
model
{
# Set up data
    for(i in 1:N) {
```

```
            for(j in 1:T) {
# risk set = 1 if obs.t > = t
                        Y[i,j] <- step(obs.t[i] - t[j] + eps)
# counting process jump = 1 if obs.t in [t[j], t[j+1])
#           i.e. if t[j] < = obs.t < t[j+1]
                    dN[i, j] <- Y[i, j] * step(t[j + 1] -
                    obs.t[i] - eps) * fail[i]
            }
    }
# Model
    for(j in 1:T) {
        for(i in 1:N) {
            dN[i, j] ~ dpois(Idt[i, j])  # Likelihood
            Idt[i, j] <- Y[i, j] * exp(beta * x1[i]) *
            dL0[j]  # Intensity
        }
        dL0[j] ~ dgamma(mu[j], c)
        mu[j] <- dL0.star[j] * c # prior mean hazard

# Survivor function = exp(-Integral{l0(u)du})^exp(beta*z)
            S.treat[j] <- pow(exp(-sum(dL0[1 : j])),
            exp(beta * -0.5));
            S.placebo[j] <- pow(exp(-sum(dL0[1 : j])),
            exp(beta * 0.5));

    }
    c <- 0.001
    r <- 0.1
    for (j in 1 : T) {
            dL0.star[j] <- r * (t[j + 1] - t[j])
    }
    beta ~ dnorm(0.0,0.000001)
    # hazard ratio for group 1 versus group 2
    HR<-exp(beta)
}

# x1 is the group indicator

list(N = 42, T = 17, eps = 1.0E-10,
    obs.t = c(1, 1, 2, 2, 3, 4, 4, 5, 5, 8, 8, 8, 8, 11,
        11, 12, 12, 15, 17, 22, 23, 6,
        6, 6, 6, 7, 9, 10, 10, 11, 13, 16, 17, 19, 20, 22, 23,
        25, 32, 32, 34, 35),
    fail = c(1, 1, 1, 1, 1, 1, 1, 1, 1, 1, 1, 1, 1, 1, 1,
        1, 1, 1, 1, 1, 1, 1, 1, 0, 1, 0, 1, 0, 0, 1, 1, 0,
        0, 0, 1, 1, 0, 0, 0, 0, 0),
    x1 = c(0.5, 0.5, 0.5, 0.5, 0.5, 0.5, 0.5, 0.5, 0.5, 0.5,
        0.5, 0.5, 0.5, 0.5, 0.5, 0.5, 0.5, 0.5, 0.5, 0.5,
        -0.5, -0.5, -0.5, -0.5, -0.5, -0.5, -0.5, -0.5,
        -0.5, -0.5, -0.5, -0.5, -0.5, -0.5, -0.5, -0.5,
        -0.5, -0.5, -0.5, -0.5, -0.5),
```

```
    t = c(1,  2,  3,  4,  5,  6,  7,  8,  10,  11,  12,  13,  15,  16,
        17,  22,  23,  35))
# initial values
list(beta = 0.0,
dL0 = c(1.0,1.0,1.0,1.0,1.0,1.0,1.0,1.0,1.0,
        1.0,1.0,1.0,1.0,1.0,1.0,  1.0,1.0))
```

Notice the first list statement that contains the following data vectors: (1) N is the total number of patients; (2) t is the number of distinct ordered times the events occur; (3) obs.t is the 42 by 1 vector of times that the events (recurrence or censoring) happen, where the placebo values are given first; (4) fail is the 42 by 1 vector that denotes the type of event, a 1 for a recurrence and a 0 for censoring; (5) x1 is a 42 by 1 vector of covariate values, where a .5 designates the placebo group, and a −.5 the treatment group, and the t vector is 17 by 1 and denotes the distinct times at which the events occur. Refer to Tables 6.2, 6.4, and 6.5 for the above values when the Freireich et al.[5] study was first described.

Based on BUGS CODE 6.6, the Bayesian analysis is executed with 65,000 observations for the simulation, with a burn-in of 5000 and a refresh of 100, and the results are reported in Table 6.17. HR is the hazard ratio, and the S.placebo vector is the estimated proportion of the number of recurrences at the 17 distinct ordered times for the placebo group, while the 17 estimated recurrence times for the treatment group are components of the vector S.treat. Finally, the beta coefficient measures the effect of the group (placebo vs. treatment) on the recurrence times. The recurrence probabilities S.placebo[i] and S.treat[i] can be plotted against the ordered times when the events occur and compared to Figure 6.1, the times to recurrence for the two groups.

The Bayesian analysis for the hazard ratio (Equation 6.32) has a posterior mean of 5.093, and a median of 4.621 with (2.133, 10.84) as the 95% credible interval. Note that the posterior mean of the regression coefficient has posterior median 1.521 and exp(1.521) = 4.576, which is approximately the posterior median of the hazard ratio. How is this estimate interpreted? Select a placebo patient and suppose the patient has not experienced a recurrence. The hazard of experiencing a recurrence for the placebo patient is 4.576 times that of a comparable treatment patient in the next instant.

Recall from the Weibull analysis of the leukemia data, Table 6.12 reports the median time to recurrence for the placebo group as 7.541 (using the posterior mean) weeks and as 28.26 weeks for the treatment group. A casual look at Table 6.17 shows that the posterior distributions of the HR and beta are skewed to the right. Also, it appears that the MCMC errors are sufficiently small so that one has confidence that the posterior means are accurate estimates of the actual posterior means.

The estimated survival probabilities are based on estimates of the survival function

$$S(t,x) = [S_0(t)]^{\exp(\sum_{i=1}^{i=p} \beta_i x_i)} \tag{6.33}$$

TABLE 6.17

Bayesian Analysis Comparing Recurrence of Placebo versus Treatment (the Cox Model)

Parameter	Mean	SD	Error	2½	Median	97½
HR	5.093	2.289	.0151	2.133	4.621	10.84
S.placebo[1]	.9282	.04863	<.0001	.8094	.9387	.9907
S.placebo[2]	.8538	.06843	<.0001	.6926	.8639	.9571
S.placebo[3]	.8161	.0756	<.0001	.6422	.8244	.9362
S.placebo[4]	.7432	.0853	<.0001	.5586	.7503	.8892
S.placebo[5]	.6703	.0925	<.0001	.4749	.6755	.835
S.placebo[6]	.5633	.0974	<.0001	.3666	.5661	.7477
S.placebo[7]	.5304	.0977	<.0001	.338	.5366	.7148
S.placebo[8]	.4142	.0938	<.0001	.2374	.4119	.6037
S.placebo[9]	.3812	.0932	<.0001	.2086	.3779	.5701
S.placebo[10]	.32	.0894	<.0001	.1583	.315	.509
S.placebo[11]	.2583	.0845	<.0001	.111	.2511	.4395
S.placebo[12]	.2257	.0810	<.0001	.0870	.2181	.402
S.placebo[13]	.1956	.0772	<.0001	.0686	.1873	.3668
S.placebo[14]	.1656	.0732	<.0001	.0488	.1305	.3298
S.placebo[15]	.1398	.0678	<.0001	.036	.0766	.2593
S.placebo[16]	.0867	.0545	<.0001	.0130	.0334	.22
S.placebo[17]	.0445	.0391	<.0001	.0025	.9866	.1484
S.treat[1]	.983	.0137	<.0001	.9473	.9866	.9982
S.treat[2]	.9643	.0217	<.0001	.9115	.9692	.9922
S.treat[3]	.9544	.0253	<.0001	.8918	.9598	.9884
S.treat[4]	.9343	.0321	<.0001	.8573	.9398	.9797
S.treat[5]	.9125	.0391	<.0001	.821	.9185	.9701
S.treat[6]	.8772	.0489	<.0001	.7654	.8838	.9521
S.treat[7]	.8652	.0523	<.0001	.745	.8717	.947
S.treat[8]	.8178	.0645	<.0001	.6736	.8246	.9229
S.treat[9]	.8024	.0687	<.0001	.6528	.8099	.9151
S.treat[10]	.771	.0761	<.0001	.6064	.7786	.8976
S.treat[11]	.7339	.0846	<.0001	.5522	.7409	.8774
S.treat[12]	.7114	.0889	<.0001	.5224	.7174	.8659
S.treat[13]	.6882	.0932	<.0001	.4913	.6937	.8528
S.treat[14]	.6619	.097	<.0001	.4641	.6669	.8355
S.treat[15]	.636	.1007	<.0001	.4318	.6406	.8191
S.treat[16]	.5662	.111	<.0001	.3453	.5688	.773
S.treat[17]	.4761	.1189	<.0001	.2502	.4747	.7085
beta	1.538	.4176	<.0001	.7718	1.521	2.384

where $S_0(t, x)$ is the baseline survival function. From a Bayesian viewpoint, the posterior distribution of each survival function is determined by the posterior distribution of beta. The corresponding code is

```
# Survivor function = exp(-Integral{l0(u)du})^exp(beta*x1)
S.treat[j] <- pow(exp(-sum(dL0[1 : j])), exp(beta * -0.5));
S.placebo[j] <- pow(exp(-sum(dL0[1 : j])), exp(beta * 0.5));
```

where −0.5 is the group indicator for placebo and 0.5 is the indicator for the treatment group.

For the leukemia study, $p = 1$, x_1 is the 42 by 1 vector of group identification in the list statement of BUGS CODE 6.7, and there are 17 ordered times at which the survival function is determined.

6.4.4 Cox Model with Covariates

In order to assess the effect of group (placebo versus treatment) on the time to recurrence, it is important that all available complete patient information be included in the analysis. For example, with the study by Freireich et al.[5], an important indicator of leukemia is the white blood cell count, which should be included in the comparison between the treatment and placebo groups, and in the estimation of the survival probabilities of the two groups. One would expect that the blood cell count would have an effect on the time to recurrence and thus on the comparison between treatment and placebo.

Recall that in the previous section, the focus was on the comparison between the two groups without including the blood cell count in the analysis, and it was found that the hazard ratio for the group effect was estimated as 4.621 with the posterior median and that the median time to recurrence for the treatment time to recurrence was 28.26 weeks for the treatment group versus 7.541 for the placebo. When the blood cell count is included, one would expect the hazard ratio of the group effect to be modified and the estimated survival probabilities to also change.

To include the patient blood cell count as a covariate, BUGS CODE 6.7 is a revision of BUGS CODE 6.6.

The code

```
Idt[i, j] <- Y[i, j] * exp(beta[1] * x1[i]+beta[2]*x2[i]) *
dL0[j]
```

includes the covariate x2 (log white blood cell) in the model, and the code

```
S.treat[j] <- pow(exp(-sum(dL0[1 : j])), exp(beta[1] *
-0.5+beta[2]*2.63));
S.placebo[j] <- pow(exp(-sum(dL0[1 : j])), exp(beta[1]*
0.5+beta[2]*3.22));
```

includes both the group ID and the log white blood cell count to compute the posterior distribution of the survival proportions for both groups. Note that the group indicator x1 is +0.5 for the placebo and is −0.5 for the treatment group, while the coefficient beta is the effect of the group on the recurrence times, and beta[2] is the effect of the log of the white blood cell count on the recurrence time. Also, 2.63 is the average blood cell count for the placebo, and for the treatment group, 3.22 is the average.

BUGS CODE 6.7

```
model
    {
    # Set up data
        for(i in 1:N) {
            for(j in 1:T) {
    # risk set = 1 if obs.t > = t
                        Y[i,j] <- step(obs.t[i] - t[j] + eps)
    # counting process jump = 1 if obs.t in [t[j], t[j+1])
    #            i.e. if t[j] < = obs.t < t[j+1]
                    dN[i, j] <- Y[i, j] * step(t[j + 1] -
                        obs.t[i] - eps) * fail[i]
        }
    }
    # Model
        for(j in 1:T) {
            for(i in 1:N) {
                dN[i, j] ~ dpois(Idt[i, j])  # Likelihood
                # With covariate log white blood cell count
                Idt[i, j] <- Y[i, j] * exp(beta[1] * x1[i]+
beta[2]*x2[i]) * dL0[j]  # Intensity
            }
        dL0[j] ~ dgamma(mu[j], c)
        mu[j] <- dL0.star[j] * c # prior mean hazard

    # Survivor function = exp(-Integral{l0(u)du})^exp(beta*x1)
    # adjusted for covariates at mean value of covariate

        S.treat[j] <- pow(exp(-sum(dL0[1 : j])), exp(beta[1]*
            -0.5+beta[2]*2.63));
        S.placebo[j] <- pow(exp(-sum(dL0[1 : j])), exp(beta[1]*
            0.5+beta[2]*3.22));
    }
    c <- 0.001
    r <- 0.1
    for (j in 1 : T) {
      dL0.star[j] <- r * (t[j + 1] - t[j])
    }
    beta[1] ~ dnorm(0.0,0.000001)
    beta[2] ~ dnorm(0.0,0.000001)

    # hazard ratio for group 1 versus group 2
    HR.beta1<-exp(beta[1])
    # HR for log wbc
    HR.beta2<-exp(beta[2])
  }

  # x1 is the group indicator
  # x2 is the log wbc count

list(N = 42, T = 17, eps = 1.0E-10,
```

```
x2 = c(2,8,5,4.91,4.48,4.01,4.36,2.42,3.49,3.97,3.52,3.05,2.32,
     3.26,3.49,2.12,1.5,3.06,2.30,2.95,2.73,1.97,
   2.31,4.06,3.28,4.43,2.96,2.88,3.60,2.32,2.57,3.20,2.80,2.70,
   2.60,2.16,2.05,2.01,1.78,2.20,2.53,1.47,1.45),
                   obs.t = c(1,  1,  2,  2,  3,  4,  4,  5,  5,  8,  8,
                        8,  8,  11,  11,  12,  12,  15,  17,  22,  23,  6,
                        6,  6,  6,  7,  9,  10,  10,  11,  13,  16,  17,  19,
                        20,  22,  23,  25,  32,  32,  34,  35),
                   fail = c(1,  1,  1,  1,  1,  1,  1,  1,  1,  1,  1,
                        1,  1,  1,  1,  1,  1,  1,  1,  1,  1,  1,  1,
                        0,  1,  0,  1,  0,  0,  1,  1,  0,  0,  0,  1,  1,
                        0,  0,  0,  0,  0),
                   x1 = c(0.5,  0.5,  0.5,  0.5,  0.5,  0.5,  0.5,
                        0.5,  0.5,  0.5,  0.5,  0.5,  0.5,  0.5,  0.5,
                        0.5,  0.5,  0.5,  0.5,  0.5,  0.5,
                        -0.5,  -0.5,  -0.5,  -0.5,  -0.5,  -0.5,
                        -0.5,  -0.5,  -0.5,  -0.5,  -0.5,  -0.5,
                        -0.5,  -0.5,  -0.5,  -0.5,  -0.5,  -0.5,
                        -0.5,  -0.5,  -0.5),
                   t = c(1,  2,  3,  4,  5,  6,  7,  8,  10,  11,  12,
                        13,  15,  16,  17,  22,  23,  35))
                   # initial values
                   list(beta = c(0,0),
dL0 = c(1.0,1.0,1.0,1.0,1.0,1.0,1.0,1.0,1.0,
     1.0,1.0,1.0,1.0,1.0,1.0,  1.0,1.0))
```

The analysis is executed with 65,000 observations, a burn-in of 5000, and a refresh of 100. Of course, the analysis is similar to that reported in Table 6.17, except that the covariate log white blood cell (lwbc) has been included. There are four parameters, delta (the effect of the white blood cell), beta (the effect of group on the time to recurrence), and the estimated survival proportions for the treatment and placebo groups.

The hazard ratio for beta[1] (the difference in the two groups) is estimated as 3.384 with the posterior mean and as 3.048 with the posterior median, while the hazard ratio for beta[2] (the effect of the log white blood cell count) is estimated as 2.766 with the posterior mean and as 2.649 with the posterior median. Note how the inclusion of the covariate x2 (log white blood cell count) changes the estimate of the hazard ratio for beta[1].

Comparing Table 6.18 with 6.17, without the covariate, the hazard ratio for beta[1] is 4.621 (estimated with the posterior median) compared to 3.048 when the covariate is included!

However, the change is even more dramatic with the posterior distributions of the survival probabilities of both groups. For example with the covariate, the estimated probability of survival at the 17th time point without the covariate is .0445 for the treatment group, but is estimated as .0146 with the covariate included. In a similar fashion, for the treatment group at the 17th time point, the estimated survival is .4012 compared

TABLE 6.18

Bayesian Analysis Comparing Times to Recurrence for Placebo versus treatment (the Cox Model with Covariates)

Parameter	Mean	SD	Error	2½	Median	97½
HR.beta[1]	3.384	1.591	.0228	1.35	3.048	7.39
HR.beta[2]	2.766	.7758	.0415	1.66	2.649	4.596
S.placebo[1]	.9828	.0196	<.0001	.9283	.9896	.9994
S.placebo[2]	.9325	.0434	.0011	.8232	.9419	.9885
S.placebo[3]	.8987	.05114	.0012	.7658	.9805	.9764
S.placebo[4]	.8256	.0733	.0015	.658	.8349	.9404
S.placebo[5]	.7399	.08593	.00155	.5447	.7483	.8897
S.placebo[6]	.617	.103	.0016	.404	.6218	.8032
S.placebo[7]	.5776	.105	.0015	.3645	.5809	.7716
S.placebo[8]	.4336	.103	.0012	.2382	.4318	.6376
S.placebo[9]	.3871	.1032	.0010	.1949	.3839	.595
S.placebo[10]	.3029	.098	<.0001	.1301	.2971	.5079
S.placebo[11]	.2123	.0902	<.0001	.0651	.2033	.4106
S.placebo[12]	.1748	.0839	<.0001	.0443	.1642	.3653
S.placebo[13]	.143	.0764	<.0001	.0306	.1316	.3221
S.placebo[14]	.1123	.0684	<.0001	.0179	.0997	.2762
S.placebo[15]	.0864	.0598	<.0001	.0099	.0731	.2347
S.placebo[16]	.0421	.0397	<.0001	.0018	.0301	.1489
S.placebo[17]	.0146	.0211	<.0001	<.00001	.0067	.0756
S.treat[1]	.9965	.0046	<.0001	.9835	.9981	.9999
S.treat[2]	.9862	.0111	<.0001	.9569	.9892	.9983
S.treat[3]	.9791	.0147	<.0001	.941	.9827	.9966
S.treat[4]	.9631	.0217	<.0001	.9088	.9677	.9914
S.treat[5]	.943	.0292	<.0001	.8714	.9485	.9834
S.treat[6]	.9106	.0394	<.0001	.8177	.9169	.9688
S.treat[7]	.899	.0430	<.0001	.7978	.9055	.9633
S.treat[8]	.8502	.0572	<.0001	.7185	.8574	.9386
S.treat[9]	.8316	.0623	<.0001	.6891	.8389	.9302
S.treat[10]	.7924	.0727	<.0001	.6299	.8002	.9107
S.treat[11]	.738	.0856	<.0001	.5517	.7456	.8821
S.treat[12]	.7096	.0911	<.0001	.5131	.717	.8653
S.treat[13]	.6811	.0966	<.0001	.4766	.6882	.8491
S.treat[14]	.6476	.1019	<.0001	.4335	.6534	.8283
S.treat[15]	.6126	.1073	<.0001	.3909	.6174	.8071
S.treat[16]	.522	.1218	<.0001	.279	.5248	.7493
S.treat[17]	.4012	.1308	.0010	.1571	.3979	.6616
beta[1]	1.124	.4322	.0065	.3021	1.114	2
beta[2]	.9824	.2598	.0136	.507	.974	1.525

to .4761 without the covariate being included! Based on Table 6.18, the 95% credible interval for beta[2] implies that the inclusion of the covariate does indeed have an important impact on the recurrence time and consequently on the estimate of the beta[1] coefficient and the hazard ratio for the beta[1] coefficient.

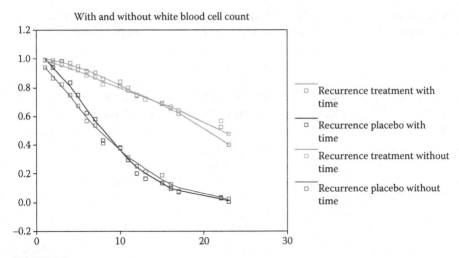

FIGURE 6.6
Times to recurrence: treatment versus placebo.

In order to see the effect of log blood cell count on survival, refer to Figure 6.6, where the upper two curves are the adjusted (for the log white blood cell count) survival proportions for the treatment group and the lower two are the survival proportions for the placebo group. The moderate effect of the covariate is clearly revealed with the upper two curves (purple and green) for the treatment group and is also revealed with the lower two curves (red and blue) for the placebo subjects.

Recall that the survivor function for this example with the covariate log white blood cell count is

$$S(t,x) = [S_0(t)]^{\exp(\sum_{i=1}^{i=p} \beta_i x_i)} \tag{6.33}$$

where p = 2, x_1 is the group indicator, and x_2 is the vector of log white blood cell counts of the list statement in BUGS CODE 6.7.

6.4.5 Testing for Proportional Hazards in the Cox Model

A crucial aspect in performing a survival analysis with the Cox model is checking the assumption of proportional hazards. Recall that the hazard function of the Cox model is

$$h(t,X) = h_0(t)e^{\sum_{i=1}^{i=p} \beta_i X_i}$$

where $h_0(t)$ is the baseline hazard function and $\beta = (\beta_1, \beta_2, ..., \beta_p)$ are unknown parameters that measure the effect of the corresponding predictors

$$X = (X_1, X_2, ..., X_p)$$.

It can be shown that the ratio of the hazard of one individual with covariate $X^* = (X_1^*, X_2^*, ..., X_p^*)$ to the hazard of another individual with covariate $X = (X_1, X_2, ..., X_p)$ is

$$HR = \frac{h(t, X^*)}{h(t, X)}$$

which reduces to

$$HR = e^{\sum_{i=1}^{i=p} \beta_i (X_i^* - X_i)} \tag{6.30}$$

Thus, the hazard ratio is independent of t and is equal to a constant! In a similar fashion, the survival function for the Cox model is

$$S(t, x) = [S_0(t)]^{\exp(\sum_{i=1}^{i=p} \beta_i x_i)} \tag{6.33}$$

where $S_0(t)$ is the baseline survival function corresponding to the baseline hazard. Note that the exponent of the baseline survival function is independent of time t.

In order to test for the proportional hazards assumption for two individuals with covariates $X^* = (X_1^*, X_2^*, ..., X_p^*)$ and $X = (X_1, X_2, ..., X_p)$, consider the survival function for the first and second individuals, namely

$$^* S(t, x^*) = [S_0(t)]^{\exp(\sum_{i=1}^{i=p} \beta_i x_i^*)} \tag{6.34}$$

and

$$S(t, x) = [S_0(t)]^{\exp(\sum_{i=1}^{i=p} \beta_i x_i)}$$

In order to test for the proportional hazards assumption, the survival function for the first individual is assumed to be

$$^* S(t, x^*) = [S_0(t)]^{\exp(\sum_{i=1}^{i=p} \beta_i x_i^* + \delta_i x_i t)} \tag{6.35}$$

and for the second individual it is assumed to be

$$S(t, x) = [S_0(t)]^{\exp(\sum_{i=1}^{i=p} \beta_i x_i + \delta_i x_i t)} \tag{6.36}$$

where t is an ordered time to an event. Note that if at least one of the δ_i is not zero, Equations 6.35 and 6.36 violate the proportional hazards assumption.

If the proportional hazard (PH) assumption is valid, the $\delta_i = 0$ for $i = 1,2,...,p$, because Equations 6.35 and 6.36 reduce to Equations 6.34 and 6.33, respectively. If the PH assumption is not true, the exponent of the baseline survival function is a function of t. In practice, the PH assumption is tested one at a time using the appropriate adjusted survival curve.

As the first illustration for testing the proportional hazard assumption, consider the study by Freireich et al.,[5] the leukemia study with two groups and including the log white blood cell count as a covariate. Suppose we want to see if the group effect follows the proportional hazard assumption, adjusted for the white blood cell account. The following code will implement the test for proportional hazard. The relevant code is the following:

```
S.treat[j] <- pow(exp(-sum(dL0[1 : j])),
  exp(beta[1]*-0.5+beta[2]*2.63 +delta*-.5*t[j]));
S.placebo[j] <- pow(exp(-sum(dL0[1 : j])),
  exp(beta[1]* 0.5+beta[2]*3.22+delta*.5*t[j]));
```

where the first statement is for the adjusted survival curve for the treatment group, and the second is for the adjusted survival for the placebo group, and the PH assumption is based on the posterior distribution of delta, which is reported in Table 6.19.

With a 95% credible interval of (−1983, 1964), one may safely conclude that delta is zero and the PH assumption for the group effect (placebo vs. treatment) is satisfied.

Analysis is executed with 55,000 observations for the simulation, with 5000 for the burn-in and a refresh of 100.

BUGS CODE 6.8

```
model
    {
    # Set up data
        for(i in 1:N) {
            for(j in 1:T) {
    # risk set = 1 if obs.t > = t
                    Y[i,j] <- step(obs.t[i] - t[j] + eps)
    # counting process jump = 1 if obs.t in [t[j], t[j+1])
    #            i.e. if t[j] < = obs.t < t[j+1]
                    dN[i, j] <- Y[i, j] * step(t[j + 1] -
```

TABLE 6.19

Test for Proportional Hazards of the Group

	Adjusted for log White Blood Cell Count					
Parameter	Mean	SD	Error	2½	Median	97½
delta	−4.397	1008	4.482	−1983	−1.001	1964

```
                              obs.t[i] - eps) * fail[i]
                  }
           }
      # Model
         for(j in 1:T) {
              for(i in 1:N) {
                      dN[i, j] ~ dpois(Idt[i, j])  # Likelihood
                      # With covariate log white blood cell
                      count
                      Idt[i, j] <- Y[i, j] * exp(beta[1] *
                      x1[i]+beta[2]*x2[i]) * dL0[j]  # Intensity
              }
              dL0[j] ~ dgamma(mu[j], c)
              mu[j] <- dL0.star[j] * c # prior mean hazard
      # Survivor function = exp(-Integral{10(u)du})^exp(beta*z)
      # adjusted for covariates at mean value of covariate
      # test for PH of group adjusted for wbc
                  S.treat[j] <- pow(exp(-sum(dL0[1 : j])),
                   exp(beta[1]* -0.5+beta[2]*2.63 +delta*-.5*t[j]));
                  S.placebo[j] <- pow(exp(-sum(dL0[1 : j])),
                   exp(beta[1]* 0.5+beta[2]*3.22+delta*.5*t[j]));
      }
      c <- 0.001
      r <- 0.1
      for (j in 1 : T) {
              dL0.star[j] <- r * (t[j + 1] - t[j])
      }
      beta[1] ~ dnorm(0.0,0.000001)
         delta ~ dnorm(0.0,0.000001)
         beta[2] ~ dnorm(0.0,0.000001)

      # hazard ratio for group 1 versus group 2
      HR.beta1<-exp(beta[1])
      HR.beta2<-exp(beta[2])

      # x1 is the group indicator (placebo.5 and treatment -.5)
      # x2 is the log wbc
  }
list(N = 42, T = 17, eps = 1.0E-10,
x2 = c(2,8,5,4.91,4.48,4.01,4.36,2.42,3.49,3.97,3.52,3.05,2.32,
3.26,3.49,2.12,1.5,3.06,2.30,2.95,2.73,1.97,
2.31,4.06,3.28,4.43,2.96,2.88,3.60,2.32,2.57,3.20,2.80,2.70,2.
60,2.16,2.05,2.01,1.78,2.20,2.53,1.47,1.45),
      obs.t = c(1, 1, 2, 2, 3, 4, 4, 5, 5, 8, 8, 8, 8, 11,
         11, 12, 12, 15, 17, 22, 23, 6,
         6, 6, 6, 7, 9, 10, 10, 11, 13, 16, 17, 19, 20, 22, 23,
         25, 32, 32, 34, 35),
      fail = c(1, 1, 1, 1, 1, 1, 1, 1, 1, 1, 1, 1, 1, 1, 1,
         1, 1, 1, 1, 1, 1, 1, 1, 1, 0, 1, 0, 1, 0, 0, 1, 1, 0,
         0, 0, 1, 1, 0, 0, 0, 0, 0),
```

```
x1 = c(0.5,  0.5,  0.5,  0.5,  0.5,  0.5,  0.5,  0.5,  0.5,
    0.5,  0.5,  0.5,  0.5,  0.5,  0.5,  0.5,  0.5,  0.5,  0.5,  0.5,
      -0.5,  -0.5,  -0.5,  -0.5,  -0.5,  -0.5,  -0.5,  -0.5,
        -0.5,  -0.5,  -0.5,  -0.5,  -0.5,  -0.5,  -0.5,  -0.5,
        -0.5,  -0.5,  -0.5,  -0.5,  -0.5),
t = c(1,  2,  3,  4,  5,  6,  7,  8,  10,  11,  12,  13,  15,  16,
    17,  22,  23,  35))
# initial values
list(beta = c(0,.0), delta = 0,
dL0 = c(1.0,1.0,1.0,1.0,1.0,1.0,1.0,1.0,1.0,
    1.0,1.0,1.0,1.0,1.0,1.0,  1.0,1.0))
```

It should be pointed out that when testing for the proportional hazard assumption for the log white blood cell count (which is a continuous variable), the counts should be partitioned into, say, three groups.

The Freireich et al.[5] study of leukemia has been covered in some detail. There is indeed a difference in the two groups and the log white blood cell count makes an impact on the time to recurrence; however, it has also been determined that gender is not an important factor in the model (see Exercise 14 and BUGS CODE 6.7 Ext). Therefore, I suggest that the appropriate model for the leukemia study includes the log white blood cell count and the group membership. The proportional hazard assumption was checked for the group effect and it appears that that factor does indeed satisfy the assumption. See Table 6.19.

The literature of survival analysis involving the Cox proportional hazards model is extensive and diverse, and the Bayesian approach presented here is a brief introduction. Of historical interest is the original publication of Cox[10] followed by Bayesian treatments of the subject authored by Kalbflseich,[11] Clayton,[12] and Chen et al.[13]

Advanced topics for survival studies include techniques that are appropriate when the PH assumption is not valid, which are not introduced in this chapter.

6.5 Comments and Conclusions

The chapter begins with the basic notation used for life tables and using this notation and referring to the format of the life table, the survival and hazard functions are defined. The use of the survival and hazard functions are illustrated in the context of the Kaplan–Meier survival curves using the leukemia study of Freireich et al.[5] A Bayesian approach to the Kaplan–Meier curve involves assuming the number of deaths and number of censored observations have a binomial distribution, then estimating the various survival probabilities by the relevant posterior distribution. If there are two or more groups, it is of interest to compare the survival times between the two. The log-rank test is very popular method by which the overall survival of two groups are compared, and a Bayesian version of the test mimics the conventional log-rank test, but the interpretation of the Bayesian test is quite different.

The latter part of the chapter is devoted to techniques that model the survival experience of each individual, that is, the actual time to the event (recurrence, time of death, etc.) or the actual time the subject is censored is taken into consideration by the model. First parametric models are considered, including the Weibull distribution. Weibull models allow one to estimate the survival time of each individual and the models allow a large number of survival scenarios to be modeled, which is displayed by Figure 6.3. When analyzing the treatment and placebo groups of the Freireich et al.[5] study, the median time to recurrence is estimated by its posterior distribution. It was shown that the time to recurrence of the treatment group was 28.26 weeks compared to 7.541 weeks for placebo. See Table 6.12 for additional details. A second example involving time to infection for kidney patients on dialysis further illustrates the power of the Bayesian approach and the analysis is reported in Table 6.14. In both cases, the P–P plot shows the Weibull distribution is appropriate as a model for the dependent variable. The last feature involving the Weibull distribution is for estimating the effects of various covariates on the time to recurrence. The effects of age, gender, and type of kidney disease on the time to infection is estimated through a Bayesian analysis reported in Table 6.15. It was found that age did not have an impact on time to infection, but that gender (male compared to female) does. This implies for a particular time t, that in the next instant, the chance of a female experiencing an infection is 80% less than that of a male.

Finally, the Cox proportional hazards model is defined in terms of the hazard function and survival function, and that the ratio of the hazards of two individuals with different covariate values is independent of time. The model is first illustrated with the leukemia study of two groups, where there is only one independent variable, namely the group (placebo versus treatment), and it was found that the hazard ratio for the group effect had a posterior mean of 5.093, implying that the hazard of a placebo patient was five times that of a patient receiving treatment (see Table 6.17). When covariates are included in the Cox model, the analysis is illustrated with the leukemia study, where the log of the white blood cell count is included as the covariate. Table 6.18 reveals the analysis, where it is reported that the hazard ratio for the white blood cell count, adjusted for group, has a posterior mean of 2.766.

Chapter 6 is concluded with a Bayesian analysis for the proportional hazards assumption, where the test for the assumption is based on the survival function of the Cox model. If the PH assumption holds, the exponent of the baseline survival does not depend on time to the event; thus the Bayesian approach is to include time in the exponent and determine the posterior distribution of the coefficient delta in Equations 6.35 and 6.36. It was found that the PH assumption is valid for the group effect, adjusted for the log of the white blood cell account (see Table 6.19).

The Freireich et al.[5] study of leukemia has been covered in detail. It was found that there is indeed a difference in the two groups and that the log white blood cell count makes an impact on the time to recurrence; however, it was also determined that gender was not an important factor in the model (see problem 14 and BUGS CODE 6.8 Ext). Therefore, I suggest that the appropriate model for the leukemia study includes the log white blood cell count and the group membership. The proportional hazard assumption was checked for the group effect and it appears that that factor does indeed satisfy the assumption (see Table 6.19).

Advanced topics for survival studies with the Cox model were not introduced, including the following two situations when the PH assumption is not valid: (1) a stratified analysis and (2) the time-dependent Cox model; the reader is referred to Kleinbaum[1] for additional information.

Exercises

1. Validate Table 6.8, the Bayesian analysis for the treatment group of 21 leukemia patients. Use 55,000 observations for the simulation, with a burn-in of 5000 and a refresh of 100. Use the information in Table 6.6 and BUGS CODE 6.1.

   a. What are the MCMC errors for the conditional probabilities and the survival probabilities?

   b. Plot the posterior density of $P[4]$, the conditional probability of survival at week 13.

   c. Is the posterior distribution of $P[4]$ skewed? If so, explain.

2. Validate Table 6.9, the Bayesian analysis for the placebo group of 21 leukemia patients. Use 55,000 observations for the simulation, with a burn-in of 5000 and a refresh of 100. Use the information in Table 6.7 and BUGS CODE 6.1.

   a. What are the MCMC errors for the conditional probabilities and the survival probabilities?

   b. Plot the posterior density of the survival probability $S[8]$ at week 12.

   c. Is the posterior distribution of $S[8]$ skewed? If so, explain.

3. Using BUGS CODE 6.2 and the information of Table 6.10 validate the Bayesian analysis of Table 6.11. Use 55,000 observations for the simulation, a burn-in of 5000, and a refresh of 100.

4. What are the MCMC errors for all the parameters of Table 6.11?

   a. Is there a difference in the overall recurrence pattern of the treatment group compared to placebo? Explain your answer.

   b. Plot the posterior density of diff (3). Is it skewed?

   c. What does Figure 6.1 imply? Figure 6.2? What is the connection between the two figures?

5. The following is the Evans County, Illinois Survival Data for 1967–1980 (Table 6.20).

With two groups and 25 subjects each in the following categories: (1) those with no history of chronic disease and (2) those with a history of chronic disease. The main end point is time to death and the censored times. This is only a subset of the number of subjects in the study, but for additional information, see Lackland et al.[14] The data can be accessed at http://stat.ethz.ch/education/semesters/ss2011/seminar/homework/solution1.pdf. I put the information in a combined format similar to Table 6.10.

TABLE 6.20

Combined Survival Experience of Two Groups of Evans County Patients

i	$t_{(i)}$	d_{1i}	d_{2i}	$R1(i)$	$R2(i)$	a_{1i}	a_{2i}	$d_{1i} - a_{1i}$	$d_{2i} - a_{2i}$
1	1.4	0	1	25	25	.50	.5	−.50	.50
2	1.6	0	1	25	24	.51	.49	−.51	.51
3	1.8	1	1	25	23	.52	.48	.48	−.48
4	2.2	1	0	24	22	.52	.48	.48	−.48
5	2.4	0	1	23	22	.51	.49	−.51	.51
6	2.5	1	0	23	21	.52	.48	.48	−.48
7	2.6	1	0	22	21	.51	.49	.49	−.49
8	2.8	0	1	21	21	.50	.50	−.50	.50
9	2.9	0	1	21	20	.51	.49	−.51	.51
10	3.0	1	0	21	19	.53	.48	.48	−.48
11	3.1	0	1	20	19	.51	.49	−.51	.51
12	3.5	1	1	20	18	1.05	.95	−.05	.05
13	3.6	0	1	19	17	.53	.47	−.53	.53
14	3.8	1	0	19	16	.54	.46	.46	−.46
15	3.9	0	1	18	16	.53	.47	−.53	.53
16	4.1	0	1	18	15	.55	.45	−.55	.55
17	4.2	0	1	18	14	.56	.44	−.56	.56
18	4.7	0	1	18	13	.58	.42	−.58	.58
19	4.9	0	1	18	12	.60	.40	−.60	.60
20	5.2	0	1	18	11	.62	.38	−.62	.62
21	5.3	1	0	18	10	.64	.36	.36	−.36
22	5.4	1	0	17	10	.63	.37	.37	−.37
23	5.7	1	0	16	10	.62	.38	.38	−.38
24	5.8	0	1	15	10	.60	.40	−.60	.60
25	5.9	0	1	15	9	.63	.38	−.63	.63
26	6.5	0	1	15	8	.65	.35	−.65	.65
27	6.6	1	0	15	7	.68	.32	.32	−.32
28	7.8	0	1	14	7	.67	.33	−.67	.67
29	8.2	1	0	14	6	.70	.30	.30	−.30
30	8.3	0	1	13	6	.68	.32	−.68	.68
31	8.4	0	1	12	5	.71	.29	−.71	.71
32	8.7	1	0	12	4	.75	.25	.25	−.25

TABLE 6.20 (Continued)

Combined Survival Experience of Two Groups of Evans County Patients

i	$t_{(i)}$	d_{1i}	d_{2i}	R1(i)	R2(i)	a_{1i}	a_{2i}	$d_{1i} - a_{1i}$	$d_{2i} - a_{2i}$
33	8.8	0	1	11	4	.73	.27	−.73	.73
34	9.1	0	1	11	3	.79	.21	−.79	.79
35	9.2	1	0	11	2	.85	.15	.15	−.15
36	9.8	1	0	10	2	.83	.17	.17	−.17
37	9.9	0	1	9	2	.82	.18	−.82	.82
38	10.0	1	0	9	1	.90	.10	.10	−.10
39	10.2	1	0	8	1	.89	.11	.11	−.11
40	10.7	1	0	7	1	.88	.13	.13	−.13
41	11.0	1	0	6	1	.86	.14	.14	−.14
42	11.1	1	0	5	1	.83	.17	.17	−.17
43	11.4	0	1	4	1	.80	.20	−.80	.80
44	11.7	1	0	4	0	1.00	0	0	0

a. Based on BUGS CODE 6.2 and the above information on the two groups, perform a Bayesian analysis with 60,000 observations, a burn-in of 5000, and a refresh of 100. Note that the data and initial values are a part of the last two list statements of BUGS CODE 6.2.

b. For each group, determine the posterior distribution of the difference between the observed minus the expected number of deaths.

c. Determine the posterior distribution of the sum (some1, see [6.17]) of the differences between the observed number of deaths and expected number of deaths of Group 1.

d. What is the 95% credible interval for some1?

e. Based on the 95% credible interval for some1, is there sufficient information to declare a difference in the survival experience between the two groups?

   Note that the 95% credible interval (−17.76, −2.107) for the posterior distribution of some1 implies the survival curves are not the same for the two groups.

f. What is the standard deviation of the posterior distribution of some1?

g. Find the posterior distribution of the likelihood ratio.

   For the likelihood ratio, I computed a posterior mean of 8.89 with a 95% credible interval (5.413, 15.37)

h. What is the standard deviation of the posterior distribution of the likelihood ratio?

i. What is the MCMC error for estimating the posterior mean of some1?

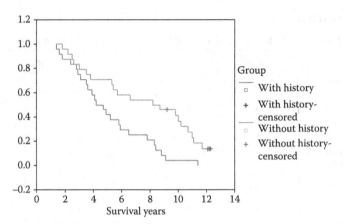

FIGURE 6.7
Kaplan–Meier plot for survival.

6. Verify Figure 6.7. This is a plot of two survival curves for the Evans County data, where the above line indicates subjects with no history of chronic disease, and the below line indicates a history of heart disease.

   a. Verify the graph of the two survival curves.

   b. Are the survival curves the same? Justify your answer from the information of problem.

7. Refer to Table 6.2, the recurrence times for the treatment group of the Freireich et al.[5] study. Using a P–P plot, does it appear reasonable that the recurrence times follow a Weibull distribution?

8. On the basis of the data from Table 6.2 and BUGS CODE 6.3, perform a Bayesian analysis that produces Table 6.12. Note the data from Table 6.2 appears in the third list statement of BUGS CODE 6.3. Use 65,000 observations for the simulation, with a burn-in of 5000 and a refresh of 100.

   a. What is the MCMC error for estimating median[1], the median of the recurrence times for the treatment group?

   b. What is the 95% credible interval for median[2], the median recurrence for the placebo group?

   c. What posterior distributions are skewed?

   d. Plot the posterior distribution of diff. Does the distribution appear skewed?

   e. What is the prior distribution of the beta parameters?

   f. What is the prior distribution for the $k[i]$ parameters?

9. Refer to the analysis for the leukemia study reported in Table 6.12. Using Formulas 6.26 and 6.27 for the mean and median survival times, estimate the average and median survival for the two groups.

10. Based on Table 6.13 and BUGS CODE 6.4, perform a Bayesian analysis of the McGilchrist and Aisbett study. Use 110,000 observations for the simulation, a burn-in of 5000, and a refresh of 100.

   a. What is the posterior mean of the median survival of times to infection?

   b. What is the posterior mean of the shape parameter $k[1]$? Is the distribution skewed?

   c. Refer to Figure 6.5. Is the Weibull distribution appropriate for the times to infection? Why?

   d. Plot the posterior density of median[1].

   e. What is the MCMC error of estimation for estimating the posterior mean of lamda[1], the scale parameter of the Weibull?

11. Portrayed below are the times to second infection of 38 patients who are on dialysis for kidney disease. This is part of the McGilchrist and Aisbett[9] study, where the median time to first infection is reported in Table 6.14 and where the Bayesian analysis is executed with BUGS CODE 6.4 and reported in Table 6.14. Using BUGS CODE 6.4 and the times to infection in the Table 6.21, execute a Bayesian analysis with 55,000 observations for the simulation, a burn-in of 5000, and a refresh of 100.

TABLE 6.21

Time to Second Infection for Kidney Patients

Patient	Recurrence Time	Event	Gender
1	16	1	1
2	108	0	2
3	318	1	2
4	66	1	2
5	9	1	1
6	201	1	2
7	154	1	1
8	25	1	1
9	38	1	2
10	43	1	2
11	4	0	1
12	8	1	2
13	78	1	2
14	13	0	2
15	12	1	1
16	30	1	2
17	333	1	2
18	70	0	2
19	117	1	2
20	159	0	2
21	28	1	1
22	24	0	2
23	245	1	1

(Continued)

TABLE 6.21 (Continued)

Time to Second Infection for Kidney Patients

Patient	Recurrence Time	Event	Gender
24	40	1	1
25	196	1	2
26	30	1	2
27	8	0	2
28	58	1	2
29	25	0	2
30	5	0	2
31	114	1	2
32	8	0	1
33	562	1	1
34	46	0	2
35	156	1	2
36	26	1	2
37	30	1	2
38	16	0	2

Source: McGilchrist, C.A., and Aisbett, C.W., *Biometrics*, 47, 461–466, 1991.

    a. Verify the posterior analysis reported in the Table 6.22.

    b. What is your estimate of the median time to second infection?

    c. Plot the posterior density of median[1].

    d. Is the posterior distribution of lamda[1] skewed?

    e. Compare the results to those reported in Table 6.14.

    f. Why is the median time to a second infection greater than that for the median time to the first infection?

12. Based on Table 6.13 and using BUGS CODE 6.3, compare the median time to first infection of males to that of females. Use 55,000 observations for the simulation with a burn-in of 5000, and a refresh of 100.

    a. What is the posterior mean of the median time to first infection for males?

    b. What is the posterior mean of the median time to first infection for females?

TABLE 6.22

Posterior Analysis for Time to Second Infection

Parameter	Mean	SD	Error	2½	Median	97½
beta	−5.017	.7683	.04185	−6.607	−4.979	−3.581
k[1]	1.02	.1439	.007606	.7843	1.015	1.316
lamda[1]	.008767	.007203	<.0001	.00135	.00688	.02786
median[1]	96.45	21	.6639	60.73	95.48	142.5

c. What is the posterior mean of the difference in the median time to first infection for males minus the median time to first infection for females?

d. Is the posterior distribution of the difference skewed? Why?

e. On the basis of the 95% credible interval for the difference, is it reasonable to believe that the median time to infection for males is different from that of females?

13. Consider Exercise 11 that involved information about the time to second infection for kidney patients undergoing dialysis. Suppose a pilot study was done with 16 patients with the results listed in Table 6.23.

Assuming a Weibull distribution is appropriate for the times to second infection for the pilot study, use the information from the pilot study as prior information for estimating the median time to second infection. Assume the pilot study was done in preparation for doing an additional study reported in the table of Exercise 11. Also assume that prior to the pilot study, the information about the parameters beta and k for the Weibull is vague, similar to that given in BUGS CODE 6.4.

Generate 65,000 observations for the simulation with a burn-in of 5000 and a refresh of 100.

TABLE 6.23

Times to Second Infection for Kidney Patients Pilot Study

Patient	Recurrence Time	Event	Gender
1	40	1	1
2	73	0	1
3	274	1	2
4	86	0	2
5	17	1	1
6	155	1	2
7	179	1	1
8	21	1	1
9	42	0	2
10	39	1	2
11	11	0	1
12	17	1	1
13	29	1	2
14	11	1	2
15	22	1	1
16	22	1	2

    a. What is the posterior mean and median of the median time to the second infection?

    b. What is the 95% credible interval for the median time to second infection?

    c. Compare the posterior distribution of the median time (using both data sets) to second infection with the posterior distribution for the same parameter reported in Exercise 11. Is there a discernible difference in the posterior means?

    d. Plot the posterior density of the median time to the second infection with both data sets (the prior information given earlier and the times reported in the table of Exercise 11).

    e. Explain the role Bayes' theorem plays in utilizing the prior information of the pilot study.

14. Using BUGS CODE 6.7, verify Table 6.18 with 55,000 observations for the simulation, 5000 for the burn-in, and 100 for the refresh. The analysis is for the Freireich et al.[5] study with two groups of patients, one is the placebo group and the other is the treatment group, each with 21 patients. The analysis is focused on the covariate, namely the log white blood cell count and its effect on the time to recurrence. Also important is the effect of the covariate on the hazard ratio for the difference in the two groups and its effect on the estimated survival probabilities of the two groups.

    a. How does the covariate affect the posterior mean of beta and the hazard ratio for beta? Compare the posterior mean and median of beta with the covariate and without the covariate. The analysis without the covariate is reported in Table 6.17.

    b. Should the log white blood cell count be used in the analysis? Why?

    c. What is the effect of the covariate on the survival probabilities of the treatment group? Compare to the survival probabilities reported in Table 6.17.

    d. What is the effect of the covariate on the survival probabilities of the placebo group? Compare the survival probabilities to those reported in Table 6.17.

    e. Are the MCMC errors small enough for you to have confidence in the posterior mean of the parameters reported in Table 6.18?

15. Consider the leukemia example of Freireich et al.[5] with two groups, one of which is a treatment group and the other placebo. Below is BUGS CODE 6.7 Ext where the list statement contains the data for the time to recurrence, the censoring times, the log of the white blood cell count, an indicator for the group identification (0 for treatment and 1 for placebo), and a vector x3 of gender identification

(a 1 for males and a 0 for females). The white blood cell count is an indication of the disease (leukemia) and its effect is examined in Exercise 14. Does gender also have an effect on the time to recurrence? BUGS CODE 6.7 Ext is a slight extension of BUGS CODE 6.7 in that the covariate x3 gender is added to assess its effect on the time to recurrence and its effect on the hazard ratio of the other variables, group and white blood cell count.

BUGS CODE 6.7 Ext

```
model
  {
# Set up data
     for(i in 1:N) {
          for(j in 1:T) {
# risk set = 1 if obs.t > = t
                     Y[i,j] <- step(obs.t[i] - t[j] + eps)
# counting process jump = 1 if obs.t in [t[j], t[j+1])
#          i.e. if t[j] < = obs.t < t[j+1]
                     dN[i, j] <- Y[i, j] * step(t[j + 1]-
                     obs.t[i] - eps) * fail[i]
          }
     }
# Model
     for(j in 1:T) {
          for(i in 1:N) {
               dN[i, j] ~ dpois(Idt[i, j]) # Likelihood
               # With covariate log white blood cell count
               Idt[i, j] <- Y[i, j] * exp(beta[1] * x1[i]+beta
               [2]*x2[i]+beta[3]*x3[i]) * dL0[j] # Intensity
               }
          dL0[j] ~ dgamma(mu[j], c)
          mu[j] <- dL0.star[j] * c # prior mean hazard
# Survivor function = exp(-Integral{10(u)du})^exp(beta*z)
# adjusted for covariates at mean value of covariate
               S.treat[j] <- pow(exp(-sum(dL0[1 : j])),
                  exp(beta[1] * -0.5+beta[2]*2.63+beta[3]*.47));
               S.placebo[j] <- pow(exp(-sum(dL0[1 : j])),
                  exp(beta[1] * 0.5+beta[2]* 3.22+beta[3]*.52));
          }
          c <- 0.001
          r <- 0.1
          for (j in 1 : T) {
               dL0.star[j] <- r * (t[j + 1] - t[j])
          }
     beta[1] ~ dnorm(0.0,0.000001)
beta[2] ~ dnorm(0.0,0.000001)
beta[3] ~ dnorm(0.0,0.000001)
```

```
      # hazard ratio for group 1 versus group 2
      HR.beta1<-exp(beta[1])
      # hazard ratio for log wbc
      HR.beta2<-exp(beta[2])
      # HR for gender
      HR.beta3<-exp(beta[3])
   }
 # x1 is the group indicator
 # x2 is log wbc
 # x3 is gender (1 male, 0 female)

list(N = 42, T = 17, eps = 1.0E-10,

x3 = c(1,1,1,1,1,1,0,1,0,0,0,1,1,1,0,0,0,0,0,0,1,
  0,1,0,0,0,0,1,1,1,0,0,0,0,0,0,1,1,1,1,1,1),
x2 = c(2,8,5,4.91,4.48,4.01,4.36,2.32,3.49,3.97,3.52,3.05,2.42
  ,3.26,3.49,2.12,1.5,3.06,2.30,2.95,2.73,1.97,
2.31,4.06,3.28,4.43,2.96,2.88,3.60,2.32,2.57,3.20,2.80,2.70,2.
  60,2.16,2.05,2.01,1.78,2.20,2.53,1.47,1.45),
      obs.t = c(1, 1, 2, 2, 3, 4, 4, 5, 5, 8, 8, 8, 8, 11, 11,
            12, 12, 15, 17, 22, 23, 6,
         6, 6, 6, 7, 9, 10, 10, 11, 13, 16, 17, 19, 20, 22, 23, 25,
            32, 32, 34, 35),
      fail = c(1, 1, 1, 1, 1, 1, 1, 1, 1, 1, 1, 1, 1, 1, 1, 1,
         1, 1, 1, 1, 1, 1, 1, 1, 0, 1, 0, 1, 0, 0, 1, 1, 0, 0, 0,
         1, 1, 0, 0, 0, 0, 0),
      x1 = c(0.5, 0.5, 0.5, 0.5, 0.5, 0.5, 0.5, 0.5, 0.5, 0.5,
         0.5, 0.5, 0.5, 0.5, 0.5, 0.5, 0.5, 0.5, 0.5, 0.5,
         -0.5, -0.5, -0.5, -0.5, -0.5, -0.5, -0.5, -0.5, -0.5,
          -0.5, -0.5, -0.5, -0.5, -0.5, -0.5, -0.5, -0.5, -0.5,
           -0.5, -0.5, -0.5),
      t = c(1, 2, 3, 4, 5, 6, 7, 8, 10, 11, 12, 13, 15, 16, 17,
       22, 23, 35))

      # initial values
      list(beta = c(0,0,0),
dL0 = c(1.0,1.0,1.0,1.0,1.0,1.0,1.0,1.0,1.0,
  1.0,1.0,1.0,1.0,1.0,1.0, 1.0,1.0))
```

Based on BUGS CODE 6.7 Ext, I executed the analysis with 55,000 observations, a burn-in of 5000, and a refresh of 100, and the results are reported as Table 6.24.

a. Verify the results. Use BUGS CODE 6.7 Ext with 55,000 observations for the simulations with a burn-in of 5000 and a refresh of 100.

b. Does the addition of gender affect the hazard ratios for beta[1] (the group effect on the recurrence time) and beta[2] (the effect of log white blood cell count on the time to recurrence)?

TABLE 6.24

A Bayesian Analysis with the Cox Model

Parameter	Mean	SD	Error	2½	Median	97½
HR.beta1	3.465	1.635	.0273	1.372	3.119	7.438
HR.beta2	2.758	.8793	.0510	1.582	2.607	5.089
HR.beta3	1.272	.5446	.0090	.5075	1.169	2.606
S.placebo[1]	.9831	.0200	<.0001	.9272	.99	.9995
S.placebo[2]	.9339	.0433	.0011	.8252	.9434	.9893
S.placebo[3]	.9004	.0547	.0013	.7679	.9101	.977
S.placebo[4]	.8264	.0730	.0015	.6584	.8356	.9407
S.placebo[5]	.7396	.0896	.0015	.5419	.7482	.8897
S.placebo[6]	.6151	.1046	.0014	.3993	.62	.8035
S.placebo[7]	.5755	.1069	.0014	.3591	.579	.7717
S.placebo[8]	.4319	.106	.0011	.2312	.4305	.6422
S.placebo[9]	.3863	.106	<.0001	.1902	.3833	.6
S.placebo[10]	.3018	.1008	<.0001	.1233	.2959	.5127
S.placebo[11]	.2105	.0930	<.0001	.0588	.2007	.4156
S.placebo[12]	.173	.0867	<.0001	.0388	.1613	.3721
S.placebo[13]	.1416	.0793	<.0001	.0260	.1289	.3291
S.placebo[14]	.1112	.0709	<.0001	.0154	.0971	.2826
S.placebo[15]	.0856	.0617	<.0001	.00879	.0716	.2408
S.placebo[16]	.0152	.0217	<.0001	.0017	.0299	.1523
S.placebo[17]	.0152	.0217	<.0001	<.00001	.0071	.0770
S.treat[1]	.9965	.0048	<.0001	.9832	.9982	.9999
S.treat[2]	.9866	.0111	<.0001	.957	.9897	.9985
S.treat[3]	.9797	.0147	<.0001	.9414	.9833	.9968
S.treat[4]	.9638	.0215	<.0001	.9089	.9684	.9916
S.treat[5]	.9438	.0288	<.0001	.8731	.9492	.9838
S.treat[6]	.9115	.0391	<.0001	.8184	.9177	.9696
S.treat[7]	.9	.0429	<.0001	.7989	.9066	.9646
S.treat[8]	.8517	.0570	<.0001	.7211	.859	.9414
S.treat[9]	.8337	.0621	<.0001	.6928	.8413	.9324
S.treat[10]	.7946	.0727	.0010	.6313	.8028	.9131
S.treat[11]	.7398	.08603	<.0001	.5507	.7482	.8838
S.treat[12]	.7112	.0916	<.0001	.5121	.7188	.8677
S.treat[13]	.6829	.0968	<.0001	.4764	.69	.8508
S.treat[14]	.6495	.1023	<.0001	.4346	.6556	.8304
S.treat[15]	.6148	.1007	<.0001	.393	.6192	.8097
S.treat[16]	.5273	.1213	<.0001	.2844	.5301	.7532
S.treat[17]	.4112	.1324	.0012	.1617	.4088	.6721
beta[1]	1.146	.4349	.0075	.3164	1.137	2.007
beta[2]	.9704	.2914	.00167	.4587	.9583	1.627
beta[3]	.1553	.4159	.0068	−.6784	.1562	.958

Source: Freireich, E.J., Gehan, E., Frei, E., and Schroeder, L.R., *Blood*, 21, 699–716, 1963. Study with Covariates.

c. Is the gender covariate needed in the model? Why or why not? Refer to the 95% credible interval for alpha.

d. Refer to Table 6.18 and compare the difference in the survival probabilities for the treatment and placebo groups with those reported in the table. Is there much of a difference in the two sets of survival proportions?

e. Plot the posterior density of the hazard ratio for beta[1]

f. Are the MCMC errors sufficiently small so that the posterior means are good approximations to the actual posterior means?

16. Consider the Freireich et al.[5] study with two groups (placebo and treatment), two genders (males and females), and the log white blood cell count for each patient. Vector x1 contains the group indicator (0.5 placebo and −0.5 treatment), vector x2 contains the log white blood cell counts, and vector x3 identifies sex (1 male and 0 female). Below is BUGS CODE 6.9, which tests for the proportional hazard assumption of gender.

BUGS CODE 6.9

```
model
  {
  # Set up data
    for(i in 1:N) {
      for(j in 1:T) {
  # risk set = 1 if obs.t > = t
        Y[i,j] <- step(obs.t[i] - t[j] + eps)
  # counting process jump = 1 if obs.t in [t[j], t[j+1])
  #              i.e. if t[j] < = obs.t < t[j+1]
                    dN[i, j] <- Y[i, j] * step(t[j + 1]-
                    obs.t[i] - eps) * fail[i]
                }
      }
  # Model
        for(j in 1:T) {
                for(i in 1:N) {
                    dN[i, j] ~ dpois(Idt[i, j]) # Likelihood
                    # With covariate log white blood cell count
                    Idt[i, j] <- Y[i, j] * exp(beta[1] * x1
                    [i]+beta[2]*x2[i]+beta[3]*x3[i])
                        * dL0[j] # Intensity
                }
        dL0[j] ~ dgamma(mu[j], c)
        mu[j] <- dL0.star[j] * c # prior mean hazard
  # Survivor function = exp(-Integral{l0(u)du})^exp(beta*z)
  # adjusted for covariates at mean value of covariate
        S.treat[j] <- pow(exp(-sum(dL0[1 : j])), exp(beta[1] *
          -0.5+beta[2]*2.63+beta[3]*.47+delta*.47*t[j]));
```

```
        S.placebo[j] <- pow(exp(-sum(dL0[1 : j])), exp(beta[1] *
          0.5+beta[2]* 3.22+beta[3]*.52+delta*.52*t[j]));
    }
    c <- 0.001
    r <- 0.1
    for (j in 1 : T) {
      dL0.star[j] <- r * (t[j + 1] - t[j])
    }
  beta[1] ~ dnorm(0.0,0.000001)
  beta[2] ~ dnorm(0.0,0.000001)
  beta[3] ~ dnorm(0.0,0.000001)
  delta ~ dnorm(0.0,0.000001)
    # hazard ratio for group 1 versus group 2
    HR.beta1<-exp(beta[1])
    # hazard ratio for log wbc
    HR.beta2<-exp(beta[2])
    # HR for gender
    HR.beta3<-exp(beta[3])   }
  # x1 is the group indicator
  # x2 is log wbc
  # x3 is gender (1 male, 0 female)

list(N = 42, T = 17, eps = 1.0E-10,

x3 = c(1,1,1,1,1,1,0,1,0,0,0,1,1,1,0,0,0,0,0,0,1,
  0,1,0,0,0,0,1,1,1,0,0,0,0,0,0,1,1,1,1,1,1),
x2 = c(2,8,5,4.91,4.48,4.01,4.36,2.32,3.49,3.97,3.52,3.05,2.42
  ,3.26,3.49,2.12,1.5,3.06,2.30,2.95,2.73,1.97,
2.31,4.06,3.28,4.43,2.96,2.88,3.60,2.32,2.57,3.20,2.80,2.70,2.
  60,2.16,2.05,2.01,1.78,2.20,2.53,1.47,1.45),
    obs.t = c(1, 1, 2, 2, 3, 4, 4, 5, 5, 8, 8, 8, 8, 11, 11,
      12, 12, 15, 17, 22, 23, 6,
      6, 6, 6, 7, 9, 10, 10, 11, 13, 16, 17, 19, 20, 22, 23, 25,
      32, 32, 34, 35),
    fail = c(1, 1, 1, 1, 1, 1, 1, 1, 1, 1, 1, 1, 1, 1, 1, 1,
      1, 1, 1, 1, 1, 1, 1, 1, 0, 1, 0, 1, 0, 0, 1, 1, 0, 0, 0,
      1, 1, 0, 0, 0, 0, 0),
    x1 = c(0.5, 0.5, 0.5, 0.5, 0.5, 0.5, 0.5, 0.5, 0.5, 0.5,
      0.5, 0.5, 0.5, 0.5, 0.5, 0.5, 0.5, 0.5, 0.5, 0.5, 0.5,
      -0.5, -0.5, -0.5, -0.5, -0.5, -0.5, -0.5, -0.5, -0.5,
      -0.5, -0.5, -0.5, -0.5, -0.5, -0.5, -0.5, -0.5, -0.5,
      -0.5, -0.5, -0.5),
    t = c(1, 2, 3, 4, 5, 6, 7, 8, 10, 11, 12, 13, 15, 16, 17,
      22, 23, 35))

    # initial values
    list(beta = c(0,0,0), delta = 0,
dL0 = c(1.0,1.0,1.0,1.0,1.0,1.0,1.0,1.0,1.0,
  1.0,1.0,1.0,1.0,1.0,1.0, 1.0,1.0))
```

TABLE 6.25

Test for PH of Gender

Parameter	Mean	SD	Error	2½	Median	97½
delta	−5.057	997	4.209	−1971	−4.587	1952

The analysis is executed using the BUGS CODE 6.9 with 55,000 observations, with a burn-in of 5000 and a refresh of 100. The code below gives the relevant operations for testing the PH.

```
S.treat[j] <- pow(exp(-sum(dL0[1 : j])), exp(beta[1] *
    -0.5+beta[2]*2.63+beta[3]*.47+delta*.47*t[j]));
```

This is the statement for the posterior distribution of delta for the placebo group. Note that delta is the relevant parameter.

```
S.placebo[j] <- pow(exp(-sum(dL0[1 : j])), exp(beta[1]
    * 0.5+beta[2]* 3.22+beta[3]*.52+delta*.52*t[j]));
```

a. Verify Table 6.25.
b. Is the proportional hazard assumption valid for gender? Why or why not?
c. Plot the posterior density of delta.
d. What does delta measure?
e. Is the posterior distribution of delta skewed?

References

1. Kleinbaum, D.G. *Survival Analysis: A Self-Learning Text*, Springer Verlag, 1996, New York.
2. Congdon, P. *Bayesian Statistical Modelling*, John Wiley & Sons Inc., 2001, New York.
3. Congdon, P. *Applied Bayesian Modelling*, John Wiley & Sons Inc., 2003, New York.
4. Newman, S.C. *Biostatistical Methods in Epidemiology*, John Wiley & Sons Inc., 2001, New York.
5. Freireich, E.J., Gehan, E., Frei, E., and Schroeder, L.R. The effect of 6-Mercaptopurine on the duration of remission in acute leukemia: a model for the evaluation of other potentially useful therapies, *Blood*, 21, 699–716, 1963.
6. Kahn, H.A. and Sempos, C.T. *Statistical Methods in Epidemiology*, Oxford University Press, 1989, Oxford, UK.
7. Peto, R., Pike, M.C., Armitage, P., Breslow, N.E., Dox, D.R., Howard, S.V., Mantel, N., McPherson, K., Peto, J., and Smith, P.G. Design and analysis of randomized clinical trials regarding prolonged observation of each patient. II. Analysis and examples, *Br. J. Cancer*, 35(1), 1–35, 1977.
8. Ibrahim, J.G, Chen, M.-H., and Sinha, D. *Bayesian Survival Analysis*, Springer, 2001, New York.

9. McGilchrist, C.A. and Aisbett, C.W. Regression with frailty in survival analysis, *Biometrics*, 47, 461–466, 1991.

10. Cox, D.R. Regression models and life tables (with discussion), *J. Roy. Stat. Soc., B*, 34, 187–220, 1972.

11. Kalbfleisch, J. Non parametric Bayesian analysis of survival time data, *J. Stat. Soc., B.*, 40, 214–221, 1978.

12. Clayton, D. A Monte Carlo method for Bayesian inference in frailty models, *Biometrics*, 47, 467–485, 1991.

13. Chen, M., Ibrahim, J., and Shoa, Q. *Monte Carlo Methods in Bayesian Computation*, Springer, 2000, New York.

14. Lackland, D.T., Egan, B.M., Mountford, W.K., Boan, A.D., Evans, D.A., Gilbert, G., McGee, D.L., Thirty-year survival for black and white hypertensive individuals in the Evans Count heart study and the hypertension detection and follow-up program, *J. Am. Soc. Hypertension*, 2(6), 446–454, 2008.

7

Screening for Disease

7.1 Introduction

Chapter 7 introduces an important topic that is part of the experience of many epidemiologists, and the topic is screening for disease among individuals who do not exhibit any symptoms of the disease. For example, the Morrison[1] book presents descriptive and analytical methods necessary to design and analyze screening programs, and the Shapiro et al.[2] analysis of the Health Insurance Plan of Greater New York (HIP) illustrate epidemiological methods for estimating the lead time and survival of the trial participants.

Chapter 7 continues by describing the fundamentals of a screening program and measures of test accuracy of the various modalities for screening. A modality is an instrument (e.g., an imaging device) that measures the health status of the participants of the screening program, whereas the several measures of test accuracy are statistical techniques that estimate the accuracy of the modality. For example, Miller, Chamberlain, Day, Hakama, and Prorok[3] describe the estimation of the specificity and sensitivity of the U.K. screening trial for breast cancer trial as reported by Chamberlain et al.[4] The positive and negative predictive values (NPVs) also estimate the test accuracy and are illustrated with a study to diagnose coronary artery disease (CAD) with the exercise stress test (EST). For screening programs with two groups, a study group and a control, the validity of the modality is often determined by estimating the lead time and the survival experience of the study group compared to that of the control.

The HIP study is one of the earlier screening programs, and consisted of two groups, where participants were randomized into the study and control groups, each with about 30,000 subjects. It was well designed and analyzed, but will be analyzed for the present with Bayesian techniques, where the analysis is comprised of estimating the lead time (the time between the time the disease is detected with the screening device and the time the disease would have been clinically detected without screening) and of estimating the survival times (from diagnosis) to death. Survival will be compared between the two groups with life table techniques and with survival models such as the Cox proportional hazards model, whereas lead time will be estimated by the so-called method of differences.

Bayesian methods for estimating lead time and survival will be fully explained and illustrated with the aid of WinBUGS. The student should be able to appreciate the importance of screening programs and the statistical methods that are used to evaluate them. The difference method to estimate lead time and the life table to compare survival between the study and control groups are unique to epidemiology, but will be interpreted with a Bayesian approach.

7.2 Principles of Screening

Screening programs attempt to identify disease among individuals who do not display any symptoms of that disease. For example, the recently completed national lung cancer screening program attempted to identify individuals with lung cancer among high-risk candidates (smokers and ex-smokers) who did not display any symptoms of the disease. There were two groups, where with one, the modality was computed tomography (CT) and with the other the modality was x-ray. Aberle et al.[5] report reduced mortality with low-dose CT compared to the standard imaging modality of x-ray. A total of 53,454 persons at high risk for the disease were studied, with 26,722 assigned to the CT and 26,732 to x-ray, and the study took place at 33 U.S. medical centers between August 2002 and April 2004. Of course, the participants are still being followed to estimate mortality and incidence of the disease as diagnosed by the two modalities. Deaths include those who participated to some extent in the three annual screening exams and those that refused screening. The disease could be detected as a result of screening and detected between screenings, and of course lung cancer continues to be diagnosed after the exams terminated.

Because a screening program is expensive and involves a large number of medical personnel over a long period of time, Mausner and Bahn[6] state that screening should only be attempted under certain conditions, including the following:

1. The disease should be an important public health problem.
2. There should be an efficacious treatment for the disease.
3. The screening modality needs to be accurate for detecting the disease and facilities for detection, diagnosis, and treatment should be available.
4. There should be a recognizable preclinical period for the disease, during which the modality can detect the disease with reasonable accuracy.
5. The modality should be acceptable to the population to be screened.

6. The natural history of the disease should be understood, from the latent period where the disease cannot be detected, to the preclinical phase, where it can be detected, to the clinical manifestation of the disease.

The usual situation with a disease is to diagnose it when it becomes clinical, which is opposed to the situation where the disease is detected in the preclinical phase for an asymptomatic patient. The interval between the time the disease is detected by the screening modality and the time it would have been detected in the absence of screening is called the lead time.

In a later section of this chapter, the Bayesian approach to estimating the lead time will be described. Note that detecting the disease in the preclinical phase is more difficult than when the clinical symptoms appear.

Once the disease is detected the patient is referred to a gold standard with two possibilities: (1) the gold standard confirms that the modality did indeed detect the disease, or (2) the gold standard is negative for disease when the modality was positive, which is called a false positive. The proportion of cases where the modality correctly identifies disease is called the sensitivity of the test (modality). For those subjects that test negative for disease through the modality, they are usually not subject immediately to the gold standard.

There are various ways to design a screening study, but for our purposes, the following scenario is assumed, namely, that the subjects are randomized to a study group and to a control group. For the study group, the patients are screened several times at equally spaced intervals, thus, the subject can be detected for disease at each screening exam by the modality, or the disease can be detected clinically between the exams; this is called an interval diagnosis. It is important to know that a certain subset of the study group can refuse to be screened, but can be diagnosed clinically during the screening experiment. Subjects in the control group receive medical care as usual, and their time of disease from diagnosis is recorded for the record. Our main emphasis for analysis will be to determine the accuracy of the screening modality, estimate the lead time, and compare the survival between the study and control groups.

7.3 Evaluation of Screening Programs

7.3.1 Introduction

Screening programs involve modalities that measure the health status of the individual. For example, mammography is employed in screening programs for breast cancer, while for lung cancer, CT is used to image the lungs in order to measure the extent (if any) of the disease. The basic measures of test

TABLE 7.1

Classification Table

	Disease	
Test	**$D = 0$**	**$D = 1$**
$X = 0$	(n_{00}, θ_{00})	(n_{01}, θ_{01})
$X = 1$	(n_{10}, θ_{10})	(n_{11}, θ_{11})

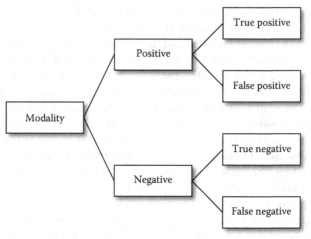

FIGURE 7.1
Four outcomes of a binary test.

accuracy are defined below and are essential in the evaluation of screening programs.

This section will introduce Bayesian techniques to estimate and test hypotheses about the basic measures of test accuracy. The measures of test accuracy are (1) classification probabilities, (2) predictive measures, and (3) diagnostic likelihood ratios. The classification probabilities are the false positive fraction (FPF) and true positive fraction (TPF), while there are two predictive values, the positive predictive value (PPV) and the NPV. Finally, there are two diagnostic likelihood ratios, the positive PDLR and the negative NDLR. These measures will be defined in the next section in the context of a cohort study. Thus, there is a random sample of size n selected from the target population and a gold standard, thus each patient is classified into the four categories of Table 7.1.

The n_{ij} are the number of subjects with test score $i = 0$ or 1 and disease status $j = 0$ or 1, while θ_{ij} is the corresponding probability.

Figure 7.1 depicts the four outcomes for a binary test. When the modality produces a positive test, the result is verified by the gold standard as a true positive, but if the gold standard does not verify the test result, a false positive occurs.

Also, if the modality produces a negative test, which is verified by the gold standard, the result is a true negative, otherwise a false negative occurs. When referring to Figure 7.1, also refer to Table 7.1, because they go hand in hand.

7.3.2 Classification Probabilities

The basic measures of test accuracy are the TPF (sensitivity) and the FPF (1 − specificity) where

$$TPF(\theta) = \frac{\theta_{11}}{(\theta_{11} + \theta_{01})} = P(X = 1 | D = 1) \tag{7.1}$$

and

$$FPF(\theta) = \frac{\theta_{10}}{(\theta_{00} + \theta_{10})} = P(X = 1 | D = 0) \tag{7.2}$$

It is important to know that the TPF and FPF are unknown parameters and are functions of θ. The Bayesian analysis determines the posterior distribution of these quantities, from which the parameters are estimated and certain inferences are performed. Assume the prior information is based on a previous study, with results given in Table 7.2
where m subjects have been classified in the same way as those in Table 7.1.
The density based on prior information is

$$\xi(\theta) \propto \theta_{00}^{m_{00}} \theta_{01}^{m_{01}} \theta_{10}^{m_{10}} \theta_{11}^{m_{11}} \tag{7.3}$$

Thus, the likelihood function for $\theta = (\theta_{00}, \theta_{01}, \theta_{10}, \theta_{11})$ is

$$L\left(\frac{\theta}{n}\right) \propto \theta_{00}^{n_{00}} \theta_{01}^{n_{01}} \theta_{10}^{n_{10}} \theta_{11}^{n_{11}} \tag{7.4}$$

and the posterior distribution is Dirichlet,

$$\frac{\theta}{(n,m)} \sim Dir(n_{00} + m_{00} + 1, n_{01} + m_{01} + 1, n_{10} + m_{10} + 1, n_{11} + m_{11} + 1) \tag{7.5}$$

TABLE 7.2

Classification Table of Prior Information

Test	Disease	
	$D = 0$	$D = 1$
$X = 0$	(m_{00}, θ_{00})	(m_{01}, θ_{01})
$X = 1$	(m_{10}, θ_{10})	(m_{11}, θ_{11})

Note, if there is no prior information, the m_{ij} are zero, and one in effect is assuming a uniform prior distribution for the θ_{ij}.

Monte Carlo Markov chain (MCMC) sampling from the Dirichlet distribution, using WinBUGS, will determine the posterior distribution of these classification probabilities.

As an example, consider an example examined by Pepe[7] and based on the study by Weiner et al.[8] This is a cohort study of 1465 subjects, where each is classified as to disease status (CAD via an angiogram) and a diagnostic test, the EST, which is a nuclear medicine procedure, where the data can be found in the list statement of BUGS CODE 7.1. Note that for the EST data, a one is added to each cell frequency. The code is documented with a # sign. By adding a one to each cell, one is in effect assuming a uniform prior for the cell probabilities.

The analysis is based on the following code.

BUGS CODE 7.1

```
# Measures of accuracy
# Binary Scores

Model;
{
# Dirichlet distribution for cell probabilities
g00~dgamma(a00,2)
g01~dgamma(a01,2)
g10~dgamma(a10,2)
g11~dgamma(a11,2)

h<-g00+g01+g10+g11
# the theta have a Dirichlet distribution
theta00<-g00/h
theta01<-g01/h
theta10<-g10/h
theta11<-g11/h
# the basic test accuracies are below
tpf<-theta11/(theta11+theta01)
se<-tpf
sp<-1-fpf
fpf<-theta10/(theta10+theta00)
tnf<-theta00/(theta00+theta10)
fnf<-theta01/(theta01+theta11)
ppv<-theta11/(theta10+theta11)
npv<-theta00/(theta00+theta01)
pdlr<-tpf/fpf
ndlr<-fnf/tnf
# sensitivity mammography for Chamberlain et al.⁴
tpfm<-(theta11+theta10)/(theta11+theta10+theta01+theta00)
```

```
# sensitivity of physical opinion
tpfpe<-(theta01+theta11)/(theta11+theta10+theta01+theta00)
}

# Exercise Stress Test Pepe
# Uniform prior (add one to each cell of the table frequencies!)
list(a00 = 328,a01 = 209,a10 = 116,a11 = 819)
# Table 7.12, Chamberlain et al.
# Uniform prior
list(a00 = 30051,a01 = 14,a10 = 1959,a11 = 181)
# Chamberlain et al.⁴ uniform prior
# add one to each cell
list(a00 = 3,a01 = 9,a10 = 50,a11 = 123)
# Problem 7, improper prior
list(a00 = 70205,a01 = 25,a10 = 2908,a11 = 262)

# Problem 8, improper prior
list(a00 = 67116,a01 = 41,a10 = 3216,a11 = 76)
# initial values
list(g00 = 1,g01 = 1,g10 = 1,g11 = 1)
```

The notes of interest for the code are headed by #. There are three list statements; the first gives the information necessary to generate a Dirichlet distribution for the four cell probabilities. The entries are the cell frequencies plus 1. In this way, a uniform prior is assumed. The third list statement gives the initial values for the MCMC procedure, and the second list statement will be used later.

A uniform prior is assumed resulting in a posterior distribution that is Dirichlet (328,209,116,819). The analysis is executed with 55,000 observations generated from the joint posterior distribution of the cell probabilities, using 5,000 as a burn-in and 100 as a refresh, resulting in Table 7.3 for the posterior analysis for the accuracy of the EST.

The sensitivity or TPF is estimated as .7967 with an associated posterior standard deviation of .0125 and (.7716, .8208) as a 95% credible interval. Note the analysis also includes 5.84×10^{-5} as the MCMC error for estimating the TPF, which implies the estimate of .7967 is within 5.84×10^{-5} units of the "true" posterior TPF. The WinBUGS output also includes plots of the marginal posterior distribution of the parameters and Figure 7.2 portrays that for the sensitivity of the EST.

TABLE 7.3

Bayesian Analysis for Exercise Stress Test

Parameter	Mean	SD	Error	Lower 2½	Median	Upper 2½
TPF	.7967	.0125	5.84×10^{-5}	.7716	.7968	.8208
FPF	.2612	.0208	9.22×10^{-5}	.2215	.2608	.3033

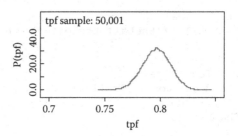

FIGURE 7.2
Posterior density of the true positive fraction.

With an estimated false positive rate of .2612 and an estimated true positive rate of .79, the EST provides good to fair accuracy, but other measures of accuracy should be considered.

7.3.3 Predictive Values

The second set of measures for test accuracy are the PPV and the NPV, which are defined as follows:

$$\text{PPV}(\theta) = \frac{\theta_{11}}{(\theta_{01} + \theta_{11})} = P(D = 1 \mid X = 1) \tag{7.6}$$

and

$$\text{NPV}(\theta) = \frac{\theta_{00}}{(\theta_{00} + \theta_{01})} = P(D = 0 \mid X = 0) \tag{7.7}$$

Because these two quantities depend on disease incidence, it is important that the patients be selected at random from the target population, so that when estimating the predictive values, the estimated disease incidence is estimated without bias. Returning to the previous example, the posterior distribution of the predictive values is provided in Table 7.4. They answer the question of primary interest to the patient: Do I have the disease? This to some extent is answered by the following posterior analysis.

The distribution of the PPV appears to be symmetric with a mean of .8759, which implies that the chance for heart disease among those patients that test positive is .87, which gives me some confidence in the EST to detect disease. On the other hand, for those who test negative, the chance of not having coronary heart disease is only .61. My confidence is somewhat lowered in the ability of the test to discriminate between diseased and nondiseased patients. If the test is negative, I am not sure whether I have the disease or not! Note that a perfect test occurs when PPV = NPV = 1. I did not give the exact figure for the MCMC error, only noting that it is quite small for these

TABLE 7.4

Distribution of Predictive Values

Parameter	Mean	SD	Error	Lower 2½	Median	Upper 2½
PPV	.8759	.0108	<.0001	.8538	.8762	.8961
NPV	.6109	.0211	<.0001	.5693	.611	.6517

two measures of test accuracy. In executing the analysis, one should vary the MCMC sample size to see its effect on the posterior distribution and the error of estimation.

7.3.4 Diagnostic Likelihood Ratios

The diagnostic likelihood ratios are a third group of test accuracy measures and are given as follows:

$$\text{PDLR}(\theta) = \frac{P(X=1|D=1)}{P(X=1|D=0)}$$

$$= \frac{\left[\dfrac{\theta_{11}}{(\theta_{11}+\theta_{01})}\right]}{\left[\dfrac{\theta_{10}}{(\theta_{10}+\theta_{00})}\right]} \tag{7.8}$$

$$= \frac{\text{TPF}(\theta)}{\text{FPF}(\theta)}$$

and

$$\text{NDLR}(\theta) = \frac{P(X=0|D=1)}{P(X=0|D=0)}$$

$$= \frac{\left[\dfrac{\theta_{01}}{(\theta_{11}+\theta_{01})}\right]}{\left[\dfrac{\theta_{00}}{(\theta_{10}+\theta_{00})}\right]} \tag{7.9}$$

$$= \frac{\text{FNF}(\theta)}{\text{TNF}(\theta)}$$

With regard to the PDLR, the more accurate the diagnostic test becomes, the larger the numerator (TPF) tends to become and the smaller the denominator (FPF) tends to become, but for the NDLR, the opposite is true. The numerator (FNF) tends to become smaller and the denominator (TNF) tends to become larger. The range of both is $(0,\infty)$.

TABLE 7.5

Distribution of Diagnostic Likelihood Ratios

Parameter	Mean	SD	Error	Lower 2½	Median	Upper 2½
PDLR	3.07	.2526	.0011	2.616	3.055	3.609
NDLR	.2755	.0187	<.0001	.2399	.275	.3135

For example, the characteristics of the posterior distribution for the likelihood ratios are given in Table 7.5.

Note the estimated simulation error for estimating PDLR is .0011, which implies that the estimate of 3.07 is within .0011 of the "true" posterior positive diagnostic likelihood ratio and that the test is positive about three times more often among the diseased, compared to those without CAD. On the other hand, among those who have the disease, the test is negative much less often compared to those without the disease. Both measures indicate an accurate test. The larger the PDLR and the smaller the NDLR, the more accurate the test.

In summary, three types of measures of accuracy have been computed for the EST. For the sensitivity and specificity, I am somewhat confident that the test is informative, but with regard to the predictive values, the NPV did not give me high confidence in the test to measure accuracy. For additional information about these basic measures of accuracy, Pepe (p. 20)[7] provides a summary.

7.3.5 ROC Curve

Consider the results of mammography given to 60 women, of which 30 had the disease, which is presented in Zhou et. al (p. 21).[9] and reported in the list statement of BUGS CODE 7.2. The a_{1j} give the data (plus 1) for each of the five outcomes for the diseased group, whereas the a_0 give the frequencies for the nondiseased group. A one is added to each cell frequency that in effect assumes a uniform prior distribution for the 10 cell frequencies.

The radiologist assigns a score from 1 to 5 to each mammogram, where 1 indicates a normal lesion, 2 a benign lesion, 3 a lesion which is probably benign, 4 indicates suspicious, and 5 malignant. How would one estimate the accuracy for mammography from this information? When the test results are binary, the observed TPF and FPF are calculated, but here there are five possible results for each image. The scores could be converted to binary by designating 4 as the threshold, then scores 1–3 as negative, and 4–5 as positive test results. Then estimate the TPF as tpf = 23/30 and the specificity (1-FPF) as (1-fpf) = 21/30. Another approach would be to use each test result as a threshold and calculate the tpf and fpf, which are depicted in Table 7.6.

Of the 30 diseased, 30 had at score of at least 1, whereas 23 had a score of at least 4. On the other hand, of the 30 without cancer, 30 had a score of at least 1, 8 had a score of at least 4, and so on.

TABLE 7.6

TPF versus FPF for Mammography

				Test Result	
Status	Normal 1	Benign 2	Probably Benign 3	Suspicious 4	Malignant 5
TPF	30/30 = 1.00	30/30 = 1.00	29/30 = .966	23/30 = .766	12/30 = .400
FPF	30/30 = 1.00	21/30 = .700	19/30 = .633	8/30 = .266	0/30 = 0.000

The area under the ROC gives the intrinsic accuracy of a diagnostic test and can be interpreted in several ways, see Zhou et. al (p. 28).[9]: the area under the ROC can be (1) the average sensitivity for all values of specificity, (2) the average specificity for all values of sensitivity, or (3) the probability that the diagnostic score of a diseased patient is more of an indication of disease than the score of a patient without the disease or condition. The problem is in determining the area under ROC. There are five points corresponding to the five threshold values. If the diagnostic score can be considered continuous (e.g., the coronary artery calcium score of the Shields Heart Study), then the curve through the points becomes more discernible and the area easier to determine.

In the case of discrete data, the area under the curve as determined by a linear interpolation of the points on the graph, including (0,0) and (1,1), has the following interpretation:

$$\text{AUC} = P(Y > X) + \left(\frac{1}{2}\right)P(Y = X) \tag{7.10}$$

See Pepe (p. 92),[7] where it is assumed that one patient is selected at random from the population of diseased patients, with a diagnostic score of Y, while another patient, with a score of X, is selected from the population of nondiseased patients. Note that the AUC depends on the parameters of the model. Let us return to the mammography example and estimate the area under the curve through a Bayesian method.

For the mammography example, the area is defined as

$$\text{AUC}(\theta, \varphi) = P(Y > X \mid \theta, \varphi) + \left(\frac{1}{2}\right)P(Y = X \mid \theta, \varphi) \tag{7.11}$$

where Y (= 1,2,3,4,5) is the diagnostic score for a person with breast cancer and X (= 1,2,3,4,5) is the score for a person without breast cancer. It can be shown that

$$\text{AUC}(\theta, \varphi) = \sum_{i=2}^{i=5} \sum_{j=1}^{j=i-1} \theta_i \phi_j + (1/2) \sum_{i=1}^{i=5} \theta_i \phi_i \tag{7.12}$$

It is assumed the Y and X are independent, given the parameters, and that $P(Y = i) = \theta_i$ and $P(X = j) = \varphi_j$, i, j = 1,2,3,4,5. AUC is a parameter that depends on θ and φ. Their posterior distributions are θ/data~Dir(2,1,7,12,13) and independent of φ/data~Dir(10,3,12,9,1), assuming a uniform prior for the parameters; see Table 7.6.

Samples from the posterior distribution of the AUC are generated by sampling from the posterior distributions of θ and φ. This is accomplished with WinBUGS, where 55,000 observations are generated from the posterior distribution of all the parameters, with a burn-in of 5000 and a refresh of 100. The code for the operation is given below as BUGS CODE 7.2 and the notes indicated by a # identify the important parts of the program. For example, the statements that follow the note "# generate Dirichlet distribution" generate the posterior distribution of the cell probabilities of Table 7.6. The first list statement is the information used to generate the gamma variables that generate the Dirichlet distribution of the cell probabilities. A one is added to each cell frequency of Table 7.6, which induces a uniform prior distribution for the cell probabilities.

BUGS CODE 7.2

```
# Area under the curve
# Ordinal values
# Five values

Model;
{

# generate Dirichlet distribution
g11~dgamma(a11,2)
g12~dgamma(a12,2)
g13~dgamma(a13,2)
g14~dgamma(a14,2)
g15~dgamma(a15,2)

g01~dgamma(a01,2)
g02~dgamma(a02,2)
g03~dgamma(a03,2)
g04~dgamma(a04,2)
g05~dgamma(a05,2)

g1<-g11+g12+g13+g14+g15

g0<-g01+g02+g03+g04+g05

# posterior distribution of probabilities for response of
diseased patients
```

```
theta1<-g11/g1
theta2<-g12/g1
theta3<-g13/g1
theta4<-g14/g1
theta5<-g15/g1

# posterior distribution for probabilities of response of
non-diseased patients
ph1<-g01/g0
ph2<-g02/g0
ph3<-g03/g0
ph4<-g04/g0
ph5<-g05/g0

# auc is area under ROC curve
# A1 is the P[Y>X]
# A2 is the P[Y = X]
auc<- A1+A2/2

A1<-theta2*ph1+theta3*(ph1+ph2)+theta4*(ph1+ph2+ph3)+
theta5*(ph1+ph2+ph3+ph4)
A2<- theta1*ph1+theta2*ph2+theta3*ph3+theta4*ph4
+theta5*ph5
}
# Mammography Example Zhou et al.[8]
# Uniform Prior
# a one is added to each cell frequency
list(a11 = 2,a12 = 1,a13 = 7,a14 = 12,a15 = 13,a01 = 10,a02 = 3,
  a03 = 12,
a04 = 9,a05 = 1)

# initial values
list(g11 = 1,g12 = 1,g13 = 1,g14 = 1,g15 = 1,g01 = 1,g02 = 1,
  g03 = 1,
g04 = 1,g05 = 1)
```

The posterior analysis is given in Table 7.7.

Notice that mammography gives fair to good accuracy based on the ROC area, which is estimated as .7811(.0514) with the posterior mean and by (.6702, .8709) using a 95% credible interval. The MCMC error for the parameter based on 50,000 observations is less than .001, but the reader should vary the simulation sample size to see its effect on the MCMC error and posterior mean. The parameter $A1$ is $P[Y>X]$ and estimated as .688(.06350) and the probability of a tie, $P[Y = X]$, given by $A2$, is estimated as .1861(.0307). The estimated area of .7811 is similar to that computed by Zhou et al (p. 30).[9]

Finally, the mammography example is concluded with a test for the usefulness of the procedure. Obviously, a perfect test has an ROC area of 1, and a useless test an area of .5. Thus, consider a Bayesian test of H: AUC <.5 versus

TABLE 7.7

Posterior Distribution of Area under ROC Curve: Mammography Example

Parameter	Mean	SD	Error	Lower 2½	Median	Upper 2½
AUC	.7811	.0514	<.0001	.6702	.7848	.8709
A1	.688	.0635	<.0001	.5564	.6909	.8036
A2	.1861	.0307	<.0001	.128	.1854	.2484

the alternative A: AUC ≥ .5. How is this performed with WinBUGS? Based on BUGS CODE 7.2, the statement `T<-step(AUC-.5)` will provide a test of the null hypothesis. The mean of T is the probability of the alternative hypothesis, and one can verify that

$$P[\text{AUC}(\theta, \varphi) \geq .5 \,|\, \text{data}] = .99999\,(.0044) \tag{7.13}$$

Therefore, the null hypothesis is rejected and one may conclude that mammography is a useful diagnostic procedure for detecting breast cancer.

7.3.6 UK Trial for Early Detection

The following examples will illustrate Bayesian inference techniques for estimating the accuracy for various modalities of screening tests. For example, Chamberlain et al.[4] provide the results for a trial for the early detection of breast cancer. Trial enrollment was between 1979 and 1981 and enrolled women aged 45–64 years living in eight locations in the United Kingdom. Annual screening by clinical examination of the breast, with mammography in alternative years was provided over 7 years for 45,841 women. An additional 63,636 were provided instruction in breast physical examination, and 127,117 served as a control and no additional services were available. Thus, there were three groups comprising the study: two study groups, one receiving both physical examination and mammography, one receiving instruction for self examination of the breast, and a control group receiving standard medical care.

Relative sensitivity and specificity are estimated from Table 7.8 based on results of the initial screening.

Each of 181 patients received one of three scores as to the status of the disease: (1) normal, (2) the lesion is localized and benign, and (3) the lesion is suspicious. Assuming that a normal rating signifies a negative test and that a localized benign or suspicious lesion indicates a positive status for breast cancer, Chamberlain et al.[4] report the relative sensitivity of mammography as (18+153)/181 = .945, whereas that for physical opinion as (35+950)/181 = .718. It should be noted that the denominator is the total number of screened detected cases.

Table 7.8 reduces to Table 7.9 when combining the localized benign and suspicious as one category.

There are 181 patients with cancer, and among those, the physical examination, erroneously classified 51 as negative, while 10 were erroneously classified as negative with mammography.

How would the Bayesian estimate the relative sensitivity and specificity of the two modalities? A uniform prior is assumed for the four cell probabilities of Table 7.9, and it is also assumed that 181 patients are selected at random from the relevant population of patients; thus, the four cell frequencies have a multinomial distribution and the four cell probabilities have a posterior distribution that is Dirichlet with parameter(3,9,50,123); thus, each cell probability has a beta distribution.

Referring to Table 7.9, then executing BUGS CODE 7.1 with 65,000 observations for the simulation, a burn-in of 5000, and a refresh of 100, the posterior distribution of the TPF (sensitivity) for mammography and physical opinion are displayed in Table 7.10.

It is seen that the TPF of mammography is estimated as .9325 with the posterior mean, and for physical opinion, the posterior mean is .7136, and both estimates agree with the usual estimates of .945 and .718, respectively, as reported by Chamberlain et al (p. 4).[4] The MCMC errors are

TABLE 7.8

Mammography and Physical Examination, UK Detection of Breast Cancer, at the Initial Screening

		Physical Examination		Total
Mammography	Normal	Localized Benign	Suspicious	
Normal	2	5	3	10
Localized benign	10	6	2	18
Suspicious	39	24	90	153
Total	51	35	95	181

Source: Miller et al., *Cancer Screening*, Cambridge University Press, New York, 1991.

TABLE 7.9

Mammography and Physical Examination: UK Detection of Breast Cancer

	Physical Opinion		Total
Mammography	Negative	Positive	
Negative	2	8	10
Positive	49	122	171
Total	51	130	181

TABLE 7.10

Posterior Distribution of the True Positive Fraction for Mammography and Physical Opinion: Initial Screening

Parameter	Mean	SD	Error	2½	Median	97½
θ_{00}	.01623	.00930	<.00001	.00335	.01448	.38825
θ_{01}	.04861	.01586	<.00001	.02228	.04703	.08391
θ_{10}	.2702	.0325	<.0001	.2087	.2695	.3361
θ_{11}	.665	.0345	<.0001	.5956	.6656	.7308
TPF mammography	.9325	.0181	<.00001	.8956	.9367	.9659
TPF physical examination	.7136	.03306	<.0001	.6467	.7144	.7758

quite small implying the 65,000 observations generated for the simulation is adequate. The Bayesian analysis implies that mammography is more accurate than physical examination, which is not surprising, and agrees with other studies that compare the two modalities. Remember that the 181 cases have been verified to be cancer patients. Also displayed in Table 7.10 is the posterior distribution of the four cell probabilities, and for example, the posterior mean of the cell where both tests are positive is .665 implying that approximately 66.5% of the total of 181 patients test positive for breast cancer by both modalities. A Bayesian analysis is continued for the UK Trial of Early Detection by considering the results of screening at round 1, displayed below in Figure 7.3. Round 1 consists of a screening by both mammography and clinical examination of the breast. In particular, the specificity, PPV, NPV, positive diagnostic likelihood ratio, and the negative diagnostic likelihood ratio will be estimated. See also Table 7.11.

Using the information in Table 7.12 and executing BUGS CODE 7.1 with 55,000 observations, a burn-in of 5000, and a refresh of 100, the Bayesian analysis is revealed in Table 7.12.

The FPF is estimated as .0612, which implies a specificity of .94, while sensitivity is estimated as .9283, which together implies that screening is accurate; however, the predictive values must also be assessed.

The PPV is

$$P[D = 1 \mid X = 1] = \frac{\theta_{11}}{(\theta_{01} + \theta_{11})} \tag{7.14}$$

and is estimated as .0845 by the posterior mean and by (.0731, .0967) for the 95% credible interval. This implies that when the test is positive, the chance of actually having the disease is only .0845! This is because the prevalence of disease is only 195/32,205 = .00605. On the other hand, the NPV

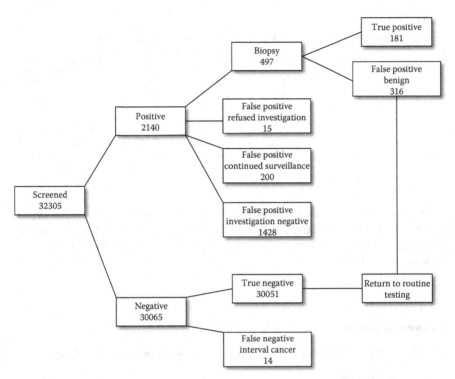

FIGURE 7.3
Results of screening for round 1. UK Early Detection Trial.

TABLE 7.11

UK Early Detection Trial: Mammography and Physical Examination at Round 1 Screening

Mammography and Physical Examination	Breast Cancer	
	$D = 0$ (negative)	$D = 1$ (positive)
Negative $X = 0$	30,051	14
Positive $X = 1$	1959	181

$$P[D = 0 \mid X = 0] = \frac{\theta_{00}}{(\theta_{00} + \theta_{01})} \tag{7.15}$$

is estimated as .9995 with the posterior mean and (.9993, .9997) with the 95% credible interval, implying that for those patients that test negative through mammography, 99.95% will not have breast cancer.

Therefore in summary, the PPV does not give one confidence in screening for those that test positive; however, for those that test negative, one has

TABLE 7.12

Bayesian Analysis for UK Early Detection Trial: Screening at Round 1

Parameter	Mean	SD	Error	2½	Median	97½
FPF	.0612	.00134	<.000001	.0586	.0611	.0638
NDLR	.07641	.0197	<.00001	.0427	.0747	.1192
NPV	.9995	.0001244	<.0000001	.9993	.9995	.9997
PDLR	15.18	.4499	.0020	14.28	15.18	16.04
PPV	.0845	.0060	<.00001	.0731	.08446	.0967
TPF	.9283	.0185	<.00001	.8881	.9298	.9601

confidence the patient does not have disease. Thus of the four measures of test accuracy, the PPV implies that the screening tests are not accurate in the diagnosis of breast cancer. For more information about the UK trial, see Chamberlain et al.[10]

7.4 HIP Study (Health Insurance Plan of Greater New York)

7.4.1 Introduction

Screening for chronic disease gained interest in the 1960s as a result of clinical experience that showed that disease detected at an earlier stage had better prognosis than disease detected at a later stage. Thus, interest in screening is based on the hypothesis that it would shift the diagnosis to an earlier stage and treatment would have a better chance to make an impact on the development of diseases such as breast cancer. According to Shapiro[2] and Ventet, Strax, and Venet,[2] reports began to appear in the early 1960s from many periodic examination programs on the detection of breast cancer by palpation. For example, Holleb, Venet, Day and Hoyt[11] and Gilbertsen and Krelsberg[12] Venet[13]uniformly indicated that a larger fraction of patients were diagnosed with a localized disease than that of the general population of patients. Of course there was considerable debate about and doubt about the impact examinations could have toward the reduction of deaths (due to breast cancer) in the more general population. It was difficult to generalize the results of these early reports because they were not designed randomized studies, but instead were based on patients who volunteered for the examinations. The selection factors associated with these groups made a meaningful comparison difficult to perform.

In the early 1960s, advances in mammography played an important role in the emergence of screening for breast cancer. It is interesting to note that just after the discovery of x-ray by Roentgen, Saloman[14] reported the use of x-ray in order to examine the breast, and he indeed recognized the potential of the modality in visualizing mass densities, mass irregularities, and micro-calcifications. Not much progress was made until after WWI when Warren[15] explored the application of radiography with emphasis on the potential in helping the clinician diagnose the disease in asymptomatic women. It was Egan[16] at the MD Anderson Hospital and Tumor Institute whose acceptance of mammography as a valuable device for the diagnosis of breast cancer had an important impact on screening for the disease. He reported mammography studies involving 2000 patients with the objective of preserving the maximum detail in the image with the lowest kilovolts for penetration, compensation for increased exposure of the radiation, and proper focus of the x-ray beam. He also emphasized careful attention to the technical details in production a good image of the breast. See Shapiro et al (pp. 710).[2] for additional information about the technical advancement in mammography and the early hospital investiga-tions, including the 15-hospital study conducted by MS Anderson Hospital and Tumor Institute in the early 1960s. As reported by Clark, Copeland, Egan, et al.,[17] the latter study showed that mammography was highly replicated among radiologists following standardized training. For an interesting his-tory of the use of x-ray for diseases of the breast, see Strax.[18]

Between 1999 and 2008, the rate of women dying from breast cancer has varied, depending on their race and ethnicity. Figure 7.4 shows that in 2008,

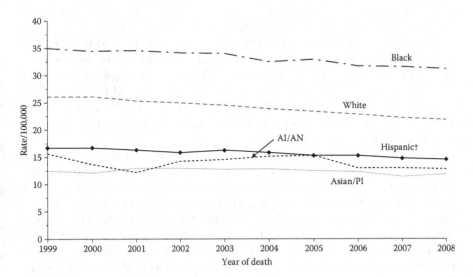

FIGURE 7.4
Female breast cancer death rates* by race and ethnicity, U.S., 1999–2008. (From U.S. Mortality Files, National Center for Health Statistics, CDC. See www.cdc.gov/cancer/breast/statistics/race.htm.)

---

* Rates are per 100,000 persons and are age-adjusted to the 2000 U.S. standard population (19 age groups—Census P25-1130). Death rates cover 100% of the U.S. population.

black women were more likely to die of breast cancer than any other group. White women had the second highest rate of deaths from breast cancer, followed by women who are Hispanic, American Indian/Alaska Native, and Asian/Pacific Islander. This illustrates quite well the impact of breast cancer. For example, for black women the death rate in 1999 was approximately 35 per 100,000, which implies that there were approximately 10,500 deaths for that year. I assumed the number of black females in 1999 was 20% of 150,000,000 females = 30,000,000. Now 30,000,000/100,000 = 300, and 300 times 35 = 10,500 approximate black female deaths in 1999. I assume the number of females in the United States in 1999 was approximately 150,000,000 and that of those 20% were black (see Figure 7.4).

The developments at MD Anderson proved very impressive with the result that the National Cancer Institute initiated a long-term study of the value of mammography in reducing breast cancer mortality, which resulted in the first randomized study involving the HIP. HIP was a good choice to execute a screening study because it included an experienced research group, where many projects were financed by NIH and various private foundations. HIP was a prepaid comprehensive medical care plan with 31 affiliated medical groups located in New York City and Long Island. About 700,000 members were group enrollments among city, state, and federal government employees. Other research projects showed that the membership constituted a wide range of socioeconomic, ethnic, and religious groups in New York City; however, there were some areas that included members with very high income. HIP also had other important assets including an electronic information system that included (1) a file with each member's ID number, name, sex month and year of birth, size of covered family, medical group membership, source of enrollment, date of enrollment, and date of termination from enrollment and (2) a reporting system of all services provided by physicians in each medical group, from which diagnostic and therapeutic services received by each patients were easy to determine.

On the basis of the incidence rates of breast cancer, during 1958–1961, it was decided to enroll women over the range from 40 to 64 years of age. Total enrollment in the 23 medical groups was about 490,000 among which 80,300 were aged 40 to 64. Within each of the medical groups, two systematic random samples were selected where every nth women was placed in the study group, and the $(n + 1)$st in the control group, resulting in about 31,000 in each. The scheduled date for the initial exam became the entry date to the study, where each control group woman was assigned the same date as the corresponding study group woman. The HIP was the first randomized study to determine the efficacy of mammography and clinical examination for screening of breast cancer. The main question was: Does screening for disease increase the survival for those that were screened compared to a comparable control group of women?

Women entered the study beginning December 1963 through June 1966 with approximately 31,000 assigned to each group. For those in the study group, about 20,220 or 61% appeared for the initial examination and a high proportion of those participated in the following three annual examinations (mammography, clinical examination, or both). Of the 20,200, close to 80% participated in the first annual examination, 73% in the second, and 69% in the third.

It is assumed the progression of disease follows three states:

$$S_0 \rightarrow S_p \rightarrow S_c$$

S_0 is the disease free state, where the subject is either disease-free or the disease is in such an early phase, it cannot be detected by mammography, while the preclinical phase is denoted by S_p when the subject is asymptomatic but breast cancer can be detected. Finally S_c is the clinical state, where the symptoms of breast cancer become apparent. It should be noted that that time in the disease-free state, and in the preclinical phase are random, as is the time from the disease-free phase to the preclinical phase, and from the preclinical to the clinical phase.

In order to illustrate the statistical procedures for this screening study, first the descriptive statistics will be provided, followed by a Bayesian method to estimate the lead time (the time from when breast cancer was detected, via screening with clinical exam and mammography, and the time the disease would have been detected in the absence of screening), and finally a comparison of the mortality between the two groups using a Bayesian life table method explained in Chapter 5. The HIP data set was provided by the National Cancer Institute, Division of Cancer Prevention, and Diane Erwin kindly provided the technical expertise needed to download the dataset and associated documentation.

7.4.2 Descriptive Statistics

The dataset consisted of three cohorts: (1) 20,177 of the study group that participated in screening, (2) 9984 of the study group that refused screening, and (3) 30,565 in the control group. Those who participated in screening of the control group had at least one examination (consisting of clinical examination [palpitation of the breast] and mammography), where there were four examinations: the initial, followed by three annual examinations. Of those patients participating in screening, breast cancer could be diagnosed by screening or by finding the disease between examinations or after all screening was completed. Note that after the fourth examination (the third annual exam, given in 1966) screening was not provided. For those patients in the control group and those in the study group not participating in screening, usual medical care was provided by the health plan.

The following variables were measured on patients:

1. Date of entry
2. Date of last follow-up
3. Subscriber status (female or spouse)
4. Study cohort (study group, group that refused, and control group)
5. Age at entry to study
6. Age at initial exam
7. Age at first annual exam
8. Age at second annual exam
9. Age at third annual exam
10. Detection classification (clinical only, radiology only, mass on image, radiology only, microcalcifications, radiology, mass and microcalcifications, clinical plus radiology both mass, clinical plus radiology mass and radiology microcalcification, interval diagnosis less than 12 months after last screening, interval diagnosis at least 12 months since last screening, study refused screening, and control)
11. Survival status (living, deceased, and lost to follow-up)
12. Date of diagnosis
13. Age at diagnosis
14. Age at death or last follow-up if alive
15. Interval in months from entry to diagnosis
16. Interval from entry to death or last follow-up
17. Interval from diagnosis to death or last follow-up in months
18. Size of lesion for hospital report (various categories)
19. Extent of disease (stage I, II, III, IV, metastasis to lymph nodes, unknown)
20. Race (white, black, other, and unknown)
21. Age at death
22. Interval from entry to death in months
23. Interval from entry to diagnosis in days
24. Interval form diagnosis to death or last follow-up in days
25. Cancer flag (breast cancer, interval from entry to diagnosis less than 15 years, other, status unknown)
26. Death flag (dead, interval from entry to death < 15 years, dead, interval from entry to death > 15 years, other, status unknown)
27. ID number of subject

Of primary interest is the frequency of age at entry that is depicted in Table 7.13 where 60,696 patients are partitioned by age over the range from 40 to 64 years of age.

The corresponding histogram is portrayed in Figure 7.5, where the least frequent age is 40 with 1087 patients and the most frequent age is 43 with 3172.

The distribution of age at entry is similar among the three cohorts, which is deduced from the box plots in Figure 7.6. Note the distribution of age at entry is slightly skewed for the study group, but appears symmetric for the refused and control groups. Also it appears that the inter-quartile range is about the same for the three cohorts and that the minimum (40 years) and maximum (age 64) are the same for the study, refused, and control groups. By using randomization

TABLE 7.13

Frequency at Age of Entry

Age	Frequency	Percent	Valid Percent	Cumulative Percent
40	1,087	1.8	1.8	1.8
41	2,620	4.3	4.3	6.1
42	3,084	5.1	5.1	11.2
43	3,172	5.2	5.2	16.4
44	3,023	5.0	5.0	21.4
45	3,049	5.0	5.0	26.4
46	2,907	4.8	4.8	31.2
47	2,918	4.8	4.8	36.0
48	2,761	4.5	4.5	40.6
49	2,860	4.7	4.7	45.3
50	2,830	4.7	4.7	49.9
51	2,834	4.7	4.7	54.6
52	2,756	4.5	4.5	59.1
53	2,660	4.4	4.4	63.5
54	2,785	4.6	4.6	68.1
55	2,651	4.4	4.4	72.5
56	2,656	4.4	4.4	76.9
57	2,372	3.9	3.9	80.8
58	2,232	3.7	3.7	84.4
59	1,945	3.2	3.2	87.7
60	1,785	2.9	2.9	90.6
61	1,652	2.7	2.7	93.3
62	1,557	2.6	2.6	95.9
63	1,312	2.2	2.2	98.0
64	1,188	2.0	2.0	100.0
Total	60,696	100.0	100.0	

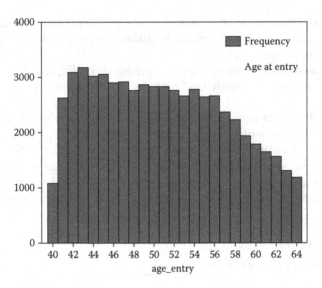

FIGURE 7.5
Age at entry.

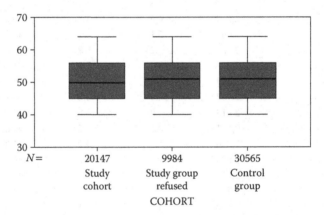

FIGURE 7.6
Box plot by cohort.

to assign patients, one would expect the distribution of age to be similar among the control and study groups, and surprisingly for those patients of the study group who refused screening, the age at entry has the same or almost the same distribution as the other two groups. See Figure 7.6, where the median (51) age at entry for the refused group is the same as the control group and is somewhat less (by a small amount) than that for the study group.

Next, the age distribution is determined for the study group for age at entry, age at the initial examination, and ages at the first, second, and third annual examinations (Table 7.14).

One sees that the average age at entry was 50.51 years with a median of 50 years, while the average age of a patient in the study group at the initial exam was 50.56 followed by an average of 53.93 at the third annual examination. There is a decline of 6244 patients from 20,147 at the initial exam to 13,903 at the third or final annual examination. From age at entry to age at the third examination, the range is, on the average, 3.42 years. After the third examination screening (clinical examination and mammography) ceased, thus, its effect will dissipate after the third examination. One sees that the dispersion as measured by the standard deviation is remarkably consistent over the 3.42 years. The manner by which a patient can participate in screening can be quite complicated. For example, one can participate in the initial examination, but not the initial annual, but can again participate in the final examination. There are too many scenarios to enumerate, but one can imagine the possibilities.

Some idea of the extent of disease diagnosed for the three cohorts is given by Table 7.15. The time to diagnosis is assumed to be less than or equal to 60 months (5 years) to see the effect of screening in the study group.

The percent diagnosed with stage I breast cancer is 62% for the study group compared to 45.5% for the control group implying that screening is detecting an earlier stage of the disease. It should be stressed that the

TABLE 7.14

Age Distribution at Entry and Four Screening Examinations

Exam	Mean	SD	Median	N
Entry	50.51	6.411	50	60,696
Initial	50.56	6.44	50	20,147
First	51.67	6.44	51	15,932
Second	52.79	6.43	52	14,763
Third	53.93	6.43	53	13,903

TABLE 7.15

Extent of Disease by Cohort: Time to Diagnosis is ≤5 years

	Frequency and (Percent)		
Stage	Study Group	Refused	Control Group
I	142(62)	32(39.5)	135(45.5)
II	65 (28.7)	27(33.3)	106(35.7)
III	10(4.4)	8(9.9)	34(11.4)
IV	3(1.3)	10(12.3)	7(2.4)
I or II with no lymph nodes excised	8(3.5)	4(4.9)	14(4.7)
Unknown	0(0)	0(0)	1(.1)

extent of disease can vary by changing the time to diagnosis from ≤5 years to some other range.

7.4.3 Estimating the Lead Time

What is the lead time? Consider the following progression of breast cancer, where S_0 is the disease free phase; S_p is the preclinical phase, where mammography and clinical exam can detect the disease; T_d is the time where the disease is in fact detected by screening; and S_c is the clinical phase, where breast cancer becomes apparent (in the absence of screening) to the patient.

$$S_0 \to S_p \to T_d \to S_c$$

The lead time is $S_c - T_d$, the time by which the diagnosis is advanced by screening. Note the beginning of the preclinical phase S_p and the beginning of the clinical phase S_c are random, as is the lead time $S_c - T_d$.

Our approach to estimating the lead time is called the method of differences and is based on the mean time of entry to diagnosis for the study group minus the mean time from entry to diagnosis for the control group. This somewhat presents a dilemma in that one must determine how to take into account the effect of screening (clinical examination and mammography) of the study group. From the time of entry to the last annual examination is about 4 years, where screening begins to have an effect at the initial examination and the last time screening can be effective are for those diagnosed by the last examination. When the time of entry to diagnosis is taken as no more than 5 years, the situation for the study group is depicted by Figure 7.7.

In order to estimate the lead time for the HIP study, consider the diagnostic profile of the study group of Figure 7.7. I have assumed the time from entry to the study to the time of diagnosis is less than or equal to 60 months. Note that of the 229 study cohort patients, 59 were diagnosed with breast cancer by clinical (breast examination), whereas 44 were diagnosed with mammography. Of the 44, 27 were diagnosed by observing a lesion (mass) and 17 by observing microcalcifications on the image. To continue, 29 were diagnosed with breast cancer by both mammography and breast examination, where 26 were observed with a tumor mass and 3 with both a mass and microcalcifications. Lastly, 97 were interval (diagnosed between screening examinations) diagnoses, and of those 45 were diagnosed within 12 months of the last examination, and 52 identified after 12 months from the last examination. In order to estimate the lead time, I will ignore interval diagnosis (breast cancer diagnosed between examinations or after the patient's last examination). This last restriction is somewhat controversial because a patient's participation make the patients more aware of the clinical symptoms of the disease; thus, an interval diagnosis can be the effect of screening. For the control group, the effect to screening is taken to be nil and the time to diagnosis from time to entry will also be assumed to be no more than 60 months.

Descriptive statistics for estimating the lead time are portrayed in Table 7.16. One sees that the median time from entry to diagnosis is 436 days with a mean of 555 days and the corresponding entries for the control group are 972 days and 945 days, respectively. On the basis of the sample median, the lead time is estimated as $972 - 436 = 536$ days or 1.468 years. On the basis of the sample mean, the lead time is estimated as $945 - 555 = 390$ days or 1.06 years. Curiously, this estimate of 1.06 years for the lead time is quite similar to that reported by Wu, Kafadar, Rosner, and Broemeling.[19] For the study group, the lead time is skewed to the right, but for the control, there is a very small left skewness. I would advise using the medians to estimate the lead time.

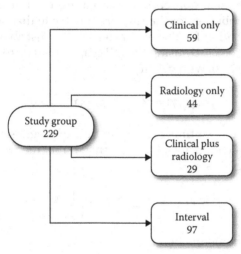

FIGURE 7.7
Diagnostic profile of the study group.

TABLE 7.16

Lead Time: Time from Entry to Diagnosis by Cohort in Days

	Time to diagnosis is ≤ 60 months			
Cohort	Mean	SD	N	Median
Study	555	473	132	436
Control	945	558	303	972

TABLE 7.17

Lead Time: Time from Entry to Diagnosis by Cohort in Days

	Time to diagnosis is ≤48 months			
Cohort	Mean	SD	N	Median
Study	548	467	131	433
Control	726	445	232	680

The effect of screening on the lead time can be demonstrated by assuming that the time from entry to diagnosis is 4 years, which is depicted in Table 7.17.

Now, with the sample median, the lead time is estimated as 680–433 = 247 days or .676 years and with the sample mean as 726–548 = 178 days or .4876 years. By assuming a shorter time to diagnosis, the lead time estimate is reduced.

What is the Bayesian approach to estimating the lead time? One must assume some distribution for the time from entry to diagnosis and assume some prior distribution for the parameters of the distribution for the time from entry to diagnosis. By using the P-P graph, it is safe to assume that this distribution is Weibull, with density

$$f(x \mid v, \gamma) = v \lambda x^{v-1} \exp(-\lambda x^v), \, x > 0 \tag{7.16}$$

where x is the time from time of entry to time of diagnosis and v and λ are positive unknown parameters. It can be shown that the mean of X, the time from entry to diagnosis, is

$$E(X \mid v, \lambda) = \left(\frac{1}{\lambda}\right)^{1/v} \Gamma\left(\frac{1+1}{v}\right) \tag{7.17}$$

where Γ is the gamma function.

The Bayesian analysis will consist of determining the posterior distribution of v and λ and thus of $E(X \mid v, \lambda)$ for the study group and for the control group, then finding the posterior distribution of

$$\text{diff} = E(X \mid v_2, \lambda_2) - E(X \mid v_1, \lambda_1) \tag{7.18}$$

where v_1, λ_1 are the parameters of the Weibull distribution (Equation 7.16) for the study group and v_2, λ_2 for the control group. Estimation of the lead time is based on the posterior distribution of diff (Formula 7.18).

BUGS CODE 7.3 will execute the Bayesian analysis and the code uses similar notation to that of Formulas 7.16 through 7.18. Note that the remarks indicated by # explain the essential parts of the program.

BUGS CODE 7.3

```
model;
# Weibull distribution for time of entry to diagnosis
# study group
{for(i in 1:n1){x1[i]~dweib(v1,lamda1)}

v1~dgamma(.001,.001)
lamda1~dgamma(.001,.001)
lme1<- (1/v1)*log(lamda1)+loggam(1+1/v1)
# me1 is the mean of time from entry to diagnosis for the
  study group
me1<-exp(lme1)
# Weibull distribution for time to diagnosis
# control group

for(i in 1:n2){x2[i]~dweib(v2,lamda2)}

v2~dgamma(.001,.001)
lamda2~dgamma(.001,.001)
lme2<- (1/v2)*log(lamda2)+loggam(1+1/v2)
# me2 is the mean of time from entry to diagnosis for the
  control group

me2<-exp(lme2)
# diff is the estimate of the lead time
diff<-me2-me1
}
# study and control groups HIP
# x1 is the vector of times from entry to diagnosis for the
  study group
# x2 is the vector of times from entry to diagnosis for the
control group
list(n1 = 132, x1 = c(81,118,428,1265,530,468,1200,496,127,29,
  599,48,751,1197,1375,1343,1279,575,37,595,406,27,412,
  126,75,568,993,1277,192,123,12,236,1358,79,384,64,439,
  47,29,845,743,1473,414,152,13,1315,549,1320,1244,1191,
  1063,448,67,392,23,959,34,739,18,808,1341,1101,345,378,
  875,1398,1177,53,34,1383,343,21,14,386,927,439,544,
  600,347,88,987,495,792,160,16,1274,750,420,147,38,730,
  110,1051,81,1335,939,871,183,490,79,404,79,404,971,123,
  1260,298,59,182,964,1520,1308,1097,16,85,300,517,22,
  1456,69,16,881,73,41,1265,1322,358,673,470,217,561,
  409,433,29), n2 = 303,
x2 = c(1468,1779,601,29,1617,1227,1470,141,751,1297,1109,1562,
  1285,345,1429,972,345,1853,1554,490,1478,1799,1791,614,
  118,50,1446,1353,510,1466,1177,146,1623,1715,70,1098,
  1352,1470,728,534,193,1446,1818,1551,14,1474,197,1037,
  1728,230,696,440,491,576,1771,1539,1025,1616,575,330,604,
  1576,528,1118,1038,1527,77,1413,1240,358,110,349,214,1096,
  1495,1143,1543,729,412,1674,105,16,684,434,7,330,1707,1433,
```

```
87,1028,1237,1729,274,682,679,703,755,505,966,1595,209,1273,
1361,495,1368,258,800,1519,532,878,1333,802,390,1482,647,
1730,1537,609,622,1472,1041,1343,1754,1505,1765,600,1632,
16,1149,415,309,1745,1797,164,68,1734,919,619,1185,1044,
1335,546,93,832,1681,1742,168,989,550,1295,99,366,1806,559,
756,1734,822,139,724,446,511,1553,1304,388,1328,1467,1526,
810,1616,252,392,1684,331,1820,1746,789,22,1562,1525,53,1507,
426,1125,719,255,1098,1567,1261,981,44,1568,1590,1810,601,
973,1539,549,1237,274,1309,1513,520,1478,1778,572,985,1665,797,
513,568,1590,1631,912,1815,185,764,1320,1633,1619,1407,779,
313,1322,1623,896,1734,151,1427,463,511,296,1362,1131,1175,
807,1154,79,415,921,659,588,1727,163,198,1829,417,839,1306,
1192,1747,778,670,1066,1630,1098,1270,133,1119,423,912,379,
734,155,1732,1460,287,1812,555,41,1081,811,1155,1559,1147,
1635,1766,722,1011,1189,1534,558,236,1301,1405,266,932,407,
428,1216,208,657,518,379,864,1051,1037,510,77,1105,319,
83,169,464))
# initial values for the Weibull distribution
list(v1 = 1,lamda1 = 1,v2 = 1,lamda2 = 1)
```

The simulation is based on 55,000 observations with a burn-in of 1000 and a refresh of 100, and the posterior analysis is portrayed in Table 7.18.

One sees that the posterior mean of the average time from entry to diagnosis is 526.6 days for the study group and 937.1 days for the control group. The main parameter of interest is the lead time diff with a posterior mean of 374.5 days with a 95% credible interval of 247.6 days to 490.3 days. Thus, based on the posterior mean, the estimated lead time is 374.5 days or 1.026 years, an estimate that approximately agrees with Wu, Kafadar, Rosner, and Broemeling;[19] however, it should be noted that the latter employed a model-based approach to estimate the lead time, an approach that will be described in Chapter 8. The estimate of 1.026 years is almost the same as that reported in Table 7.16, an estimated lead time based on descriptive statistics.

TABLE 7.18

Posterior Distribution of the Lead Time

Parameter	Mean	SD	Error	2½	Median	97½
diff	374.5	61.88	.4252	247.6	376	490.3
lamda1	.002398	.001185	<.00001	.000815	.00216	.00533
lamda2	.0000201	.0000128	<.0000001	.0000052	.0000170	.00005
me1	526.6	51.16	.1754	471	559.7	671.7
me2	937.1	34.85	.388	870.6	936.6	1008
v1	.9735	.0707	.0022	.8397	.9719	1.115
v2	1.581	.0815	.0038	1.421	1.58	1.742

7.4.4 Estimating and Comparing Survival

There are three components of a screening trial that need to be explained: (1) estimation of the lead time, (2) comparing mortality between the study group and the control cohort, and (3) estimating the accuracy of the modality (in this case, clinical examination of the breast and mammography).

7.4.4.1 Life Tables

Recall Chapter 5, where the Bayesian approach to life tables is introduced. First to be developed is the case where the number of deaths and the number of censored observations are known for each time period. In the tables to follow, a period is 6 months of which there are 40; thus, there are 20 years of data for patients who are diagnosed with breast cancer (see Table 7.19).

TABLE 7.19

Survival and Censored Observations for Study Cohort of HIP Study

Interval t	People Living at Beginning of Interval $O[t]$	Last Report Dead $m[t]$	Last Report Living $w[t]$
1	304	11	0
2	293	8	0
3	285	12	0
4	273	11	0
5	262	5	0
6	257	9	0
7	248	8	0
8	240	3	0
9	237	5	0
10	232	7	0
11	225	7	0
12	218	7	0
13	211	5	0
14	206	2	0
15	204	12	0
16	192	3	0
17	189	6	0
18	183	2	0
19	181	4	0
20	177	10	0
21	167	1	0
22	166	7	0
23	159	3	0
24	156	2	0

(Continued)

TABLE 7.19 (Continued)

Survival and Censored Observations for Study Cohort of HIP Study

Interval t	People Living at Beginning of Interval O[t]	Last Report Dead m[t]	Last Report Living w[t]
25	154	2	0
26	152	3	0
27	149	0	4
28	145	2	5
29	138	3	6
30	129	1	12
31	116	3	17
32	96	3	16
33	77	2	11
34	64	2	13
35	49	1	14
36	34	1	7
37	26	0	12
38	15	0	9
39	51	0	4
40	1	0	1

Source: Shapiro, S., Venet, W., Strax, P., and Venet, L., *Periodic Screening for Breast Cancer, The Health Insurance Project and Its Sequelae, 1963–1986*, table 1, appendix, page 154, Johns Hopkins University Press, Baltimore, 1988.

Let $O[t]$ be the number of patients alive at the beginning of period t, $m[t]$ the number that die during that period, and $w[t]$ the number who were alive at last contact during that period. The latter are lost to follow-up. Suppose the distribution of the number that die in the interval t is binomial with parameters $q[t]$ and $O[t]-w[t]$, where $q[t]$ is the probability that a patient will die in the interval t. Assume that the prior distribution of $q[t]$ is beta(1,1), that is, is uniform, namely

$$m[t] \sim \text{binomial}\big(q[t], O[t]-w[t]\big) \qquad (7.19)$$

for $t = 2,...,40$.

Suppose that the number alive at time t, $O[t] + L[t-1]$, follows a binomial distribution with parameters sr[t] and $O[t]$, where sr[t] is the survival rate at time t and $L[t-1]$ is the cumulative number alive at last contact immediately before time t. Thus

$$O[t]+L[t-1] \sim \text{dbinom}\big(\text{sr}[t], O[t]\big) \qquad (7.20)$$

Table 7.20 portrays the survival experience of 304 subjects of the study group who were diagnosed with breast cancer over a 20-year period, where each period represents 6 months. The first column gives the interval number, the second gives the number $O[t]$ who are alive at the beginning of the interval, the third gives the number $m[t]$ who die in the tth interval, and $w[t]$ gives the number of subjects in the tth interval who were alive when contact was lost with the subject. A similar interpretation follows for the 295 patients of the control group depicted in Table 7.20. Note the 304 subjects of the study group included those that refused screening.

TABLE 7.20

Survival and Censored Observations for Control Cohort of HIP Study

Period t	People Living at Beginning of Interval $O[t]$	Last Report Dead $m[t]$	Last Report Living $w[t]$
1	295	8	0
2	287	14	0
3	373	21	0
4	252	18	0
5	234	14	0
6	220	9	0
7	211	14	0
8	197	8	0
9	189	9	0
10	180	4	0
11	176	3	0
12	173	4	0
13	169	6	0
14	163	4	0
15	159	4	0
16	155	5	0
17	150	5	0
18	145	4	0
19	141	3	0
20	138	1	0
21	137	1	0
22	136	2	0
23	134	6	0
24	128	1	0
25	127	2	0
26	125	2	0
27	123	2	2

(Continued)

TABLE 7.20 (*Continued*)

Survival and Censored Observations for Control Cohort of HIP Study

Period *t*	People Living at Beginning of Interval *O[t]*	Last Report Dead *m[t]*	Last Report Living *w[t]*
28	119	2	5
29	112	0	9
30	103	3	16
31	84	2	13
32	69	1	5
33	63	3	6
34	54	0	9
35	45	0	8
36	37	0	14
37	23	0	12
38	11	0	7
39	4	0	3
40	1	0	1

Source: Shapiro, S., Venet, W., Strax, P., and Venet, L, *Periodic Screening for Breast Cancer, The Health Insurance Project and its Sequelae, 1963–1986,* table 11, appendix, page 164, Johns Hopkins University Press, Baltimore, 1988.

The code below is similar to BUGS CODE 5.4 of Chapter 5 and will be employed to execute the Bayesian analysis that computes the posterior mean and standard deviation of the probability of survival.

BUGS CODE 7.4

```
Model;
{
# group 1
for(i in 1:a){m1[i]~dbin(q1[i], Op1[i])}
# prior distribution of mortality
for(i in 1:a){q1[i]~dbeta(1,1)}

for(i in 1:a){Op1[i]<-O1[i]-w1[i]}
# below is applied when only half the withdrawals are used
# for(i in 1:a){Op1[i]<-O1[i]-w1[i]/2}
# survival probabilities
for(i in 2:a){L1[i]<-L1[i-1]+w1[i]}
L1[1]<-w1[1]
# sr1 survival rate for group 1
for(i in 1:a){s1[i]~dbin(sr1[i], O1[1])}
# prior distribution for survival rate
for(i in 1:a){sr1[i]~dbeta(1,1)}
for(i in 2:a){s1[i]<- O1[i]+L1[i-1]}
```

```
# group 2
for(i in 1:a){m2[i]~dbin(q2[i], Op2[i])}
for(i in 1:a){q2[i]~dbeta(1,1)}
for(i in 1:a){Op2[i]<-O2[i]-w2[i]}
# below is applied when only half the withdrawals are used
# for(i in 1:a){Op2[i]<-O2[i]-w2[i]/2}
# survival probabilities
for(i in 2:a){L2[i]<-L2[i-1]+w2[i]}
L2[1]<-w2[1]
# sr2 survival rate for group 2
for(i in 1:a){s2[i]~dbin(sr2[i], O2[1])}
for(i in 1:a){sr2[i]~dbeta(1,1)}
for(i in 2:a){s2[i]<- O2[i]+L2[i-1]}
}
# data for HIP study and control cohorts
# group 1 is study cohort
# group 2 is control
list(a = 40, m1 = c(11,8,12,11,5,9,8,3,5,7,7,7,5,2,12,3,6,2,4,
  10,1,7,3,2,
2,3,0,2,3,1,3,3,2,2,1,1,0,0,0,0), O1 =
  c(304,293,285,273,262,257,248,240,237,232,225,218,211,206,
204,192,189,183,181,177,167,166,159,156,154,152,149,145,138,
  129,116,96,77,64,
49,34,26,14,5,1),w1 = c(0,0,0,0,0,0,0,0,0,0,0,0,0,0,0,0,0,0,0,0,0,
  0,0,0,0,0,0,0,4,5,6,
12,17,16,11,13,14,7,12,9,4,1), m2 = c(8,14,21,18,14,9,14,8,9,4
  ,3,4,6,4,4,5,5,4,
3,1,1,2,6,1,2,2,2,2,0,3,2,1,3,.1,0,0,0,0,0,0),
O2 =
c(295,287,273,252,234,220,211,197,189,180,176,173,169,163,159,
  155,150,145,
141,138,137,136,134,128,127,125,123,119,112,103,84,69,63,54,45
  ,37,23,11,4,1),
w2 = c(0,0,0,0,0,0,0,0,0,0,0,0,0,0,0,0,0,0,0,0,0,0,0,0,0,0,0,2,5
  ,9,16,13,5,6,9,8,14,12,
7,3,1))
# data for HIP screened cohort and control cohort
# group 1 is screened cohort
# group 2 is control
list(a = 40, m1 =
  c(3,6,8,6,4,6,5,3,5,6,7,6,2,1,9,2,6,1,3,8,0,3,3,2,2,3,0,2,2,
  1,2,1,2,2,1,0,0,
0,0,0), O1 = c(225,222,216,208,202,198,192,187,184,179,173,166
  ,160,158,157,148,146,140,139,136,128,128,125,122,120,118,115
  ,112,107,100,
89,77,62,51,39,27,20,10,3,1),
w1 =
c(0,0,0,0,0,0,0,0,0,0,0,0,0,0,0,0,0,0,0,0,0,0,0,0,0,0,0,3,3,5,
  10,10,14,9,10,11,7,10,7,2,1), m2 = c(8,14,21,18,14,9,14,8,
  9,4,3,4,6,4,4,5,5,4,
```

```
3,1,1,2,6,1,2,2,2,2,0,3,2,1,3,.1,0,0,0,0,0,0),
O2 = c(295,287,273,252,234,220,211,197,189,180,176,173,169,163
    ,159,155,150,145,
141,138,137,136,134,128,127,125,123,119,112,103,84,69,63,54,45
    ,37,23,11,4,1),
w2 = c(0,0,0,0,0,0,0,0,0,0,0,0,0,0,0,0,0,0,0,0,0,0,0,0,0,0,0,2,5
    ,9,16,13,5,6,9,8,14,12,
7,3,1))
# initial values
list(q1 =
c(.5,.5,.5,.5,.5,.5,.5,.5,.5,.5,.5,.5,.5,.5,.5,.5,.5,.5,.5,
    .5,.5,.5,.5,.5,.5,.5,.5,.5,.5,
.5,.5,.5,.5,.5,.5,.5,.5,.5,.5),q2 = c(.5,.5,.5,.5,.5,.5,.5,.5,
    .5,.5,.5,.5,.5,.5,.5,.5,.5,.5,.5,.5,
    .5,.5,.5,.5,.5,.5,.5,.5,.5,.5,.5,.5,.5),
sr1 = c(.5,.5,.5,.5,.5,.5,.5,.5,.5,.5,.5,.5,.5,.5,.5,.5,.5,.5,
    .5,.5,.5,.5,.5,.5,.5,.5,.5,.5,.5,.5,
.5,.5,.5,.5,.5,.5,.5,.5,.5,.5),
sr2 = c(.5,.5,.5,.5,.5,.5,.5,.5,.5,.5,.5,.5,.5,.5,.5,.5,.5,.5,
    .5,.5,.5,.5,.5,.5,.5,.5,.5,.5,.5,.5,
.5,.5,.5,.5,.5,.5,.5,.5,.5,.5))
```

A Bayesian analysis is executed with 55,000 observations from the joint posterior distribution with a burn-in of 5000 and a refresh of 100, and the results are reported in Table 7.21. Columns 2 and 3 portray the estimated probability of death at each period, while the estimated survival rates are recorded in the last two columns. Also reported are the associated posterior standard deviations. It is obvious beginning with the third interval that the posterior mean of the survival rate for the study group is greater than that for the control. Median survival for the study cohort occurs at period 26 (year 13) and at period 17 (year 8.5) for the control cohort.

Observe in Figure 7.8 a plot of the survival rates for the study and control cohorts, where the top curve corresponds to the study cohort and the green curve to the control. What are plotted are the posterior means of Table 7.22 versus the period. One might conclude that screening is indeed more effective, but the study group subjects include those patients who refused screening; therefore, one should compute the mortality and survival rates for the study group that does not include the refusals. By not including the refusals, what is the effect on Figure 7.8? Will the top curve portray a larger difference than that appearing in Figure 7.8? See problem 17 at the end of the chapter.

What is the Bayesian approach to comparing the survival experience of the study cohort to that of the control? Recall Section 6.3.4 of Chapter 6 where the Bayesian version of the log-rank test is presented.

The log-rank test is one of the conventional procedures by which the overall survival experience of two groups is compared. The emphasis is on the word *overall*, that is, the two groups are compared from the first ordered

TABLE 7.21

Posterior Distribution of Mortality and Survival Rates

	Study and Control Cohorts—Posterior Mean and (Standard Deviation)			
Period t	Mortality Study $q1[t]$	Mortality Control $q2[t]$	Survival Rate Study $sr1[t]$	Survival Rate Control $sr2[t]$
1			1(0)	1(0)
2	.03058(.0111)	.0519(.0130)	.9608(.0110)	.9697
3	.04531(.01228)	.0798(.0163)	.9346(.0141)	.9224(.0155)
4	.04354(.0123)	.0748(.0163)	.8953(.0174)	.8518(.0205)
5	.0228(.0092)	.0635(.0157)	.8594(.0199)	.7914(.0234)
6	.0386(.0119)	.0449(.0138)	.843(.0206)	.7441(.0252)
7	.0359(.0116)	.0702(.0175)	.8138(.0222)	.7138(.0262)
8	.0164(.0081)	.0451(.0146)	.7876(.0233)	.6667(.0272)
9	.0251(.0100)	.0523(.0160)	.7776(.0237)	.6397(.0277)
10	.0341(.0118)	.0275(.0121)	.7614(.0242)	.6094(.0282)
11	.0353(.0122)	.0224(.0111)	.7386(.0250)	.5961(.0284)
12	.0364(.0126)	.0285(.0125)	.7156(.0256)	.5858(.0285)
13	.0281(.0113)	.0409(.0150)	.6931(.0264)	.5725(.0287)
14	.0144(.0082)	.0303(.0132)	.6766(.0267)	.5523(.0288)
15	.0632(.0168)	.0311(.0136)	.6697(.0268)	.5388(.0289)
16	.0205(.0101)	.0381(.01530)	.6308(.0275)	.5253(.0290)
17	.0366(.0136)	.0395(.0157)	.6208(.0276)	.5084(.0289)
18	.0162(.00920	.0339(.0149)	.6013(.0279)	.4915(.029)
19	.0273(.0119)	.0279(.0137)	.5946(.0281)	.4781(.0289)
20	.0614(.0179)	.0143(.0100)	.5817(.0280)	.468(.0290)
21	.0117(.0082)	.0143(.0101)	.5489(.0283)	.4647(.0288)
22	.0475(.0163)	.0217(.01230)	.5458(.0285)	.4612(.0288)
23	.0248(.0121)	.0514(.0107)	.5229(.0286)	.4544(.0288)
24	.0189(.0108)	.0154(.0107)	.5489(.0283)	.4344(.0288)
25	.0192(.0109)	.0232(.0132)	.5065(.0285)	.4309(.0288)
26	.0259(.0127)	.0235(.0133)	.4998(.0285)	.4241(.0285)
27	.0068(.0068)	.0244(.0138)	.4901(.0285)	.4174(.0286)
28	.0211(.012)	.0259(.0147)	.4902(.02840)	.4106(.0286)
29	.0298(.0147)	.0095(.0095)	.4837(.0285)	.404(.0285)
30	.0167(.0117)	.045(.0218)	.4739(.0285)	.404(.02850)
31	.0396(.0193)	.0409(.0228)	.4707(.0285)	.394(.02830)
32	.0486(.0234)	.0305(.0211)	.4607(.0285)	.3873(.0281)
33	.0441(.0246)	.0676(.0325)	.451(.0284)	.3839(.0281)
34	.0567(.0316)	.0235(.0218)	.4446(.0285)	.3737(.0281)
35	.0537(.0364)	.0255(.0205)	.438(.0282)	.3737(.0279)
36	.0687(.04620)	.0401(.0387)	.4346(.0282)	.3737(.0279)
37	.0624(.0587)	.0767(.0711)	.4315(.0283)	.3737(.0281)
38	.1428(.1233)	.1676(.1412)	.4315(.0284)	.3737(.0281)
39	.3338(.2363)	.3339(.2349)	.4314(.0284)	.3738(.0280)
40	.4976(.2876)	.5004(.2881)	.4314(.0281)	.3739(.0280)

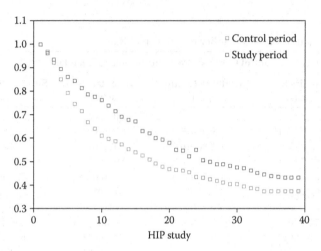

FIGURE 7.8
Survival study and control cohorts.

observation to the last ordered observation (of time to event). A Bayesian version of the test is presented in that the Bayesian approach mimics to some extent the conventional; however, the interpretation of the test is strictly Bayesian. The general idea is presented using the above example of two cohorts from the HIP study, the study cohort and the control.

Basically, the log-rank test consists of comparing the two groups by comparing the observed minus the expected number of events (deaths), normalized by the variance of the difference. Recall the HIP breast cancer screening trial of two cohorts of patients.

The ordered times of the two groups are combined and appear in the second column, the third column is the number of deaths for the study cohort, and the deaths of the control cohort are in the fourth column. The fifth and sixth columns are the number at risk for the study and control cohorts, respectively, whereas the seventh and eighth are the average number of deaths in cohorts 1 and 2 respectively, assuming there is no difference in the overall deaths between the two cohorts. The last two columns are the difference in the number of deaths minus the average number of deaths for the two groups. There are 40 periods of 6 months each covering a 20-year range.

The average or expected number of events for cohort 1 is computed as follows:

$$a_{1i} = \left[\frac{R1(i)}{(R1(i) + R2(i))} \right] (d_{1i} + d_{2i}) \qquad (7.21)$$

where it is assumed that the probability of an event (and the probability of a censored observation) is the same for the two groups at each time period. Recall that the number of events and the number of censored observations for each time period is assumed to be a binomial random variable; thus, the

TABLE 7.22

Combined Survival Experience of the Study and Control Cohorts of HIP Study

i	$t_{(i)}$	d_{1i}	d_{2i}	R1(i)	R2(i)	a_{1i}	a_{2i}	$d_{1i} - a_{1i}$	$d_{2i} - a_{2i}$
1	1	11	8	304	295	9.64	9.36	1.36	−1.36
2	2	8	14	293	287	11.11	10.89	−3.11	3.11
3	3	12	21	285	373	14.29	18.71	−2.29	2.29
4	4	11	18	273	252	15.05	13.92	−4.08	4.08
5	5	5	14	262	234	10.04	8.96	−5.04	5.04
6	6	9	9	257	220	9.70	8.30	−.70	.70
7	7	8	14	248	211	11.89	10.11	−3.89	3.89
8	8	3	8	240	197	6.04	4.96	−3.04	3.04
9	9	5	9	237	189	7.96	6.21	−2.79	2.79
10	10	7	4	232	180	6.19	4.81	.81	−.81
11	11	7	3	225	176	5.61	4.39	1.39	−1.39
12	12	7	4	218	173	6.13	4.87	.87	−.87
13	13	5	6	211	169	6.11	4.89	−1.11	1.11
14	14	2	4	206	163	3.35	2.65	−1.35	1.35
15	15	12	4	204	159	8.99	7.01	3.01	−3.01
16	16	3	5	192	155	4.43	3.57	−1.43	1.43
17	17	6	5	189	150	6.13	4.87	−.13	.13
18	18	2	4	183	145	3.35	2.65	−1.35	1.35
19	19	4	3	181	141	3.93	3.07	.07	−.07
20	20	10	1	177	138	6.18	4.82	3.82	−3.82
21	21	1	1	167	137	1.1	.9	−.10	.10
22	22	7	2	166	136	4.95	4.05	2.05	−2.05
23	23	3	6	159	134	4.88	4.12	−1.88	1.88
24	24	2	1	156	128	1.65	1.35	.35	−.35
25	25	2	2	154	127	2.19	1.81	−.19	.19
26	26	3	2	152	125	2.74	2.26	.26	−.26
27	27	0	2	149	123	1.10	.90	−1.10	1.10
28	28	2	2	145	119	2.20	1.80	−.20	.20
29	29	3	0	138	112	1.66	1.34	1.34	−1.34
30	30	1	3	129	103	2.22	1.78	−1.22	1.22
31	31	3	2	116	84	2.90	2.10	.10	−.10
32	32	3	1	96	69	2.33	1.67	.67	−.67
33	33	2	3	77	63	2.75	2.25	−.75	.75
34	34	2	0	64	54	1.08	.92	.92	−.92
35	35	1	0	49	45	.52	.48	.48	−.48
36	36	1	0	34	37	.48	.52	.52`	−.52
37	37	0	0	26	23	0	0	0	0
38	38	0	0	15	11	0	0	0	0
39	39	0	0	51	4	0	0	0	0
40	40	0	0	1	1	0	0	0	0

number at risk is also a random variable. The number at risk will decrease by the number of events and the number of censored observations.

BUGS CODE 7.5 is for calculating the difference in the observed minus expected events for Group 1, assuming there is no difference in the probability of an event between the two. Note, the observed minus expected differences for Group 2 (the control) differ only in sign to those of the first group (study cohort). The number of deaths and number of censored observations are assumed to have a binomial distribution, whose corresponding probabilities are given beta (.01, .01) prior distributions. This induces a posterior distribution to the number of expected events given by Equation 7.21 and consequently induces a posterior distribution to the observed minus expected differences of cohort 1. Of interest is testing for a difference in the overall recurrence between the two groups, which is expressed by the sum of the differences

$$\text{some1} = \sum_{i=1}^{i=m} (d_{1i} - a_{1i}) \tag{7.22}$$

where m = 17 is the number of distinct observed times.

In a similar way, the sum of the observed minus expected differences for Group 2 is

$$\text{some2} = \sum_{i=1}^{i=m} (d_{2i} - a_{2i}) \tag{7.23}$$

where

$$a_{2i} = \left[\frac{R(2i)}{(R1(i) + R2(i))} \right] (d_{1i} + d_{2i}) \tag{7.24}$$

is the expected number of recurrences for cohort 2 for the *i*th time.

If there is no difference in the event rate of cohort 1 versus cohort 2, one would expect half the differences to be positive and half to be negative; thus, on the average, one would expect the sum to average 0. If the 95% credible interval for the sum of the differences (Equation 7.22) does not contain 0, one is inclined to believe that the two groups do indeed differ with respect to the overall death rate. BUGS CODE 7.5 is well documented with statements indicated by a # sign, and the code closely follows Equations 7.21 through 7.24

BUGS CODE 7.5

```
model;
# log-rank Bayesian approach
{
# for group 1
# distribution of the number of deaths
```

```
for (i in 1:m1){d1[i]~dbin(q1[i],r1[i])}
# distribution of the number of censored
for (i in 1:m1){c1[i]~dbin(qc1[i],r1[i])}
# prior distribution of q1[i]
for (i in 1:m1){q1[i]~dbeta (1,1)}
for (i in 1:m1){qc1[i]~dbeta (1,1)}
# number at risk
r1[1]<- n1
for(i in 2:m1){r1[i]<-r1[i-1]-d1[i-1]-c1[i-1]}
# for group 2
# distribution of the number of deaths
for (i in 1:m2){d2[i]~dbin(q2[i],r2[i])}
# distribution of the number of censored
for (i in 1:m2){c2[i]~dbin(qc2[i],r2[i])}
# prior distribution of q2[i]
for (i in 1:m2){q2[i]~dbeta (1,1)}
for (i in 1:m2){qc2[i]~dbeta (1,1)}
# number at risk
r2[1]<- n2
for(i in 2:m2){r2[i]<-r2[i-1]-d2[i-1]-c2[i-1]}
# mortality group 1
for(i in 1:m1){s1[i]<- q1[i]*r1[i]}
# mortality group 2
for(i in 1:m2){s2[i]<- q2[i]*r2[i]}
# Expected number of deaths
for(i in 1:m1){e1[i]<- e11[i]*e12[i]}
for(i in 1:m2){e11[i]<- r1[i]/(r1[i]+r2[i])}
for(i in 1:m2){e12[i]<- s1[i]+s2[i]}
for(i in 1:m1){e2[i]<- e21[i]*e22[i]}
for(i in 1:m2){e21[i]<- r2[i]/(r1[i]+r2[i])}
for(i in 1:m2){e22[i]<- s1[i]+s2[i]}
# variance of difference
for(i in 1:m2){t1[i]<- r1[i]*r2[i]*(s1[i]+s2[i])*(r1[i]
+r2[i]- s1[i]-s2[i])}
for(i in 1:m2){t2[i]<-(r1[i]+r2[i])*(r1[i]+r2[i])*(r1[i]+r2[i]-1)}
# difference observed minus expected
for(i in 1:m1){ome1[i]<- d1[i]-e1[i]}
for(i in 1:m1){v1r1[i]<-t1[i]/t2[i]}
for(i in 1:m2){ome2[i]<-d2[i]-e2[i]}
# sum of observed minus expected group 1
some1<- sum(ome1[])
# sum of expected group 1
se1<-sum(e1[])
# sum of expected group 2
se2<-sum(e2[])
# sum of observed minus expected group 2
some2<- sum(ome2[])
# likelihood ratio
LR<- some1*some1/se1+some2*some2/se2
}
```

```
# HIP study
list(n1 = 304,n2 = 295, m1 = 40, d1 = c(11,8,12,11,5,9,8,3,5,7,
    7,7,5,2,12,
3,6,2,4,10,1,7,3,2,2,3,0,2,3,1,3,3,2,2,1,1,0,0,0,0), c1 =
    c(0,0,0,0,0,0,
0,0,0,0,0,0,0,0,0,0,0,0,0,0,0,4,5,6,12,17,16,11,13,1
    4,7,12,9,4,1),
m2 = 40, d2 =
c(8,14,21,18,14,9,14,8,9,4,3,4,6,4,4,5,5,4,3,1,1,2,6,1,2,2,2,2,
    0,3,2,1,3,
0,0,0,0,0,0,0),
c2 = c(0,0,0,0,0,0,0,0,0,0,0,0,0,0,0,0,0,0,0,0,0,0,0,0,0,0,0,2,5,
    9,16,13,5,6,9,
8,14,12,7,3,1))
# initial values study versus control
list(q1 =
 c(.5,.5,.5,.5,.5,.5,.5,.5,.5,.5,.5,.5,.5,.5,.5,.5,.5,.5,.5,.5,.
5,.5,.5,.5,.5,.5,.5,.5,.5,.5,.5,.5,.5,.5,.5,.5,.5,.5,.5), q2 = c(.5
,.5,.5,.5,.5,.5,.5,.5,.5,.5,.5,.5,.5,.5,.5,.5,.5,.5,.5,.5,.5,.5,
.5,.5,.5,.5,.5,.5,.5,.5,.5,.5,.5,.5,.5,.5,.5),
qc1 =
c(.5,.5,.5,.5,.5,.5,.5,.5,.5,.5,.5,.5,.5,.5,.5,.5,.5,.5,.5,.5,.5,
.5,.5,.5,.5,.5,.5,.5,.5,.5,.5,.5,.5,.5,.5,.5,.5),
qc2 =
c(.5,.5,.5,.5,.5,.5,.5,.5,.5,.5,.5,.5,.5,.5,.5,.5,.5,.5,.5,.5,.5,
.5,.5,.5,.5,.5,.5,.5,.5,.5,.5,.5,.5,.5,.5,.5,.5))
# Screened versus control
list(n1 = 225,n2 = 295, m1 = 20, d1 =
c(9,14,10,8,11,13,3,11,7,11,3,5,5,2,3,3,4,1,0,0), c1 = c(0,0,0,
0,0,0,0,0,0,0,0,0,6,15,24,19,18,17,3),
m2 = 20, d2 = c(22,39,23,22,13,7,10,9,9,4,3,7,4,4,3,3,3,0,0,0),
c2 = c(0,0,0,0,0,0,0,0,0,0,0,0,0,7,24,18,15,22,19,4))
# initial values screened vs control
list(q1 = c(.5,.5,.5,.5,.5,.5,.5,.5,.5,.5,.5,.5,.5,.5,.5,.5,.5,
    .5,.5,.5),
q2 = c(.5,.5,.5,.5,.5,.5,.5,.5,.5,.5,.5,.5,.5,.5,.5,.5,.5,.5,.5,.5),
qc1 = c(.5,.5,.5,.5,.5,.5,.5,.5,.5,.5,.5,.5,.5,.5,.5,.5,.5,.5,
    .5,.5),
qc2 = c(.5,.5,.5,.5,.5,.5,.5,.5,.5,.5,.5,.5,.5,.5,.5,.5,.5,.5,
    .5,.5))
```

The Bayesian analysis is based on Table 7.22 and executed through BUGS CODE 7.5 with 55,000 observations for the simulation, a burn-in of 5000, and a refresh of 100, where the main focus is on the posterior distribution of the sum (Equation 7.22) of the differences observed minus average number of events of the study cohort. See Table 7.23 for the Bayesian analysis.

Our focus is on some1, the sum of the observed minus expected deaths of the study cohort, which has a posterior mean of −20.73 and posterior standard deviation of 9.994, and a 95% credible interval of (−40.75, −1.548). If the

TABLE 7.23

Posterior Analysis: Log-Rank Test Comparing Study and Control Cohorts of HIP Trial

Parameter	Mean	SD	Error	2½	Median	97½
LR	5.59	1.393	.0042	4.412	5.091	9.547
ome[1]	1.353	2.174	.0094	−3.359	1.503	5.15
ome[2]	−3.124	2.327	.0095	−8.106	−2.985	.9903
ome[3]	−4.881	2.849	.0121	−10.94	−4.729	.256
ome[4]	−4.096	2.72	.0114	−9.838	−3.953	.8051
ome[5]	−5.048	2.245	.0096	−9.881	−4.886	−1.102
ome[6]	−.7089	2.234	.0087	−5.537	−.5601	3.187
ome[7]	−3.894	2.471	.0100	−9.146	−3.735	.4504
ome[8]	−3.056	1.793	.0078	−7.032	−2.884	−.0631
ome[9]	−2.789	2.035	.0088	−7.224	−2.662	.698
ome[10]	.8108	1.843	.0074	−3.293	.9871	3.889
ome[11]	1.385	1.743	.0079	−2.525	1.568	4.278
ome[12]	.8746	1.823	.0076	−3.197	1.053	3.912
ome[13]	−1.11	1.8	.0076	−5.122	−.9284	1.905
ome[14]	−1.361	1.352	.0062	−4.463	−1.184	.761
ome[15]	2.987	2.183	.0093	−1.737	3.147	6.77
ome[16]	−1.443	1.544	.0070	−4.932	−1.257	1.052
ome[17]	−.144	1.824	.0083	−4.166	.0252	2.893
ome[18]	−1.362	1.353	.0061	−4.47	−1.184	.7426
ome[19]	.0433	1.477	.0060	−3.301	.2223	2.395
ome[20]	3.793	1.818	.0074	−.201	3.959	6.863
ome[21]	−.1116	.7812	.0033	−2.081	.0652	.8634
ome[22]	2.031	1.632	.0069	−1.602	2.199	4.721
ome[23]	−1.889	1.597	.0073	−5.476	−1.726	.7279
ome[24]	.3446	.9432	.0041	−1.946	.5174	1.651
ome[25]	−.1991	1.081	.0048	−2.768	−.0246	1.389
ome[26]	.24	1.208	.0052	−2.55	.4239	2.106
ome[27]	−1.102	.759	.0032	−2.996	−.9375	=.136
ome[28]	−.2081	1.085	.0044	−2.783	−.0391	1.388
ome[29]	1.338	.9431	.0043	−.9498	1.516	2.65
ome[30]	−1.236	1.098	.0046	−3.826	−1.061	.3846
ome[31]	.0995	1.276	.0056	−2.886	.2801	2.048
ome[32]	.6579	1.15	.0047	−2.062	.8403	2.349
ome[33]	−.7611	1.2	.0050	−3.525	−.5913	1.089
ome[34]	.9031	.7563	.0031	−.982	1.07	1.861
ome[35]	.4713	.5137	.0021	−.9001	.628	.9853
ome[36]	.512	.4705	.0020	−.7595	.6521	.9864
ome[37]	−.01098	.0767	.000319	−.1071	0	0
ome[38]	−.0105	.0724	.000301	−.1008	0	0
ome[39]	−.0114	.0731	.00032	−.1156	0	0
ome[40]	−.0099	.0502	.000213	−.1479	0	0
some1	−20.73	9.994	.0247	−40.75	−20.58	−1.548
some2	19.94	8.509	.02091	2.902	20.09	36.28

survival experience of the two cohorts is the same, one would expect the 95% credible interval to contain zero, but it does not; thus, one would conclude that the survival of the study cohort is better than that of the control cohort. Of course this confirms the evidence portrayed by Figure 7.8, the plot of the two survival curves. The posterior means of the ome vector are quite similar to the next to last column of Table 7.22, as they should be, thus serving as a check on the statements of BUGS CODE 7.5.

7.4.4.2 Survival Models

Analysis of the HIP study is continued using the Cox proportional hazards model, which was introduced in Chapter 6. Of interest is comparing the survival of the study cohort with the control cohort. The Cox proportional hazards model is one of the most useful in biostatistics and appears in many of the major medical journals.

Recall, it is defined as

$$h(t, X) = h_0(t)e^{\sum_{i=1}^{i=p} \beta_i X_i} \tag{7.25}$$

where $h_0(t)$ is the baseline hazard function, the β_i are unknown regression parameters, and the X_i are known covariates or independent variables. Note that the baseline hazard function is a function of time only, but that the covariates are not functions of t. The time t is the time to the event of interest, which is usually the survival time of a group of patients or the time to recurrence, or some other event measured by time. Recall that the regression function is defined in the terms of the hazard function

$$h(t) = \lim_{\Delta t \to \infty} \frac{P(t \leq T < t + \Delta t \mid T \geq t)}{\Delta t} \tag{7.26}$$

which in turn is related to the survival function $S(t)$

$$h(t) = \frac{\left[dS(t)/dt\right]}{S(t)} \tag{7.27}$$

In addition, the survival function is

$$S(t) = P(T > t) \tag{7.28}$$

where T denotes the survival time of a subject. In survival studies, the Cox regression model is expressed as a hazard, whereas the usual way to express a regression is more directly using T as a function of unknown regression coefficients. One reason the Cox model is so popular is its versatility; for example, if the actual survival time has, say, an exponential distribution or a Weibull distribution, the Cox model will provide similar results.

With the Cox model, the time variable T is not assumed to have a specific distribution; thus, the model is quite general in that it can be applied in a large variety of time to event studies. Also note that the p covariates $X_1, X_2, ..., X_p$ are not functions of t, however, there are cases where one would have time-dependent covariates, in which case, a more general Cox model is appropriate. The most important assumption of the Cox model is that if one is comparing the survival of two groups, the corresponding hazard functions must be proportional.

The most important parameter in survival studies is the hazard ratio

$$HR = \frac{h(t, X^*)}{h(t, X)} \qquad (7.29)$$

between two individuals, one with the covariate measurement X^* and the other with the measurement X, on the p covariates. Note that it easy to show that the hazard ratio (Equation 6.24) is equivalent to

$$HR = e^{\sum_{i=1}^{i=p} \beta_i (X_i^* - X_i)} \qquad (7.30)$$

where both X^* and X are known. It is important to remember that to the Bayesian approach the HR hazard ratio (Equation 7.30) is an unknown parameter because it depends on p unknown parameters

$$\beta = (\beta_1, \beta_2, ..., \beta_p) \qquad (7.31)$$

Thus, the Bayesian analysis must specify a prior distribution for β, then, through Bayes' theorem, determine the posterior distribution of β and any function of β such as the hazard ratio.

As an example of the hazard ratio, consider the two cohorts of HIP patients, where cohort 1 is the study group and cohort 2 the control, with one covariate X, where $X^* = 0$ denotes the treatment group and $X = 1$ the placebo. Then the hazard ratio (Equation 7.29) reduces to

$$HR(\beta) = e^\beta \qquad (7.32)$$

where one individual is a patient from the treatment group and the other a patient from the placebo. Thus, Equation 7.32 expresses the effect of screening as a hazard ratio. Once one estimates β, one has an estimate of the hazard ratio, or for our interest, once one has the posterior distribution of β, one has the posterior distribution of HR.

As the first example of using the Cox model in this chapter, consider the comparison of the survival times of the screened cohort with the control group of the HIP study. The following BUGS CODE 7.6 follows the leukemia example of volume I of WinBUGS. In the list statement for the data, N is the total number of patients, 520; $T = 19$ is the number of distinct times of death or the times of censored observations; obs. t is the vector of the 520 times of death or censored observations; and the fail vector is of dimension 520, where the first 225 indicate whether the event is a death or a censored observation for the screened cohort, while the remaining 295 are indicators for the control group. A 1 denotes a death and a 0 denotes a censored observation (patient at last contact was alive). Next in the list statement is the x1 vector, the vector indicating cohort membership, where a .5 indicates the screened cohort and −.5 the control cohort. Finally the t vector is of dimension 19, indicating the number of distinct times of death (or times of a censored observation). The second list statement is a vector of initial values for beta and the dL0 vector of dimension 19. The data for the screened group are selected from table 2 of the appendix of Shapiro et al.,[2] while the information for the control group is taken from table 11 of the appendix, and it should be stressed that the 520 patients were diagnosed with breast cancer within 5 years of entry to the study. Thus, the effect of screening (if it exists) can be estimated by comparing the survival experience of the two cohorts.

BUGS CODE 7.6

```
model;
{
# Set up data
  for(i in 1:N) {
    for(j in 1:T) {
# risk set = 1 if obs.t > = t
      Y[i,j] <- step(obs.t[i] - t[j] + eps)
# counting process jump = 1 if obs.t in [t[j], t[j+1])
#    i.e. if t[j] < = obs.t < t[j+1]
      dN[i, j] <- Y[i, j] * step(t[j + 1] - obs.t[i] - eps) *
fail[i]
    }
  }
# Model
  for(j in 1:T) {
    for(i in 1:N) {
      dN[i, j] ~ dpois(Idt[i, j]) # Likelihood
      Idt[i, j] <- Y[i, j] * exp(beta * x1[i]) * dL0[j] #
Intensity
    }
    dL0[j] ~ dgamma(mu[j], c)
    mu[j] <- dL0.star[j] * c # prior mean hazard
# Survivor function = exp(-Integral{l0(u)du})^exp(beta*z)
    S.screened[j] <- pow(exp(-sum(dL0[1 : j])), exp(beta *
      -0.5));
```

```
        S.control[j] <- pow(exp(-sum(dL0[1 : j])), exp(beta * 0.5));
    }
    c <- 0.1
    r <-.1
    for (j in 1 : T) {
        dL0.star[j] <- r * (t[j + 1] - t[j])
    }
    beta ~ dnorm(0.0,0.000001)
    # hazard ratio for group 1 versus group 2
    HR<-exp(beta)
}
# HIP information
# screened group table 2 of appendix
# control group table 11 of appendix
# see Shapiro, et al.[2]
list(N = 520,T = 19, eps = 1.0E-10,
obs.t = c(6,6,6,6,6,6,6,6,6,
18,18,18,18,18,18,18,18,18,18,18,18,18,18,
30,30,30,30,30,30,30,30,30,30,
42,42,42,42,42,42,42,42,
54,54,54,54,54,54,54,54,54,54,54,
66,66,66,66,66,66,66,66,66,66,66,66,66,
78,78,78,
90,90,90,90,90,90,90,90,90,90,90,
102,102,102,102,102,102,102,
114,114,114,114,114,114,114,114,114,114,114,
126,126,126,
138,138,138,138,138,
150,150,150,150,150,
162,162,
162,162,162,162,162,162,
174,174,174,
174,174,174,174,174,174,174,174,174,174,174,174,174,174,174,
186,186,186,
186,186,186,186,186, 186,186,186,186,186, 186,186,186,186,186,
  186,186,186,186,186,
186,186,186,186,
198,198,198,198,
198,198,198,198,198, 198,198,198,198,198, 198,198,198,198,198,
  198,198,198,198,
210,
210,210,210,210,210, 210,210,210,210,210, 210,210,210,210,210,
  210,210,210,
222,222,222,222,222, 222,222,222,222,222,
  222,222,222,222,222,222,222,
234,234,234,
6,6,6,6,6,6,6,6,6,6,6,6,6,6,6,6,6,6,6,6,6,6,6,6,
18,18,18,18,18,18,18,18,18,18,18,18,18,18,18,18,18,18,18, 18,
  18,18,18,18,18,18,18,18,18,18,18,18,18,18,18,18,18,18,
18,
```

```
30,30,30,30,30,30,30,30,30,30,30,30,30,30,30,30,30,30,30,30,30
  ,30,30,
42,42,42,42,42,42,42,42,42,42,42,42,42,42,42,42,42,42,42,42,42,42,
54,54,54,54,54,54,54,54,54,54,54,54,54,
66,66,66,66,66,66,66,
78,78,78,78,78,78,78,78,78,78,
90,90,90,90,90,90,90,90,99,
102,102,102,102,102,102,102,102,102,
114,114,114,114,
126,126,126,
138,138,138,138,138,138,138,
150,150,150,150,
162,162,162,162,
162,162,162,162,162,162,162,
174,174,174,
174,174,174,174,174, 174,174,174,174,174, 174,174,174,174,174,
  174,174,174,174,174,
174,174,174,174,174,
186,186,186,
186,186,186,186,186, 186,186,186,186,186,
186,186,186,186,186, 186,186,186,
198,198,198,
198,198,198,198,198, 198,198,198,198,198, 198,198,198,198,198,
210,210,210,210,210, 210,210,210,210,210, 210,210,210,210,210,
  210,210,210,210,210,
210,210,
222,222,222,222,222, 222,222,222,222,222, 222,222,222,222,222,
  222,222,222,222,
234,234,234,234),
fail = c(
1,1,1,1,1,1,1,1,1,
1,1,1,1,1,1,1,1,1,1,1,1,1,1,1,
1,1,1,1,1,1,1,1,1,1,1,
1,1,1,1,1,1,1,1,
1,1,1,1,1,1,1,1,1,1,1,1,
1,1,1,1,1,1,1,1,1,1,1,1,1,1,
1,1,1,
1,1,1,1,1,1,1,1,1,1,1,1,
1,1,1,1,1,1,1,
1,1,1,1,1,1,1,1,1,1,1,1,
1,1,1,
1,1,1,1,1,
1,1,1,1,1,
1,1,0,0,0,0,0,0,
1,1,1,0,0,0,0,0,0,0,0,0,0,0,0,0,0,0,0,
1,1,1,0,0,0,0,0,0,0,0,0,0,0,0,0,0,0,0,0,0,0,0,0,0,0,0,
1,1,1,1,0,0,0,0,0,0,0,0,0,0,0,0,0,0,0,0,0,0,0,0,
1,0,0,0,0,0,0,0,0,0,0,0,0,0,0,0,0,0,0,0,
0,0,0,0,0,0,0,0,0,0,0,0,0,0,0,0,0,0,
0,0,0,
```

```
1,1,1,1,1,1,1,1,1,1,1,1,1,1,1,1,1,1,1,1,1,1,1,1,
1,1,1,1,1,1,1,1,1,1,1,1,1,1,1,1,1,1,1,1,1,1,1,1,1,1,1,1,1,1,1,1,
  1,1,1,1,1,1,1,1,
1,1,1,1,1,1,1,1,1,1,1,1,1,1,1,1,1,1,1,1,1,1,1,
1,1,1,1,1,1,1,1,1,1,1,1,1,1,1,1,1,1,1,1,1,1,1,
1,1,1,1,1,1,1,1,1,1,1,1,1,
1,1,1,1,1,1,1,
1,1,1,1,1,1,1,1,1,1,
1,1,1,1,1,1,1,1,1,
1,1,1,1,1,1,1,1,1,
1,1,1,1,
1,1,1,
1,1,1,1,1,1,1,
1,1,1,1,
1,1,1,1,0,0,0,0,0,0,0,
1,1,1,0,0,0,0,0,0,0,0,0,0,0,0,0,0,0,0,0,0,0,0,0,0,0,0,0,
1,1,1,0,0,0,0,0,0,0,0,0,0,0,0,0,0,0,0,0,0,0,
1,1,1,0,0,0,0,0,0,0,0,0,0,0,0,0,0,0,0,0,
0,0,0,0,0,0,0,0,0,0,0,0,0,0,0,0,0,0,0,0,0,0,0,0,
0,0,0,0,0,0,0,0,0,0,0,0,0,0,0,0,0,0,0,0,
0,0,0,0),
x1 = c(
.5,.5,.5,.5,.5,.5,.5,.5,.5,.5,.5,.5,.5,.5,.5,.5,.5,.5,.5,.5,.5,
  .5,.5,.5,.5,.5,.5,.5,.5,
.5,.5,.5,.5,.5,.5,.5,.5,.5,.5,.5,.5,.5,.5,.5,.5,.5,.5,.5,.5,.5,
  .5,.5,.5,.5,.5,.5,.5,.5,
.5,.5,.5,.5,.5,.5,.5,.5,.5,.5,.5,.5,.5,.5,.5,.5,.5,.5,.5,.5,.5,
  .5,.5,.5,.5,.5,.5,.5,.5,
.5,.5,.5,.5,.5,.5,.5,.5,.5,.5,.5,.5,.5,.5,.5,.5,.5,.5,.5,.5,.5,
  .5,.5,.5,.5,.5,.5,.5,.5,
.5,.5,.5,.5,.5,.5,.5,.5,.5,.5,.5,.5,.5,.5,.5,.5,.5,.5,.5,.5,.5,
  .5,.5,.5,.5,.5,.5,.5,.5,
.5,.5,.5,.5,.5,.5,.5,.5,.5,.5,.5,.5,.5,.5,.5,.5,.5,.5,.5,.5,.5,
  .5,.5,.5,.5,.5,.5,.5,.5,
.5,.5,.5,.5,.5,.5,.5,.5,.5,.5,.5,.5,.5,.5,.5,.5,.5,.5,.5,.5,.5,
  .5,.5,.5,.5,.5,.5,.5,.5,
.5,.5,.5,.5,.5,.5,.5,.5,.5,.5,.5,.5,.5,.5,.5,.5,.5,.5,.5,.5,.5,
  .5,.5,.5,.5,.5,.5,.5,.5,
.5,.5,.5,.5,.5,.5,.5,.5,.5,.5,.5,.5,.5,.5,.5,.5,.5,.5,.5,.5,.5,
  .5,.5,.5,.5,.5,.5,.5,.5,
-.5,-.5,-.5,-.5,-.5,  -.5,-.5,-.5,-.5,-.5,  -.5,-.5,-.5,-.5,-.5,
  -.5,-.5,-.5,-.5,-.5,
-.5,-.5,-.5,-.5,-.5,  -.5,-.5,-.5,-.5,-.5,  -.5,-.5,-.5,-.5,-.5,
  -.5,-.5,-.5,-.5,-.5,
-.5,-.5,-.5,-.5,-.5,  -.5,-.5,-.5,-.5,-.5,  -.5,-.5,-.5,-.5,-.5,
  -.5,-.5,-.5,-.5,-.5,
-.5,-.5,-.5,-.5,-.5,  -.5,-.5,-.5,-.5,-.5,  -.5,-.5,-.5,-.5,-.5,
  -.5,-.5,-.5,-.5,-.5,
-.5,-.5,-.5,-.5,-.5,  -.5,-.5,-.5,-.5,-.5,  -.5,-.5,-.5,-.5,-.5,
  -.5,-.5,-.5,-.5,-.5,
```

```
-.5,-.5,-.5,-.5,-.5,  -.5,-.5,-.5,-.5,-.5,  -.5,-.5,-.5,-.5,-.5,
 -.5,-.5,-.5,-.5,-.5,
-.5,-.5,-.5,-.5,-.5,  -.5,-.5,-.5,-.5,-.5,  -.5,-.5,-.5,-.5,-.5,
 -.5,-.5,-.5,-.5,-.5,
-.5,-.5,-.5,-.5,-.5,  -.5,-.5,-.5,-.5,-.5,  -.5,-.5,-.5,-.5,-.5,
 -.5,-.5,-.5,-.5,-.5,
-.5,-.5,-.5,-.5,-.5,  -.5,-.5,-.5,-.5,-.5,  -.5,-.5,-.5,-.5,-.5,
 -.5,-.5,-.5,-.5,-.5,
-.5,-.5,-.5,-.5,-.5,  -.5,-.5,-.5,-.5,-.5,  -.5,-.5,-.5,-.5,-.5,
 -.5,-.5,-.5,-.5,-.5,
-.5,-.5,-.5,-.5,-.5,  -.5,-.5,-.5,-.5,-.5,  -.5,-.5,-.5,-.5,-.5,
 -.5,-.5,-.5,-.5,-.5,
-.5,-.5,-.5,-.5,-.5,  -.5,-.5,-.5,-.5,-.5,  -.5,-.5,-.5,-.5,-.5,
 -.5,-.5,-.5,-.5,-.5,
-.5,-.5,-.5,-.5,-.5,  -.5,-.5,-.5,-.5,-.5,  -.5,-.5,-.5,-.5,-.5,
 -.5,-.5,-.5,-.5,-.5,
-.5,-.5,-.5,-.5,-.5,  -.5,-.5,-.5,-.5,-.5,  -.5,-.5,-.5,-.5,-.5,
 -.5,-.5,-.5,-.5,-.5,
-.5,-.5,-.5,-.5,-.5,  -.5,-.5,-.5,-.5,-.5,  -.5,-.5,-.5,-.5,-.5,
 -.5,-.5,-.5,-.5,-.5,
-.5,-.5,-.5,-.5,-.5,  -.5,-.5,-.5,-.5,-.5,
 -.5,-.5,-.5,-.5,-.5),
t = c(6,18,30,42,54,66,78,90,102,114,126,138,150,162,174,186,
198,210,222, 234))
# initial values
list(beta = 0.0,
dL0 = c(1.0,1.0,1.0,1.0,1.0,1.0,1.0,1.0,1.0,1.0,1.0,1.0,1.0,1.
0,1.0, 1.0,1.0,1.0,1.0))
```

A Bayesian analysis is executed with 5000 observations for the simulation, a burn-in of 1000, and a refresh of 100, and the results are depicted in Table 7.24, where the posterior mean, standard deviation, and 95% credible interval are computed. Note that the MCMC errors are quite small, of the order <.0001. S.control[3] indicates the third year of the study for the control group, while S.screened[3] is the corresponding parameter for the screened group. If screening (mammography and clinical examination) gives a survival advantage, one would expect the posterior means of the control group to not be larger than those of the corresponding screened entry.

When comparing the two survival curves, one sees that at each year, the screened survival estimate is not less than that for the control cohort, implying that screening does indeed have an impact on survival. Figure 7.9 confirms this assertion that screening does give women a survival advantage.

It appears from Table 7.24 and Figure 7.9 that the median survival time for the control cohort is approximately 10 months and 13 months for the screened cohort, another confirmation that screening is efficacious for the detection of breast cancer. Recall the previous section, where a Bayesian version of the log-rank test is presented. The log-rank test is a formal way to compare the survival

TABLE 7.24

Posterior Analysis for the Survival of the Screened and Control Groups (HIP Study)

Parameter	Mean	SD	Error	2½	Median	97½
S.control[1]	.9384	.0110	.0002011	.9154	.9392	.9575
S.control[2]	.8359	.0187	.000397	.7976	.8365	.8701
S.control[3]	.7709	.0219	.000448	.7275	.7716	.8128
S.control[4]	.712	.0240	.000517	.6646	.7124	.7591
S.control[5]	.6647	.0254	.000569	.6164	.6651	.7147
S.control[6]	.6253	.0267	.000632	.5733	.6253	.6783
S.control[7]	.5992	.0274	.000637	.5456	.5992	.6543
S.control[8]	.5602	.02814	.000678	.5047	.5603	.6159
S.control[9]	.5287	.0284	.000701	.4723	.5288	.5847
S.control[10]	.4992	.0287	.000691	.4428	.4994	.5574
S.control[11]	.4871	.0289	.000694	.4301	.4869	.5451
S.control[12]	.4635	.0289	.000681	.4082	.4629	.5217
S.control[13]	.4457	.0290	.000667	.3895	.445	.5307
S.control[14]	.4338	.0291	.000663	.3777	.4331	.4919
S.control[15]	.4212	.0292	.000671	.3636	.4207	.479
S.control[16]	.4059	.0291	.000673	.348	.4057	.4645
S.control[17]	.3823	.0294	.000713	.3251	.3824	.4408
S.control[18]	.3769	.0297	.000714	.3185	.3766	.4374
S.control[19]	.3758	.0299	.000708	.3171	.3756	.4368
S.screened[1]	.9455	.0098	.000160	.9246	.9463	.9631
S.screened[2]	.854	.0169	.000360	.8201	.8544	.8853
S.screened[3]	.7952	.02011	.000463	.7558	.7956	.8331
S.screened[4]	.7414	.0226	.000546	.6962	.7419	.7848
S.screened[5]	.6979	.0245	.000581	.6484	.6978	.7437
S.screened[6]	.6612	.0262	.000615	.6102	.6614	.7112
S.screened[7]	.6369	.0270	.000629	.5826	.6368	.6883
S.screened[8]	.6002	.0279	.000668	.5443	.6002	.6528
S.screened[9]	.5704	.0286	.000691	.5146	.5699	.6257
S.screened[10]	.5423	.0291	.000722	.4848	.5422	.598
S.screened[11]	.5307	.0291	.000720	.4736	.5309	.5868
S.screened[12]	.508	.0293	.000724	.4504	.5079	.5644
S.screened[13]	.4908	.0294	.000727	.4344	.4904	.5495
S.screened[14]	.4793	.0296	.000727	.4219	.4792	.5368
S.screened[15]	.4669	.0299	.000739	.4091	.4667	.5243
S.screened[16]	.4519	.0300	.000737	.394	.4516	.5106
S.screened[17]	.4287	.0307	.000737	.3686	.4287	.4882
S.screened[18]	.4234	.0309	.000752	.3622	.4234	.4831
S.screened[19]	.4223	.0311	.000759	.3606	.422	.4822

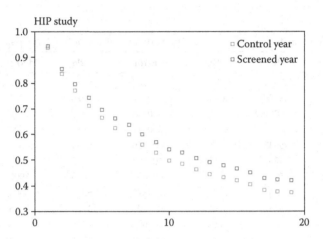

FIGURE 7.9
Survival of screened cohort versus control cohort.

experience of two groups, such as the screened and control cohorts described
earlier. Figure 7.9 and Table 7.24 in an informal way imply that screening gives
a woman a survival advantage, which also can be confirmed by the log-rank
test presented in Exercise 19.

7.5. Comments and Conclusions

Chapter 7 introduces the necessary Bayesian tools to analyze a screening
study, beginning with the definition of test accuracy and ending with the
Cox proportional hazards model to compare survival between cohorts of
the HIP study. There are three components of screening: the accuracy of the
modality, estimation of the lead time, and comparing survival between the
screened cohort and the control.

Classification probabilities for estimating test accuracy are first defined,
namely, the TPF and FPF, true and false negative fractions, PPVs and NPVs,
and positive and negative diagnostic likelihood ratios. This is followed by
Bayesian methods to estimate the classification probabilities illustrated
with an EST to detect coronary heart disease. Accuracy of a diagnostic test
is expanded to estimate the area under the ROC curve using an example
from mammography. The area under the ROC is estimated by a Bayesian
technique that computes the posterior mean, standard deviation, and 95%
credible interval and for the mammography example, the posterior mean of
the ROC area is .784, indicating fair to good accuracy of mammography to
diagnose breast cancer.

The U.K. Early Detection Trial for breast cancer is our first example of a
screening trial for chronic disease where the classification probabilities

(TPF, FPF, PPV and NPVs, and positive and negative diagnostic likelihood ratios) are estimated by a Bayesian posterior analysis. Interesting results are observed for the first round of testing with posterior means for the TPF of .9283 a FPF of .0612, but a PPV of only .0845. The latter estimate is quite small because the fraction with the disease (the proportion with breast cancer) is very small. The estimate of .0845 for the PPV implies that of those that test positive (via mammography and clinical examination) only 8.45% actually indeed have breast cancer!!

Presentation of the UK Early Detection Trial is followed by an in-depth Bayesian analysis of the HIP study that involved approximately 60,000 subjects randomly assigned to two groups, a study group that had the opportunity to have a mammography and clinical examination for breast cancer and a control group that received usual medical care. Mammography is an x-ray technique for imaging the breast, and Egan at MD Anderson Cancer Hospital and Tumor Institute in Houston showed that mammography did have enough accuracy to detect breast cancer. Based on these results the NCI showed interest in sponsoring large-scale screening trials with the result that the HIP study was initiated. The serious problem with breast cancer is revealed by Figure 7.4, which shows the mortality rates for breast cancer by race.

In order to analyze the HIP trial, I used the HIP database provided by the Biometric Research Branch of the National Cancer Institute and downloaded some 25 variables. For the study group, an initial examination followed by three examinations are available for the participants, but among the study group there is a subgroup that declined to be screened. First to be presented are the various descriptive statistics that are necessary to understand the analysis. For example, for the study cohort, the average, standard deviation, and median age at entry, the initial examination, and the three annual examinations are computed. The average age at entry was 50.51 years while that for the third annual examination was 53.93 years, which showed that on the average the time from entry to the last examination was 3.42 years. Also presented is the extent of disease by cohort where the distribution of the number of patients with Stage I, II, III, and IV disease is computed. Next, the lead time descriptive statistics are portrayed. In order to estimate the lead time, the time from entry to diagnosis of breast cancer is recorded for those patients that are diagnosed within 5 years of entry to the trial. This is done so that if screening has an effect it can be measured. After 4 years from entry, screening is not available to subjects in the study group, thus one would expect time from entry to diagnosis for the study and control cohorts to be equivalent. It is shown that the average time to diagnosis for the screened group is 555 days compared to 945 for the control giving 390 days or 1.06 years as an estimate of the lead time. For the screened group, only those diagnosed with clinical examination or mammography are included; thus, interval diagnoses are ignored.

Bayesian analysis for the HIP study is continued by modeling the time from entry to diagnosis by the Weibull distribution and the posterior mean of the median time from entry to breast cancer diagnosis for the screened group is 526.6 days, while the corresponding posterior mean for the control group is 937.1 days giving a posterior mean of the lead time as 937.1 − 526.6 = 410.5 days or 1.12 years. Finally, the survival time of the screened group is compared to that of the control group. There are 225 in the screened cohort and 295 in the control cohort and these include only those are diagnosed within 5 years of entry to the trial. Life table methods are employed to estimate the survival profile of the study group of 304 patients and the 295 control patients where all patients were diagnosed with breast cancers within 5 years of entry.

Finally, the survival profile of these two cohorts is estimated with the Cox proportional hazards model. In addition, a Bayesian log-rank test compares the survival experience of the two cohorts implying that screening does provide a survival advantage to those taking advantage of mammography and palpitation of the breast.

Screening for chronic disease is an interesting topic and only the fundamentals of the Bayesian approach are introduced. HIP was one of the first well-designed randomized trials to investigate how well mammography and clinical examination detect breast cancer and whether screening does indeed give the screened patients a survival advantage. Since the HIP trial, there have been many more that will now be briefly described.

There have been almost half a million participants from four countries involved in screening trials that examined mortality from breast cancer. For example, the Canadian National Breast Cancer Screening Study compared mammography plus clinical breast examination (CBE) with breast examination alone, but the other eight compared screening mammograms versus with or without CBE while those in the control group received usual care. Different designs were employed as well as different recruitment methods for subjects and different types of administration of the control patients. Also, there were variations in the compliance of participants to their assigned cohort (screened and control). The following lists several breast cancer screening trials:

1. HIP, United States, 1963[2]

2. Malmo, Sweden, 1976[20]

3. Ostergotland (County E of a Two-County Trial), Sweden, 1977[21]

4. Kopparberg (County W of Two-County Trial), Sweden, 1977[22]

5. Edinburg, United Kingdom, 1976[23]

6. NBSS-1, Canada, 1980[24]

7. NBSS-2, Canada, 1980[25]

8. Stockholm, Sweden, 1981[26]

9. Gothenberg, Sweden, 1982[27]

10. Age Trial[28]

For additional information about these trials see Breast Cancer Screening (PDQ), the Effect of Screening on Breast Cancer Mortality, at http://www.cancer.gov/pdq/screening/breast/healthprofessionals/page5.[29]

It is interesting to observe that the relative risk (screening versus control) varies from .71 to .97, where the lower risk is reported by the HIP trial and the higher risk by NBSS-1, the Canadian National Breast Screening Study. It should be remembered that compared to the HIP trial, mammography has increased in accuracy and that the trials vary greatly in regard to design randomization, compliance, and so on. In addition to randomized trials, there have been many population-based screening programs that examine the effectiveness of screening in the general population, not under the more ideal conditions of a well-designed randomized study. For example see Elmore et al.[30]

To some extent, Chapter 8 will continue the presentation of screening for chronic disease, but the emphasis will be on advanced modeling techniques to estimate lead time and morality.

Exercises

1. Based on BUGS CODE 7.1, verify the Bayesian analysis reported in Table 7.3. Use 55,000 observations for the simulation with a burn-in of 5000 and a refresh of 100.

   a. What is the 95% credible interval for the TPF (sensitivity)?

   b. What is the estimate of the specificity of the EST?

   c. Plot the posterior density of the FPF.

   d. Is the posterior density of the FPF skewed? Why?

2. Based on BUGS CODE 7.1, verify the Bayesian analysis of Table 7.4. Use 55,000 observations for the simulation with a burn-in of 5000, and a refresh of 100.

   a. Interpret the estimate (the posterior mean) of the PPV.

   b. For those patients that test positive, what percentage will have CAD? Explain your answer.

   c. Of those patients that test negative, what proportion will not have the disease? Based on the information of Table 7.4, explain your answer.

   d. Plot the posterior density of PPV.

   e. Is the posterior density of the PPV skewed? Why?

   f. Based on the estimated PPV and NPV, is the EST accurate for diagnosing CAD?

3. Based on BUGS CODE 7.1, verify the Bayesian analysis of Table 7.5. Use 55,000 observations for the simulation with a burn-in of 5000 and a refresh of 100.

    a. What is the 95% credible interval for the positive diagnostic likelihood ratio, PDLR?

    b. Plot the posterior density of the negative diagnostic likelihood ratio, NDLR.

    c. Is the posterior density of the NDLR skewed? Explain your answer.

    d. What is the MCMC error for estimating the posterior mean of the PDLR?

    e. Based on the estimated PDLR and the NDLR, is the EST accurate for the diagnosis of CAD?

4. Based on BUGS CODE 7.2, verify the Bayesian analysis of Table 7.7. Use 55,000 observations for the simulation with a burn-in of 5000 and a refresh of 100.

    a. What is posterior median of the area under the ROC curve?

    b. Is the posterior density of the ROC area skewed? Why?

    c. What are A1 and A2? What is their relation to the ROC area for mammography?

    d. Based on the estimated ROC area, is mammography accurate? Why?

    e. Plot the posterior density of AUC, the area under the ROC curve.

5. Based on BUGS CODE 7.1, verify the Bayesian analysis of Table 7.10. Use 55,000 observations for the simulation with a burn-in of 5000 and a refresh of 100.

    a. What is the posterior mean of the probability that mammography and physical examination are both positive?

    b. What is your estimate of the sensitivity of mammography?

    c. What is the posterior median of the physical examination?

    d. Plot the posterior density of the TPF of mammography.

    e. Is the posterior density of the TPF of physical examination skewed? Why?

    f. Use 110,000 observations for the simulation. How much is the MCMC error reduced for estimating the posterior mean of θ_{00}?

6. Based on Table 7.11 and BUGS CODE 7.1, verify the Bayesian analysis of Table 7.12. Use 55,000 observations for the simulation with a burn-in of 5000 and a refresh of 100.

    a. What is your estimate of the PPV for screening in round 1?

    b. What is your estimate of the sensitivity for screening in round 1?

TABLE 7.25

Results of Screening for Rounds 3, 5, and 7 of UK Early Detection of Breast Cancer

Mammography and Physical Examination	Breast Cancer	
	$D = 0$ (negative)	$D = 1$ (positive)
Negative $X = 0$	70,205	25
Positive $X = 1$	2,908	262

c. What is your estimate of the specificity for screening?

d. Is the posterior density of PPV skewed? Why?

e. Plot the posterior density of the PDLR.

f. Based on the results of Table 7.13, is screening accurate for detecting breast cancer in round 1?

7. Consider the UK Early Detection Trial and the results of screening for rounds 3, 5, and 7 given in Table 7.25. There are 73,401 subjects, among which 70,230 tested negative and 3171 tested positive for further investigation. Of the 70,230 who tested negative, 72,205 were true negative and 25 were false negative.

Using BUGS CODE 7.1, execute a Bayesian analysis with 55,000 observations for the simulation, with a burn-in of 5000 and a refresh of 100. The list statement for problem 7 assumes an improper prior.

a. Verify that the posterior mean for the FPF is .0377 with a 95% credible interval of (.0383, .0412). This implies that the specificity is approximately .94.

b. Verify that the posterior median of the PPV is .0825.

c. Plot the posterior density of the FPF.

d. Verify that the posterior mean of the TPF is .9129 with a 95% credible interval of (.8744, .9425).

e. Based on these results, describe the accuracy of the screening modality.

f. Is the posterior distribution of the negative predictive density skewed? Why?

8. Problem 8 is a continuation of problem 7, where Table 7.26 gives the results of rounds 2, 4, and 6 of the UK Early Detection Trial. There were a total of 70,449 subjects with 117 cases of cancer and 67,157 who tested negative for cancer.

Based on the above table and BUGS CODE 7.1, execute a Bayesian analysis with 55,000 observations, with a burn-in of 5000 and a

TABLE 7.26

Results with Physical Examination for Rounds 2, 4, and 6 of UK
Early Detection of Breast Cancer

	Breast Cancer	
Physical Examination	$D = 0$ (negative)	$D = 1$ (positive)
Negative $X = 0$	67,116	41
Positive $X = 1$	3,216	76

Source: Chamberlain et al., Sensitivity and specificity of screening in the
   UK trial of early detection of breast cancer, Article in *Cancer
   Screening*, pp. 3–17, Edited by Miller, A.B., Chamberlain, J., Day,
   N.E., Hakama, M., and Prorok, P.C., Cambridge University Press,
   Cambridge, UK, p. 14, 1990.

refresh of 100. Note: Use a list statement for problem 8 that assumes
an improper prior distribution for the cell probabilities.

a. What is the posterior mean of the FPF? I got .0457 with (.0442,
    .0472) as the 95% credible interval.

b. Compare the posterior mean of the FPF of physical examina-
    tion at rounds 2, 4, and 6, with the posterior mean of the FPF for
    rounds 3, 5, and 7 of problem 7.

9. Refer to Figure 7.4 and the 1999 death rate for white females in 1999.
   Assume the mortality rate is 26.5 per 100,000. Assume the female
   population was 150,000,000 females in 1999, of which 60% were
   white. What is your estimate of the number of deaths among white
   females during 1999?

10. Refer to Table 7.12 that gives the frequency distribution of the age of
    entry of the subjects of the HIP study.

    a. What percentage of patients, among the total number, were
        between 40 and 49 (inclusive) years of age?

    b. What percentage of patients of the HIP study were greater than
        59 years of age?

11. Refer to Table 7.14. What is the difference in years between the aver-
    age age at entry and the average age at the third annual examination?

12. Refer to Table 7.15.

    a. Compare the percentage of patients diagnosed as stage I disease
        between the study and control groups.

    b. What does this imply about the effectiveness of screening?

13. Refer to Figure 7.7, which displays the number of study patients
    diagnosed with breast cancer within 5 years of entry to the program.

    a. What percentage, among the 229, were interval diagnoses?

    b. What percentage, among the 229, were clinical only diagnoses?

c. A total of 97 of the patients had interval diagnoses, of which 45 were diagnosed within 12 months of their last examination, while the remaining 52 were diagnosed at least 12 months after their last examination. Assuming the 45 corresponded to earlier false negatives and ignoring the remaining 52, what is your estimate of sensitivity of the screening program?

14. Refer to Tables 7.16 and 7.17, where the former assumes the time to diagnosis from entry is no more than 60 months and the latter assumes the time to diagnosis from entry is no more than 48 months.

   a. Based on the means of Table 7.16, what is the estimated lead time?

   b. Based on the means of Table 7.17, what is the estimated lead time?

   c. Why are the two lead times different? Explain your answer.

15. Based on BUGS CODE 7.3, verify Table 7.18. Use 55,000 observations for the simulation, a burn-in of 1000 and a refresh of 100.

   a. What is the posterior mean of the mean times from entry to diagnosis for the study group?

   b. What is the posterior mean of the mean times from entry to diagnosis for the control group?

   c. What is your estimate of the lead time? Explain your answer?

   d. Plot the posterior density of the parameter diff, whose posterior distribution gives the estimated lead time.

16. Below is a list statement for the times from entry to diagnosis for the study and control groups. $x1$ is the vector for the study group with 131 entries, while $x2$ is the vector for the control group with 232 entries, assuming the time from entry to diagnosis is no more than 48 months (4 years).

   Use BUGS CODE 7.3 with the information below and perform a Bayesian posterior analysis with 55,000 observations for the simulation, a burn-in of 5000, and a refresh of 100.

   a. Find the posterior distribution of me1, the mean of the times from entry to diagnosis for the study cohort.

   b. Find the posterior distribution of me2, the mean of the times from entry to diagnosis for the control cohort.

   c. Determine the posterior distribution of the lead time diff.

   d. What is the posterior mean of the lead time?

   e. Compare the posterior distribution of the lead time with that given in Table 7.18. Why the difference in the two distributions?

```
list(n1 = 131, x1 = c(81,118,428,1265,530,468,1200,496,127,
   29,
599,48,751,1197,1375,1343,1279,575,37,595,406,27,412,
126,75,568,993,1277,192,123,12,236,1358,79,384,64,439,
```

```
47,29,845,743,1473,414,152,13,1315,549,1320,1244,1191,
1063,448,67,392,23,959,34,739,18,808,1341,1101,345,378,
875,1398,1177,53,34,1383,343,21,14,386,927,439,544,
600,347,88,987,495,792,160,16,1274,750,420,147,38,730,
110,1051,81,1335,939,871,183,490,79,404,79,404,971,123,
1260,298,59,182,964,1308,1097,16,85,300,517,22,
1456,69,16,881,73,41,1265,1322,358,673,470,217,561,
409,433,29), n2 = 232,
x2 = c(1468,,601,29,1227,1470,141,751,1297,1109,
1285,345,1429,972,345,490,1478,614,
118,50,1446,1353,510,1466,1177,146,70,1098,
1352,1470,728,534,193,1446,14,1474,197,1037,
230,696,440,491,576,1025,1616,575,330,604,
528,1118,1038,77,1413,1240,358,110,349,214,1096,      1143,729
,412,105,16,684,434,7,330,1433,
87,1028,1237,274,682,679,703,755,505,966,209,1273,
1361,495,1368,258,800,532,878,1333,802,390,1482,647,
609,622,1472,1041,1343,600,
16,1149,415,309,164,68,919,619,1185,1044,
1335,546,93,832,168,989,550,1295,99,366,559,
756,822,139,724,446,511,1304,388,1328,1467,
   810,252,392,331,789,22,53,
426,1125,719,255,1098,1261,981,44,601,
973,549,1237,274,1309,520,1478,572,985,797,
513,568,912,185,764,1320,1407,779,
313,1322,896,151,1427,463,511,296,1362,1131,1175,
807,1154,79,415,921,659,588,163,198,417,839,1306,
1192,778,670,1066,1098,1270,133,1119,423,912,379,
734,155,1460,287,555,41,1081,811,1155,1147,
722,1011,1189,558,236,1301,1405,266,932,407,
428,1216,208,657,518,379,864,1051,1037,510,77,1105,319,
83,169,464))
```

17. Refer to BUGS CODE 7.4, where the second list statement includes the survival experience of the 225 patients of the screened cohort and of the 295 patients of the control cohort that were diagnosed with breast cancer. Perform a Bayesian analysis with 55,000 observations generated for the simulation, a burn-in of 5000, and a refresh of 100. Compute the posterior means and standard deviations of the survival rate for the screened group and plot the survival rate of the three cohorts versus the period, and note that Table 7.21 reports the posterior means of the survival rate for the control and study groups.

    a. What is the median survival for the three cohorts?

    b. Compare the median of the screened cohort to that of the study group.

    c. Does it seem that the survival experience of the screened cohort is better than that of the control? Does this show that screening is efficacious?

18. Using the information in Table 7.22 and BUGS CODE 7.5, verify the posterior analysis of Table 7.23. Execute the analysis with 60,000

observations for the simulation with a burn-in of 5000 and a refresh of 100.

a. What is the 95% credible interval for some1, the sum of the observed minus expected deaths of the study cohort?

b. Based on the answer to part a, are the survival patterns of both cohorts the same?

c. Plot the posterior density of some1.

d. Based on Table 7.23, is screening efficacious?

e. What information is conveyed by the posterior distribution of the likelihood ratio LR?

19. Refer to BUGS CODE 7.5 and the last two list statements. One contains the data for the screened and control groups, where the screened cohort has 225 patients and the control 295 over a 20-year period. The objective here is to compare the survival profile of the two cohorts with the log-rank test, which can be executed with BUGS CODE 7.5. Perform the simulation with 55,000 observations, a burn-in of 5000, and a refresh of 100.

a. Verify the following posterior analysis.

If the two cohorts do not have similar survival profiles, one would expect the sum of the observed minus expected deaths for the screened group to not be zero (on the average) and the 95% credible interval for some1 to not include zero. In fact the 95% credible interval is (−38.83, −7.4530). See Table 7.27.

b. Is the posterior distribution of some1 skewed?

TABLE 7.27

Posterior Analysis for Bayesian Log-Rank Test (Screened versus Control Cohorts)

Parameter	Mean	SD	Error	2½	Median	97½
ome[1]	−4.418	2.334	.0107	−9.358	−4.3	−.2186
ome[2]	−9.394	3.001	.0137	−15.64	−9.26	−3.877
ome[3]	−5.293	2.546	.0112	−10.66	−5.142	−.6998
ome[4]	−6.3	2.496	.0107	−11.54	−6.176	−1.805
ome[5]	−.8617	2.344	.0104	−5.85	−.7284	3.346
ome[6]	3.08	2.145	.0091	−1.523	3.237	6.851
ome[7]	−3.32	1.721	.0075	−7.093	−3.181	−.3843
ome[8]	1.044	2.143	.0096	−3.547	1.186	4.83
ome[9]	−.8982	1.915	.00845	−5.023	−.747	2.404
ome[10]	3.542	1.857	.0091	−.541	3.692	6.735
ome[11]	.0873	1.166	.0051	−2.582	.2366	1.922
ome[12]	−.8098	1.63	.0076	−4.387	−.6649	1.957
ome[13]	.617	1.425	.0065	−2.56	.7718	2.952

(*Continued*)

TABLE 7.27 (Continued)

Posterior Analysis for Bayesian Log-Rank Test. Screened versus Control Cohorts

Parameter	Mean	SD	Error	2½	Median	97½
ome[14]	−.9191	1.168	.0048	−3.598	−.7724	.9069
ome[15]	.0502	1.177	.0048	−2.653	.1972	1.908
ome[16]	−.0819	1.231	.0053	−2.945	.0827	1.844
ome[17]	.5523	1.251	.0052	−2.24	.687	2.592
ome[18]	.5325	.4533	.0019	−.6632	.6703	.987
ome[19]	−.0090	.0625	.000288	−.0870	0	0
ome[20]	−.00726	.0462	.000195	−.0757	0	0
some1	−22.81	8.017	.0355	−38.83	−22.63	−7.453

    c. Plot the posterior density of some1.

    d. Does screening give a survival advantage to women? Explain your answer!

    e. What is the interpretation of ome[12]?

References

1. Morrison, A.S. *Screening in Chronic Disease*, Oxford University Press, 2nd edition, 1992, New York.
2. Shpiro, S., Venet, W., Strax, P., and Venet, L. *Periodic Screening for Breast Cancer, The Health Insurance Project and its Sequelae, 1963-1986*, Johns Hopkins University Press, 1988, Baltimore.
3. Miller, A.B., Chamberlain, J., Day, N.E., Hakama, M., and Prorok, P.C. *Cancer Screening*, Cambridge University Press, 1990, Cambridge, UK.
4. Chamberlain, J. et al. Sensitivity and specificity of screening in the UK trial of early detection of breast cancer, Article in *Cancer Screening*, pp. 3–17, Edited by Miller, A.B., Chamberlain, J., Day, N.E., Hakama. M., and Prorok, P.C., Cambridge University Press, 1990, Cambridge, UK.
5. Aberle, D.R., Adams, A.M., Berg, C.D., Blacl, W.C., Clapp, J.D., Fagerstrom, R.M., Gareen, I.F., et al. Reduced lung cancer mortality with low-dose computed tomographic screening, *N. Eng. J. Med.*, 365(5), 395–409, 2011.
6. Mausner, J.S. and Kramer, S. *Epidemiology-An Introductory Text*, W.B. Saunders Company, 1985, Philadelphia.
7. Pepe, M.S., *The Statistical Evaluation of Medical Tests for Classification and Prediction*, Oxford University Press, 2003, Oxford, UK.
8. Weiner, D.A., Ryan, T.J., McCabe, C.H., Kennedy, J.W., Schloss, M., Tristani, F., Chaitman, B.R., et al. Correlations among history of angina, ST-segmented response and prevalence of coronary artery disease, *N. Eng. J. Med.*, 301, 230, 1979.
9. Zhou, X.H., McClish, D.K. and Obuchowski, N.A. *Statistical Methods for Diagnostic Medicine*, John Wiley & Sons, 2002, New York.

10. Chamberlain, J. et al. First results on mortality in the UK trial of early detection of breast cancer, *The Lancet*, 332, 411–416, 1988.
11. Holleb, A., Venet, L., Day, E., and Hoyt, S. Breast cancer detected by routine examination, *N.Y. State Journal of Medicine*, 60, 823, 1960.
12. Gilbertsen, W. and Krelsberg, I. Detection of breast cancer by periodic utilization of methods of physical diagnosis, *Cancer*, 28, 1552, 1971.
13. Day, E. and Venet, L. Periodic cancer detection as a cancer control measure, *Proceedings Fourth National Cancer Conference,* Philadelphia, Lippincott, 1961.
14 .Saloman, A. Beitrage zur pathologie und klinik des mammakarzinims [Contributions to the pathology and clinic of mammakarzinims]. *Arch. F. Kun. Chir.*, 101, 573, 1913.
15. Warren, S. Roentgenologic study of the breast, *Am. J. Roentgenology*, 24, 223, 1930.
16. Egan, R. Experiences with mammography in a tumor institution, evaluation of 1000 studies, *Radiology*, 75, 894–900, 1960.
17. Clark, R., Copeland, M., Egan, R., Gallager, H.S., Geller, H., Lindsay, J.P., Robbins, L.C., and White, EC. Reproducibility of the technique of mammography for cancer of the breast, *Am. J. Surgeons*, 109, 127, 1965.
18. Strax, P. Evolution of techniques in breast screening, in *Control of Breast Cancer Through Screening*, Edited by Strax, P., PSG Publishing, 1979, Littleton, MA.
19. Wu, D., Kafadar, K., Rosner, G.L., and Broemeling, L.D. The lead time distribution, when lifetime is a random variable in periodic cancer screening, *Int. J. Biostatistics*, 8(1), 1–14. 2012.
20. Andersson, I., Aspergren, K., Janzon, I., Landberg, T., Lindholm, K., Linell, F., Ljungberg, O., et al. Mammographic screening and mortality from breast cancer: The Malmo mammographic screening trial, *British Med. J.*, 297(6654), 943–8,1988.
21. Tabar, I., Fagerberg, C.J., Gad A., Baldetorp, L., Holmberg, L.H., Gröntoft, O., Ljungquist, U., et al. Reduction in mortality from breast cancer after mass screening with mammography. Randomized trial from the Breast Cancer Screening Working Group of the Swedish National Board of Health Welfare, *Lancet*, 1(8433), 829–32, 1985.
22. Tabar, I., Fagerberg, G., Duffy, S.W., and Day NE. The Swedish two county trial of mammographic screening for breast cancer: recent results and calculation of benefit, *J. Epidemiology and Community Health*, 43(2), 107–14, 1989.
23. Roberts, M.M., Alexander, F.E., Anderson, T.J., Chetty, U., Donnan, PT., Forrest, P., Hepburn, W., et al. Edinburgh trial of screening for breast cancer: mortality at seven years, *Lancet*, 335(8684), 241–6, 1990.
24. Miller, A.B., To, T., Baines, C.J., and Wall C. The Canadian National Breast Cancer Screening Study-1: Breast cancer mortality after 11 to 16 years of follow-up. A randomized screening trial of mammography in women age 40 to 49 years, *Ann. Intern. Med.*, 137(5 part 1), 305–12, 2002.
25. Mandelblatt, J.S., Cronin, K.A., Bailey, S., Berry, D.A, de Koning, H.J., Draisma, G., Huang, H., et al. Effects of mammography screening under different screening schedules: model estimates of potential benefits and harms, *Ann. Intern. Med.*, 151(10), 738–47, 2009.
26. Frisell, J., Eklund, G., Hellstrom, I., Lidbrink, E., Rutqvist, L.E., Somell, A. Randomized study of mammography screening – preliminary report on mortality in the Stockholm trial, *Breast Cancer Res. Treat.*, 18(1), 49–56, 1991.

27. Nystrom, I., Rutqvist, L.E., Wall, S., Lindgren, A., Lindqvist, M., Rydén, S., Andersson, I., et al. Breast cancer screening with mammography: overview of Swedish trials, *Lancet*, 341(8851), 973–8, 1993.
28. Moss, S.M., Cuckle, H., Evans, S., Johns, L., Waller, M., Bobrow, L., Trial Management Group. Effect of mammographic screening from age 40 to 49 years on breast cancer mortality at 10 years follow-up: a randomized controlled trial, *Lancet*, 368(9552), 2053–60, 2006.
29. Breast Cancer Screening (PDQ). The Effect of Screening on Breast Cancer Mortality, at: http://www.cancer.gov/pdq/screening/breasr/healthprofessionals/page5.
30. Elmore, J.G., Reisch, L.M., Barton, M.B., Barlow, W.E., Rolnick, S., Harris, E.L., Herrinton, L.J., et al. Efficacy of breast cancer screening in the community according to the risk level, *J. Nat. Cancer Inst.*, 97(14) 1035–43, 2005.

8

Statistical Models for Epidemiology

8.1 Introduction

Chapter 8 presents regression models that are useful in epidemiology. Examples of some elementary models were introduced in early chapters and included simple linear and multivariate models where the dependent variable has a normal distribution or is binary. Also presented were models for survival, which included the Weibull and Cox models. The main focus of this chapter will be on models that generalize the models introduced in earlier chapters.

8.2 Review of Models for Epidemiology

In previous chapters, various models were introduced to demonstrate an association between various exposures to risk factors and disease. Regression models such as the logistic are appropriate when the dependent variable is binary (yes or no), and the independent variables are measurements of the various risk factors that represent exposures to the patient. In Chapter 4, the logistic model is defined by Equations 4.1 through 4.6, and a Bayesian analysis was used to demonstrate the association between systolic blood pressure and age for the Israeli Heart Disease Study. In this analysis, the dependent variable denotes the occurrence or nonoccurrence of a heart attack and the two independent variables are measurements of the occurrence of systolic blood pressure and age, and the analysis is executed with BUGS CODE 4.1, where the prior distribution for the regression coefficient (the intercept, the coefficient for age, and the coefficient for systolic blood pressure) is a vague normal distribution. Table 4.2 reports the Bayesian analysis that consists of the posterior distribution for the odds ratio for age and blood pressure, and the former estimate is 2.207, which means that the odds ratio for a heart attack among older patients is estimated as 2.207 times higher than that among younger patients. The logistic regression model was also used to analyze the effect of race on the occurrence of coronary artery disease. How well did the logistic model fit the data for the latter study? Table 4.6 portrays the information for the goodness-of-fit, where the posterior distribution of the predicted number of patients with coronary artery disease by race is reported. The posterior mean of the four predicted values (by race) agrees extremely well with the actual number with the disease.

Chapter 4 also introduces the student to the simple linear regression model, where the dependent variable (which measures disease status) has a normal distribution and the one independent variable measures exposure, and the model is defined by Equations 4.8 and 4.9. It should be remembered that the assumptions of a normal simple linear regression model are somewhat strict: (1) the observations should have a normal distribution, (2) the variance of all the observations are the same, and (3) the average value of the dependent variable is a linear function of the independent variable. The example of simple linear regression examines the association between systolic blood pressure and age, that is, do older people, on the average, have higher blood pressure than younger people? See Table 4.9. The Bayesian approach assumes vague normal distributions for the prior and examines the posterior distribution of the coefficient for age. BUGS CODE 4.3 is executed with 55,000 observations, and the posterior analysis reported in Table 4.10, with a posterior mean of 2.19 reported for the coefficient of age. The 95% credible interval for the age coefficient is (.9094, 3.478) and implies that for each year increase in age, the average systolic blood pressure increases by 2.19 units. The goodness-of-fit of the model is examined by plotting the predicted values (the posterior means of the predicted values) versus the actual blood pressure value, see Figure 4.5.

Simple linear regression is generalized to multiple linear regression, defined by Equation 4.10, and demonstrated with a model with two independent variables. The dependent variable is again systolic blood pressure and the independent variables are age and weight, with values given in Table 4.9. The unknown parameters are the intercept term, the coefficient for age, and that for weight, and the variance about the regression line. Emphasis will be primarily on the coefficients for age and systolic blood pressure. For multiple regression, the interpretation of the regression coefficients is somewhat restrictive. For example, the age coefficient is assumed to be the same for all values of the independent variables. Table 4.11 reports the Bayesian analysis with posterior means of −.05125 and .8173 for age and weight, respectively. This implies that for each increase of one unit of weight, the average blood pressure increases by .8173 mm of Hg (for each possible age). Figure 4.8 is a plot of the predicted values of blood pressure versus the corresponding actual values and shows that the fit is excellent. Another example of multiple linear regression is demonstrated with cigarette consumption as a function of price per pack and personal income. BUGS CODE 4.4 is executed with 55,000 observations for the simulation, and the analysis is reported in Table 4.13. As with simple linear regression, the prior distribution for the regression coefficients is assumed to be a vague normal with mean zero and precision .0001.

Chapter 4 also introduces the idea of weighted liner regression when the dependent variable has a normal distribution, but where the observations do not all have the same variance. When the variance of the observations is proportional to some function of the mean of the observations, there are special transformations (of the observations) that will stabilize the variance of the observations. For example, Table 4.16 reports that if the variance is proportional to the mean, then replacing each observation by its square root will stabilize the variance. Or, if the variance is proportional to the square root of the mean, then the log-log transformation will stabilize the variance. A weighted regression example involving state expenditures for education is analyzed with the square root transformation with the results of the Bayesian analysis portrayed in Table 4.17.

Thus, up to this point, the logistic and normal regression models have been reviewed, and now the models for analysis of survival studies will be summarized. The last class of models to be presented from previous chapters is that employed for the analysis of survival studies, namely, the Weibull and Cox models. A good example of a parametric survival model is the Weibull, where it is assumed that the distribution of the survival times follow a Weibull distribution, but on the other hand, the Cox model is a good example of a nonparametric model, because no parametric assumption is made about the distribution of survival times. Recall from Chapter 6 that survival models were introduced after explaining the Bayesian analysis of life tables. Survival models estimate the survival experience of a cohort of people based on the survival times and censoring times, as do the Weibull and Cox models. The Weibull model is defined by Equations 6.23 through 6.27, and the Bayesian analysis is illustrated using the Freireich et al.[1] data of two groups of leukemia patients. One group is the treatment group and the other placebo, and several patients have censored times, where the analysis is executed with BUGS CODE 6.3 and 55,000 observations for the simulation, and the results given by Table 6.9. Median survival is reported for both groups with a median survival of 28.26 weeks for the treatment group (as estimated by the posterior mean) compared with 7.541 weeks for the placebo. Another example also displayed the Bayesian analysis for one group of patients undergoing dialysis for kidney disease, where the main end point is the time to first infection. See Table 6.11 for the Bayesian analysis, which estimated the median time to infection as 81.46 days (as estimated by the posterior mean). In both examples, the prior distribution for the parameters of the Weibull survival model is assumed to be vague and uninformative. The Weibull model is expanded to include covariates and the dialysis is revisited using age and gender and their interaction as covariates, and the Bayesian analysis is executed with BUGS CODE 6.5 and results reported in Table 6.14. It is found that the coefficients of the model are important in predicting time to first infection.

The Cox proportional hazards model is one of the most important tools for the analysis of survival data, and the model is defined in Chapter 6 by the hazard function

$$h(t,x) = h_0(t)\exp\sum_{i=1}^{i=p}\beta_i X_i$$

where $h_0(t)$ is the baseline hazard function, X_i is the i-th of patient characteristics, and

$$\beta = (\beta_1, \beta_2, ..., \beta_p)$$

is a vector of unknown parameters. The model has the characteristic that the hazard ratio

$$HR(X^*, X) = \exp\sum_{i=1}^{i=p}\beta_i(X_i^* - X_i)$$

where X_i^* are the covariate values for one patient and X_i that for another. Note that the hazard ratio is not a function of time but only of the two covariate vectors and unknown parameters.

The Cox model is illustrated with the Freireich et al.'s study of two groups of leukemia patients, and the analysis executed with BUGS CODE 6.6 using 55,000 observations for the simulation; the posterior analysis is reported in Table 6.15. It was shown that the hazard ratio for the recurrence of the treatment group relative to the placebo group have a posterior mean of 5.093 with a 95% credible interval of (2.133, 10.84). The chapter is concluded where the Cox proportional hazards model is expanded to include multiple covariates, and Freireich et al.'s[1] study is reexamined with the inclusion of log white blood cell count as an additional covariate (group membership of treatment versus placebo is the other covariate). A Bayesian analysis based on BUGS CODE 6.7 is executed and reported in Table 6.16.

The remainder of Chapter 8 presents the reader with models that are generalizations of those presented in previous chapters and reviewed in this section.

8.3 Categorical Regression Models

Categorical and ordinal regression models are generalizations of the logistic model, where the dependent variable has two values; thus, regression models with dependent variables that assume a small number of responses will be presented. The model is referred to as an ordinal regression when the responses of the dependent variable are ordinal (can be ordered from smallest to largest).

An ordinal regression model is employed to estimate the receiver operating characteristic (ROC) area for medical tests with ordinal scores. This particular formulation of regression uses an underlying latent scale assumption. The cumulative odds model is often expressed in terms of an underlying continuous response.

The following specification of the ordinal model follows Congdon (p. 102),[2] where the observed response score Y_i with possible values 1, 2, ..., K is taken to reflect an underlying continuous part of the cumulative probability

$$\gamma_{ij} = \Pr(Y_i \le j) = F(\theta_j - \mu_i) \tag{8.1}$$

where i = 1, 2, ..., N is the number of patients and j = 1, 2, ..., K–1.
It is noted that

$$\mu_i = \beta X_i \tag{8.2}$$

expresses the regression relationship between the ordinal responses and the covariates X_i for the ith patient. F is a distribution function and θ_j are the cut points corresponding to the jth rank. For our purposes, F is usually given a logistic or probit link, where the former leads to a proportional odds model. Suppose p_{ij} is the probability that the ith patient has response j. Then

$$\gamma_{ij} = p_{i1} + p_{i2} + ... + p_{ij} \tag{8.3}$$

Of course, the above equation can also be inverted to give

$$p_{i1} = \gamma_{i1}$$

and $$p_{ij} = \gamma_{ij} - \gamma_{i,j-1} \tag{8.4}$$

$$p_{i,K} = 1 - \gamma_{i,K-1}$$

Suppose F is the logistic distribution function, and

$$C_{ij} = \log it(\gamma_{ij})$$
$$= \theta_j - \beta X_i \tag{8.5}$$

where β the vector of unknown regression coefficients is constant across response categories j. Then the θ_j are the logits of the probabilities of belonging to the categories 1, 2, ...,j as compared with belonging to the categories j + 1 ,..., K for subjects with X = 0. The difference in cumulative logits for different values of X says X_1 and X_2 are independent of j, which is called the proportional odds assumption, namely

$$C_{1j} - C_{2j} = \beta(X_1 - X_2) \tag{8.6}$$

Using the above ordinal regression model, the posterior distribution of the individual probabilities p_{ij} is determined, along with the probabilities q_j (j = 1, 2, ..., K) of the basic ordinal responses.

Once the posterior distribution of the basic responses is known for the diseased and nondiseased groups, the posterior distribution of the area under the ROC curve can also be computed. Several scenarios will be displayed for a given example of ordinal regression: (1) the ROC area induced by all covariates or selected subsets of covariates, and (2) the ROC area conditional on certain values of the covariates or subsets of covariates.

An example with ordinal scores, a study involving melanoma metastasis to the lymph nodes, is considered. A sentinel lymph node biopsy is performed on the patients to determine the degree of metastasis, where the diagnosis is made on the basis of the depth of the primary lesion, the Clark level of the primary lesion, and the age and gender of the patient. The procedure involves the cooperation of an oncologist, a surgical team that dissects the primary tumor, pathologists, and radiologists who perform the imaging aspect of the biopsy. A radiologist makes the primary determination of the degree of metastasis on a five-point ordinal scale, where 1 designates absolutely no evidence of metastasis, 2 means no evidence of a biopsy, 3 indicates very little evidence of metastasis, 4 implies there is some evidence, and 5 denotes strong evidence of metastasis.

The study is paired where each radiologist examines each patient. The reports of the hypothetical example are listed in Table 8.1.

The melanoma study has one covariate, namely the reader; thus, the study is analyzed under the following scenarios: (1) using the effect of the four radiologists simultaneously, (2) determining if the effect of the four is the same on the ROC area, (3) determining the ROC area separately for each reader, and (4) estimating the ROC area conditionally on a particular reader. The analysis is executed with the following code:

TABLE 8.1

Metastasis of Melanoma Patients

Reader	Metastasis	Rating of Metastasis					
		1	2	3	4	5	Total
1	0	12	10	5	2	1	30
1	1	3	7	11	13	16	50
2	0	15	7	4	3	1	30
2	1	2	8	10	12	18	50
3	0	11	9	2	3	5	30
3	1	8	10	6	10	16	50
4	0	13	8	6	2	1	30
4	1	10	6	8	14	12	50

BUGS CODE 8.1

```
model;
{
# 4 readers
# ROC area
# melanoma example

# code is from Congdon (p. 102)²
# non diseased
for(i in 1:30){for(j in 1:5){logit(ndgamma[i,j])<-ndtheta[j]
   -ndmu[i]}}
for(i in 1:30){ndp[i,1]<-ndgamma[i,1]}
for(i in 1:30){ndp[i,2]<-ndgamma[i,2]-ndgamma[i,1]}
for(i in 1:30){ndp[i,3]<-ndgamma[i,3]-ndgamma[i,2]}

for(i in 1:30){ndp[i,4]<-ndgamma[i,4]-
ndgamma[i,3]}
for(i in 1:30){ndy[i]~dcat(ndp[i,1:5])}
for(i in 1:30){ndp[i,5]<-1-ndgamma[i,4]}
# intercept depends on y
for(i in 1:30){
ndmu[i]<-ndb0[ndy[i]]+ ndx1[i]*ndb[1]+ndx2[i]*ndb[2]+ndx3[i]
   *ndb[3]}
for(i in 1:3){ndb[i]~dnorm(0,.001)}
for(i in 1:5){ndb0[i]~dnorm(0,.001)}
ndtheta[1]~dnorm(0,1)
ndtheta[2]~dnorm(0,1)
ndtheta[3]~dnorm(0,1)
ndtheta[4]~dnorm(0,1)
ndtheta[5]~dnorm(0,1)I(ndtheta[4],)
for(i in 1:5){ndq[i]<-mean(ndp[,i])}
# diseased population
for(i in 1:50){for(j in 1:5){logit(dgamma[i,j])<-dtheta[j]
   -dmu[i]}}
for(i in 1:50){dp[i,1]<-dgamma[i,1]}
for(i in 1:50){dp[i,2]<-dgamma[i,2]-
dgamma[i,1]}
for(i in 1:50){dp[i,3]<-dgamma[i,3]-
dgamma[i,2]}
for(i in 1:50){dp[i,4]<-dgamma[i,4]-
dgamma[i,3]}
for(i in 1:50){dy[i]~dcat(dp[i,1:5])}
for(i in 1:50){dp[i,5]<-1-dgamma[i,4]}
# intercept depends on y
for(i in 1:50){
dmu[i]<-db0[dy[i]]+ dx1[i]*db[1]+dx2[i]*db[2]+dx3[i]*db[3]}
for(i in 1:3){db[i]~dnorm(0,.001)}
for(i in 1:5){db0[i]~dnorm(0,.001)}
dtheta[1]~dnorm(0,1)
```

```
dtheta[2]~dnorm(0,1)
dtheta[3]~dnorm(0,1)
dtheta[4]~dnorm(0,1)
dtheta[5]~dnorm(0,1)I(dtheta[4],)
for(i in 1:5){dq[i]<-mean(dp[,i])}
# roc area
area<-a1+a2/2
a1<-dq[2]*ndq[1]+dq[3]*(ndq[1]+ndq[2])+
dq[4]*(ndq[1]+ndq[2]+ndq[3])+
dq[5]*(ndq[1]+ndq[2]+ndq[3]+ndq[4])
a2<-dq[1]*ndq[1]+dq[2]*ndq[2]+dq[3]*ndq[3]+
dq[4]*ndq[4]+dq[5]*ndq[5]
}


list(

ndy = c(1,1,1,1,1,1,1,1,1,1,1,1,
          2,2,2,2,2,2,2,2,2,2,
          3,3,3,3,3,
          4,4,
          5,
          1,1,1,1,1,1,1,1,1,1,1,1,1,1,1,
          2,2,2,2,2,2,2,
          3,3,3,3,
          4,4,4,
          5,
          1,1,1,1,1,1,1,1,1,1,1,
          2,2,2,2,2,2,2,2,2,
          3,3,
          4,4,4,
          5,5,5,5,5,
          1,1,1,1,1,1,1,1,1,1,1,1,1,
          2,2,2,2,2,2,2,2,
          3,3,3,3,3,3,
          4,4,
          5
),

ndx1 = c(1,1,1,1,1,1,1,1,1,1,1,1,1,1,1,1,1,1,1,1,
          1,1,1,1,1,1,1,1,1,1,

          0,0,0,0,0,0,0,0,0,0,0,0,0,0,0,0,0,0,0,0,
          0,0,0,0,0,0,0,0,0,0,

          0,0,0,0,0,0,0,0,0,0,0,0,0,0,0,0,0,0,0,0,
          0,0,0,0,0,0,0,0,0,0,

          0,0,0,0,0,0,0,0,0,0,0,0,0,0,0,0,0,0,0,0,
          0,0,0,0,0,0,0,0,0,0
),
```

```
ndx2 = c(0,0,0,0,0,0,0,0,0,0,0,0,0,0,0,0,0,0,0,0,
          0,0,0,0,0,0,0,0,0,0,
          1,1,1,1,1,1,1,1,1,1,1,1,1,1,1,1,1,1,1,1,
          1,1,1,1,1,1,1,1,1,1,
          0,0,0,0,0,0,0,0,0,0,0,0,0,0,0,0,0,0,0,0,
          0,0,0,0,0,0,0,0,0,0,
          0,0,0,0,0,0,0,0,0,0,0,0,0,0,0,0,0,0,0,0,
          0,0,0,0,0,0,0,0,0,0
),

ndx3 = c(0,0,0,0,0,0,0,0,0,0,0,0,0,0,0,0,0,0,0,0,
          0,0,0,0,0,0,0,0,0,0,
          0,0,0,0,0,0,0,0,0,0,0,0,0,0,0,0,0,0,0,0,
          0,0,0,0,0,0,0,0,0,0,
          1,1,1,1,1,1,1,1,1,1,1,1,1,1,1,1,1,1,1,1,
          1,1,1,1,1,1,1,1,1,1,
          0,0,0,0,0,0,0,0,0,0,0,0,0,0,0,0,0,0,0,0,
          0,0,0,0,0,0,0,0,0,0
),

# data for diseased

dy = c(1,1,1,
        2,2,2,2,2,2,2,
        3,3,3,3,3,3,3,3,3,3,3,
        4,4,4,4,4,4,4,4,4,4,4,4,4,
        5,5,5,5,5,5,5,5,5,5,5,5,5,5,5,5,5,

        1,1,
        2,2,2,2,2,2,2,2,
        3,3,3,3,3,3,3,3,3,3,
        4,4,4,4,4,4,4,4,4,4,4,4,
        5,5,5,5,5,5,5,5,5,5,5,5,5,5,5,5,5,5,

        1,1,1,1,1,1,1,
        2,2,2,2,2,2,2,2,2,2,
        3,3,3,3,3,3,
        4,4,4,4,4,4,4,4,4,4,
        5,5,5,5,5,5,5,5,5,5,5,5,5,5,5,5,

        1,1,1,1,1,1,1,1,1,1,
        2,2,2,2,2,2,
        3,3,3,3,3,3,3,3,
        4,4,4,4,4,4,4,4,4,4,4,4,4,4,
        5,5,5,5,5,5,5,5,5,5,5,5
),

dx1 = c(
```

```
            1,1,1,1,1,1,1,1,1,1,1,1,1,1,1,1,1,1,1,1,
            1,1,1,1,1,1,1,1,1,1,1,1,1,1,1,1,1,1,1,1,
            1,1,1,1,1,1,1,1,1,1,

            0,0,0,0,0,0,0,0,0,0,0,0,0,0,0,0,0,0,0,0,
            0,0,0,0,0,0,0,0,0,0,0,0,0,0,0,0,0,0,0,0,
            0,0,0,0,0,0,0,0,0,0,
            0,0,0,0,0,0,0,0,0,0,0,0,0,0,0,0,0,0,0,0,
            0,0,0,0,0,0,0,0,0,0,0,0,0,0,0,0,0,0,0,0,
            0,0,0,0,0,0,0,0,0,0,
            0,0,0,0,0,0,0,0,0,0,0,0,0,0,0,0,0,0,0,0,
            0,0,0,0,0,0,0,0,0,0,0,0,0,0,0,0,0,0,0,0,
            0,0,0,0,0,0,0,0,0,0
),

dx2 = c(0,0,0,0,0,0,0,0,0,0,0,0,0,0,0,0,0,0,0,0,
            0,0,0,0,0,0,0,0,0,0,0,0,0,0,0,0,0,0,0,0,
            0,0,0,0,0,0,0,0,0,0,
            1,1,1,1,1,1,1,1,1,1,1,1,1,1,1,1,1,1,1,1,
            1,1,1,1,1,1,1,1,1,1,1,1,1,1,1,1,1,1,1,1,
            1,1,1,1,1,1,1,1,1,1,
            0,0,0,0,0,0,0,0,0,0,0,0,0,0,0,0,0,0,0,0,
            0,0,0,0,0,0,0,0,0,0,0,0,0,0,0,0,0,0,0,0,
            0,0,0,0,0,0,0,0,0,0,
            0,0,0,0,0,0,0,0,0,0,0,0,0,0,0,0,0,0,0,0,
            0,0,0,0,0,0,0,0,0,0,0,0,0,0,0,0,0,0,0,0,
            0,0,0,0,0,0,0,0,0,0
),

dx3 = c(0,0,0,0,0,0,0,0,0,0,0,0,0,0,0,0,0,0,0,0,
            0,0,0,0,0,0,0,0,0,0,0,0,0,0,0,0,0,0,0,0,
            0,0,0,0,0,0,0,0,0,0,
            0,0,0,0,0,0,0,0,0,0,0,0,0,0,0,0,0,0,0,0,
            0,0,0,0,0,0,0,0,0,0,0,0,0,0,0,0,0,0,0,0,
            0,0,0,0,0,0,0,0,0,0,
            1,1,1,1,1,1,1,1,1,1,1,1,1,1,1,1,1,1,1,1,
            1,1,1,1,1,1,1,1,1,1,1,1,1,1,1,1,1,1,1,1,
            1,1,1,1,1,1,1,1,1,1,
            0,0,0,0,0,0,0,0,0,0,0,0,0,0,0,0,0,0,0,0,
            0,0,0,0,0,0,0,0,0,0,0,0,0,0,0,0,0,0,0,0,
            0,0,0,0,0,0,0,0,0,0
))
# initial values
list(ndtheta = c(0,0,0,0,0),dtheta = c(0,0,0,0,0))
```

The first list statement gives the basic information, where ndy refers to the ratings for those patients without metastasis, ndx1 gives the indicator (where 1 indicates the corresponding rating in ndy given by radiologist 1 and 0 otherwise) for the first reader for those patients without metastasis, ndx2 for the second reader, and so on. The variable dy refers to the rating for the patients with metastasis and dx1 is the column of the indicator (the numeral 0 indicates that the first rater did not give the rating and 1 indicates that reader 1 gives the rating) of the first reader, for those patients with metastasis, and so on. From Table 8.1 and the first list statement, the method for coding the data is obvious.

A Bayesian analysis is executed with 75,000 observations, a burn-in of 5000, and a refresh of 100. The output is given in Table 8.2 with the following identification for the parameters: area refers to the ROC area; db[1] refers to the effect of reader 1 for the diseased (those with metastasis) patients; and db[3] is the effect of reader 3. In addition, db0[1] is the estimate of the

TABLE 8.2

Bayesian Analysis for Melanoma Study with Four Radiologists

Parameter	Mean	SD	Error	2½	Median	97½
a1	.711	.0296	<.0001	.6501	.712	.7663
a2	.1522	.01172	<.0001	.1301	.1519	.1761
area	.7871	.0242	<.0001	.7373	.788	.8321
db[1]	.7786	5.535	.3389	−8.559	.9357	9.274
db[2]	−.0165	31.63	.1189	−61.86	.05461	61.66
db[3]	.2739	31.76	.1267	−61.73	.1926	62.59
db0[1]	−28.85	18.65	.2988	−73.45	−25.27	−1.922
db0[2]	−3.139	5.594	.3402	−11.94	−3.407	6.138
db0[3]	−1.006	5.56	.3391	−9.757	−1.276	8.071
db0[4]	1.478	5.523	.3367	−7.289	1.309	10.56
db0[5]	29.64	18.66	.2877	2.925	26.07	74.36
dq[1]	.1128	.0226	<.0001	.07577	.1103	.1628
dq[2]	.1153	.0308	<.0001	.0603	.1137	.1794
dq[3]	.1812	.0382	<.0001	.1105	.1797	.2602
dq[4]	.2108	.0370	<.0001	.1402	.2101	.2861
dq[5]	.3799	.0258	<.0001	.3387	.3769	.4385
ndb[1]	−9.859	12.43	.7639	−30.82	−6.585	8.275
ndb[2]	−.0148	31.86	.1198	−62.59	.0011	62.66
ndb[3]	.0700	31.52	.109	−61.67	−.0528	61.85
ndb0[1]	−23.55	21.04	.4894	−70.02	−21.36	14.22
ndb0[2]	8.081	12.45	.7644	−10.22	4.845	28.97
ndb0[3]	10.22	12.43	.7628	−8.036	7.048	31.21
ndb0[4]	11.62	12.4	.7588	−6.845	8.59	32.67
ndb0[5]	33.68	19.24	.5689	1.399	32.6	76.41
ndq[1]	.4828	.0381	<.0001	.4229	.4785	.5693
ndq[2]	.244	.0475	<.0001	.1508	.2438	.3376
ndq[3]	.1219	.0405	<.0001	.0531	.1185	.2113
ndq[4]	.0586	.0288	<.0001	.0148	.0546	.125
ndq[5]	.0926	.0292	<.0001	.0468	.0893	.1597

intercept≈corresponding to the ordinal response 1 for diseased patients, while db0[5] is the intercept corresponding to ordinal response 5. Continuing in a similar fashion, ndb[1] is the effect of reader 1 on the logit scale for nondiseased (those without metastasis) patients, ndb0[3] is the intercept corresponding to ordinal score 3 for nondiseased patients, and so on.

Also, dq[1] is the probability of ordinal score 1 for diseased patients, ndq[1] is the corresponding quantity for the nondiseased patients, and so on.

Note that with 75,000 observations for the simulation, the Monte Carlo Markov chain (MCMC) error for db[i] and db0[i] are relatively large and the corresponding posterior distributions are very skewed. Recall that the db[i], for i = 1, 2, 3 are the effects of readers 1, 2, and 3, respectively, for the diseased patients, whereas the db0[i], for i = 1, 2, 3, 4, 5, are the intercepts for the five regressions corresponding to the five ordinal responses of those patients where the disease has metastasized. There are five regressions of the cumulative logits on the readers. The same is observed for the nondiseased patients, that is, the MCMC errors are fairly large and the posterior distributions are skewed for the effects of the three readers and the five intercepts. Thus, the posterior median should be used for estimating the location of the skewed distributions. Although it does appear that the effects of the three readers on the cumulative logits are not the same for the diseased and the nondiseased patients, it is safe to say that the pattern of the posterior medians and the intercepts is the same for the diseased and the nondiseased. The skewness of the posterior distribution of db0[1] is exhibited in Figure 8.1.

On the other hand, the MCMC error is quite small for the ROC area, which has a posterior mean of .7871 (.0242) and a 95% credible interval of (.7373, .8321). The ROC area is "adjusted" for the simultaneous effects of the four readers. I also used **SPSS 11.5** to estimate the ROC area and got a value of .767 (.027). When the ROC area is estimated with reader 3 information only, the posterior mean is .6483 (.0213) and the 95% credible interval is (.6053, .6891). I revised BUGS CODE 8.1 and executed the analysis with 75,000 observations with a burn-in of 5000 and a refresh of 100. The MCMC error of the ROC area is <.0001. Thus, it appears that the readers do not have the same effect on the ROC area. When all four are used simultaneously, the area is .7871, but with reader 3, it is only .6484. Quite a difference!

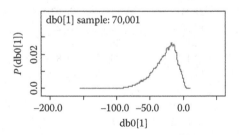

FIGURE 8.1
Posterior density of the intercept for ordinal score 1 for diseased patients.

For additional information about Bayesian categorical and/or ordinal regression, see the studies by Broemeling[3] and Johnson and Albert.[4] A good introduction to the general area of ordinal regression from a Bayesian point of view is the study by Albert and Chib.[5]

8.4 Nonlinear Regression Models

Nonlinearity is a feature of many studies involving epidemiology. A nonlinear regression model is defined as

$$Y_n = f(x_n, \theta) + e_n \tag{8.7}$$

where the nth observation of the dependent variable Y is Y_n, x_n is the corresponding observation of the q by an independent variable x, and e_n is the corresponding error term corresponding to the nth observation. It is assumed that the N error terms e_n are independent random variables with a normal distribution, mean zero, and unknown variance σ^2. The N values of Y_n and the vector x_n are known, but the p by 1 vector

$$\theta = (\theta_1, \theta_2, ..., \theta_p) \tag{8.8}$$

is assumed to be unknown.

Using a Bayesian approach, the objective is to examine the effect of exposure to q risk factors on the dependent variable Y. A prior distribution is placed on the unknown parameters θ and σ^2, then the posterior distributions of θ and σ^2 are determined via Bayes' theorem. As before, the posterior analysis will be executed using WinBUGS.

In order to illustrate the nonlinear regression, several examples will be presented as a five-step procedure:

1. Plots of the independent variables versus the dependent variable will be portrayed.
2. A model will be defined based on the plots of (1).
3. A prior distribution will be assumed for the unknown parameters θ and σ^2.
4. Based on WinBUGS, the posterior distribution of θ and σ^2 will be determined.
5. The goodness-of-fit of the model will be assessed by plotting the predicted values of Y (those predicted by the model) versus the corresponding actual Y_n values.

In order to investigate the polychlorinated biphenyls (PCB) concentration in fish from Cayuga Lake, New York, Bache et al.[6] conducted a study and

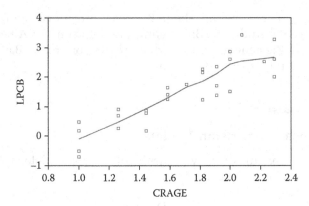

FIGURE 8.2
Log of PCB versus cube root of age.

determined the effect of the age of the fish on PCB concentration. Since the fishe are annually stocked as yearlings and distinctly marked as to year class, the ages of the fish are accurately known. The fish is mechanically chopped, ground, and mixed, and 5-g samples taken. Age is recorded in years and the PCB concentration is expressed as parts per million (ppm). PCB is a toxin, and it is important for public health to know its concentration in the environment. The data are listed in the first-list statement of BUGS CODE 8.2.

In order to investigate the effect of age on PCB, the plot of the log of the PCB values versus the cube root of age reveals a linear association as depicted in Figure 8.2.

Based on the plot, the regression model is assumed to be

$$\ln(Y_n) = \theta_1 + \theta_2 * x_n + e_n \tag{8.9}$$

where n = 1, 2, ..., 28, and 28 is the total number of observations shown in Table 8.3. The dependent variable is the natural log of PCB = Y_n and the independent variable is cube root of age = x_n, and the association is linear with unknown regression coefficients $\theta = (\theta_1, \theta_2)$; thus, it can be analyzed like a simple normal linear regression model.

However, it should be noted that the association between PCB and age can be expressed as nonlinear regression, namely,

$$Y_n = \exp\left(\beta_1 + \beta_2 * \sqrt[3]{x_n}\right) \tag{8.10}$$

A Bayesian analysis is based on Equation 8.9 and BUGS CODE 8.2.

The program statements are well documented and the code closely follows Equations 8.9 and 8.10. Note that the first list statement contains two columns, one for the PCB data and the other for the age of the fish.

TABLE 8.3

Posterior Analysis for PCB Study: Concentration of PCB in Lake Cayuga Trout

Parameter	Mean	SD	Error	2½	Median	97½
Theta[1]	−2.398	.4148	.01064	−3.21	−2.4	−1.581
Theta[2]	2.307	.2406	.0061	1.831	2.307	2.778
Sigma	.267	.08064	.000503	.1528	.2523	.4635
W[1]	−.0889	.552	.005062	−1.18	−.08778	.9973
W[2]	−.09105	.552	.005108	−1.181	−.08926	.991
W[3]	−.09073	.5507	.005058	−1.177	−.09177	.99
W[4]	−.0898	.5523	.005026	01.173	−.08976	1.003
W[5]	.5047	.5375	.003725	−.5582	.5063	1.56
W[6]	.5075	.537	.003669	−.5554	.5118	1.57
W[7]	.5078	.5346	.00373	−.5473	.5094	1.555
W[8]	.924	.5259	.00280	−.1065	.9226	1.965
W[9]	.9275	.5299	.00294	−.1233	.9334	1.969
W[10]	.9289	.5303	.00294	−.1231	.9334	1.969
W[11]	1.263	.5264	.00252	.2201	1.265	2.309
W[12]	1.264	.5302	.00240	.2336	1.264	2.315
W[13]	1.265	.527	.00217	.232	1.265	2.307
W[14]	1.541	.5297	.00225	.486	1.542	2.579
W[15]	1.79	.5288	.00239	.7396	1.791	2.834
W[16]	1.791	.5259	.00239	.7554	1.789	2.849
W[17]	1.792	.5276	.00252	.75	1.794	2.831
W[18]	2.014	.5283	.00257	.9642	2.013	3.064
W[19]	2.007	.5279	.00262	.9633	2.008	3.045
W[20]	2.012	.5316	.00261	.9637	2.011	3.062
W[21]	2.213	.5312	.00286	1.162	2.214	3.268
W[22]	2.214	.5313	.00263	1.16	2.213	3.267
W[23]	2.213	.5309	.00294	1.168	2.211	3.265
W[24]	2.396	.5352	.00330	1.339	2.392	3.454
W[25]	2.727	.5391	.00386	1.657	2.729	3.793
W[26]	2.876	.5465	.00418	1.797	2.876	3.596
W[27]	2.878	.545	.00402	1.792	2.877	3.95
W[28]	2.878	.5443	.003869	1.801	2.877	3.954

BUGS CODE 8.2

```
model;
{
# nonlinear regression of PCB on cube root of age
for(i in 1:N){y[i]~dnorm(vu[i], tauy)}
for(i in 1:N){vu[i]<-exp(beta[1]+beta[2]*x[i])}
# predicted values of PCB
for(i in 1:N){u[i]~dnorm(vu[i], tauy)}
# linear regression of log PCB on cube root of age
for(i in 1:N){z[i]~dnorm(mu[i], tau)}
# z is natural log of y
```

```
for(i in 1:N){z[i]<-log(y[i])}
for(i in 1:N){mu[i]<-theta[1]+theta[2]*x[i]}
# x is the cube root of age
for(i in 1:N){x[i]<-pow(age[i],.333)}
# prior distribution of the beta
for(i in 1:2){beta[i]~dnorm(0.000,0.0001)}
for(i in 1:2){theta[i]~dnorm(0.000,0.0001)}
# prior distribution for tau
tau~dgamma(.0001,.0001)
tauy~dgamma(.0001,.0001)
# predicted values of log PCB
for(i in 1:N){w[i]~dnorm(mu[i], tau)}
# sigma is the inverse of the precision tau
sigma<-1/tau
}

# PCB data
list(N = 28,
y = c(.6,1.6,.5,1.2,2,1.3,2.5,2.2,2.4,1.2,3.5,4.1,5.1,5.7,3.4,
   9.7,8.6,
4.0,5.5,10.5,17.5,13.4,4.5,30.4,12.4,13.4,26.2,7.4),
age = c(1,1,1,1,2,2,2,3,3,3,4,4,4,5,6,6,6,7,7,7,8,8,8,9,11,12,
   12,12))

# initial values
list(beta = c(0,0), theta = c(0,0), tauy = 1,tau = 1)
```

A Bayesian analysis is executed with 55,000 observations for the simulation, a burn-in of 5000 and a refresh of 100, and the results are reported in Table 8.3.

The simple linear regression of log PCB on the cube root of age has an estimated (posterior mean) intercept of −2.398 and 95% credible interval of (−3.21, −1.581), while the slope is estimated as 2.307 (posterior mean) with a 95% credible interval of (1.831, 2.778); thus, as the cube root of age increases by one unit, the average log PCB increases by 2.307 units. The W-values are the predicted values of PCB concentration corresponding to the observed concentration values of PCB.

Consider an alternative way of assessing the association between PCB and age directly using the nonlinear regression model

$$Y_n = \exp(\beta_1 + \beta_2 * \sqrt[3]{x_n}) + e_n \qquad (8.11)$$

where the thetas have been replaced by betas and an error term has been added.

BUGS CODE 8.2 is executed again with 55,000 observations for the simulation, a burn-in of 5000, and a refresh of 100 to give the following results for the nonlinear regression of PCB of the cube root of age.

From Table 8.4, the posterior mean of the intercept is −1.67 with a 95% credible interval of (−4.042, .05629), and the posterior mean of the slope is 1.981 with a 95% credible interval of (1.151, 3.063). From Figure 8.3, which is a plot of the posterior density of the slope, it appears that the distribution is symmetric about the posterior mean 1.981.

Note that U[2] is the predicted value of the PCB for the second observation of 1.60. The intercept and slope of the simple linear regression of log PCB on the cube root of age (Equation 8.9) are estimated (by the posterior mean) as −2.398 and 2.307, respectively. These are different from those estimated from the nonlinear regression (Equation 8.10), but not too different, and note that the two models are not the same.

TABLE 8.4

Posterior Analysis for Nonlinear Regression of PCB on the Cube Root of Age: PCB Concentration in Lake Cayuga Trout

Parameter	Mean	SD	Error	2½	Median	97½
Beta[1]	−1.67	1.047	.01268	−4.042	−1.565	.05629
Beta[2]	1.981	.4899	.00583	1.151	1.939	3.063
Tauy	.0362	.0101	.0000598	.01911	.0352	.0585
U[1]	1.55	5.543	.02417	−9.351	1.532	12.6
U[2]	1.618	5.553	.02446	−9.279	1.566	12.62
U[3]	1.559	5.521	.02608	−9.281	1.513	12.54
U[4]	1.559	5.514	.02703	−9.338	1.566	12.44
U[5]	2.475	5.552	.02601	−8.529	2.454	13.46
U[6]	2.472	5.573	.0254	−8.551	2.462	13.56
U[7]	2.476	5.581	.02662	−8.512	2.453	13.5
U[8]	3.463	5.563	.02647	−7.588	3.483	14.45
U[9]	3.488	5.588	.02711	−7.545	3.49	14.51
U[10]	3.502	5.594	.02824	−7.512	3.504	14.52
U[11]	4.544	5.609	.02973	−6.587	4.566	5.6
U[12]	4.538	5.613	.02673	−6.591	4.52	15.6
U[13]	4.547	5.593	.02744	−6.52	4.533	15.64
U[14]	5.568	5.612	.03007	−5.498	5.697	16.7
U[15]	7.031	5.615	.02692	−4.418	7.018	18.06
U[16]	7.003	5.63	.02964	−4.094	7.011	18.13
U[17]	6.992	5.624	.03191	−4.165	6.998	18.07
U[18]	8.835	5.631	.03129	−2.862	8.433	19.46
U[19]	8.411	5.656	.03055	−2.741	8.393	19.56
U[20]	8.435	5.607	.028	−2.751	8.437	19.48
U[21]	9.933	5.646	.02855	−1.201	9.91	21.07
U[22]	9.919	5.641	.03111	−1.246	9.953	20.96
U[23]	9.292	5.625	.02733	−1.31	9.953	21.02
U[24]	11.64	5.623	.02987	.5627	11.65	2.71
U[25]	15.51	5.833	.03418	4.03	15.51	27.04
U[26]	17.68	6.049	.03663	5.75	17.7	29.62
U[27]	17.7	6.048	.03822	5.756	17.72	29.56
U[28]	17.64	6.047	.03573	5.671	17.67	29.57

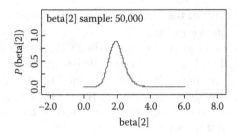

FIGURE 8.3
Posterior density of slope. Nonlinear regression.

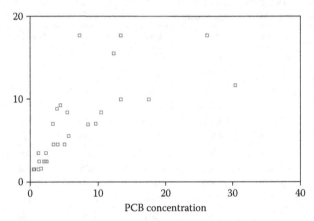

FIGURE 8.4
Predicted PCB versus PCB nonlinear regression.

How well does the nonlinear regression model fit the data? To investigate this, plot the 28 U values of Table 8.4 versus the actual PCB values (Figure 8.4). For a good fit, one would expect the plot to be linear with a slope of 1 going through the origin; thus, one might conclude that the nonlinear regression model is a "bad" fit. How would you improve the fit to the PCB study?

As a second example, consider Sredni's[7] study that investigated the transport of sulfite ions suspended in a salt solution of blood cells. The chloride concentration in percentage versus time is displayed in Figure 8.5.

Table 8.5 reports the time and concentration measurements corresponding to Figure 8.5. How does time affect the concentration?

The main objective of the study is to determine the effect of time on concentration with the model

$$Y_n = \theta_1 \left(1 - \exp\left(-\theta_2 x_n\right)\right) + e_n \tag{8.12}$$

which will be used for the analysis, where the main parameter is θ_2, which measures the decrease in minus the concentration per unit time.

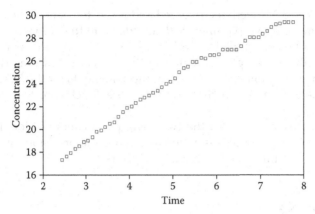

FIGURE 8.5
Chloride concentration (%) versus time (min.).

TABLE 8.5

Chlorine Concentration Values versus Time

Time (min.)	Concentration (%)	Time (min.)	Concentration (%)	Time (min.)	Concentration (%)
2.45	17.3	4.25	22.6	6.05	26.6
2.55	17.6	4.35	22.8	6.15	27.0
2.65	17.9	4.45	23.0	6.25	27.0
2.75	18.3	4.55	23.2	6.35	27.0
2.85	18.5	4.65	23.4	6.45	27.0
2.95	18.9	4.75	23.7	6.55	27.3
3.05	19.0	4.85	24.0	6.65	27.8
3.15	19.3	4.95	24.2	6.75	28.1
3.25	19.8	5.05	24.5	6.85	28.1
3.35	19.9	5.15	25	6.95	28.1
3.45	20.2	5.25	25.4	7.05	28.4
3.55	20.5	5.35	25.5	7.15	28.6
3.65	20.6	5.45	25.9	7.25	29.0
3.75	21.1	5.55	25.9	7.35	29.2
3.85	21.5	5.65	26.3	7.45	29.3
3.95	21.9	5.75	26.2	7.55	29.4
4.05	22.0	5.85	26.5	7.65	29.4
4.15	22.3	5.95	26.5	7.75	29.4

Source: Sredni, J. Problems of Design, Estimation, and Lack of Fit in Model Building, Ph.D. Thesis, University of Wisconsin Madison, 1970.

BUGS CODE 8.3 is well documented and is written to execute the analysis that estimates the two parameters θ_1 and θ_2. Note that # indicates the code that will not be executed! The code consists of two nonlinear regressions, one for the BOD example presented in the exercises and the other for the chlorine concentration program. All of the statements for the BOD program are preceded by a #, which deactivates the BOD nonlinear regression statements.

The prior distributions for the two theta parameters of Model 8.12 have normal (1, .0001) distributions, whereas that for the precision tau is gamma (.0001, .0001), noninformative vague type priors.

BUGS CODE 8.3

```
model;
{

# nonlinear regression of BOD on time
#for(i in 1:N){y[i]~dnorm(mu[i], tau)}
#for(i in 1:N){mu[i]<-theta[1]*(1-exp(-theta[2]*x[i]))}
# predicted values of BOD
#for(i in 1:N){w[i]~dnorm(mu[i], tau)}
# prior distributions
#theta[1]~ dnorm(1,.001)
#theta[2]~ dnorm(1,.001)
# tau~dgamma(.001,.001)

# regression of chlorine concentration on time
for(i in 1:N){y[i]~dnorm(mu[i], tau)}
for(i in 1:N){mu[i]<-theta[1]*(1-exp(-theta[2]*x[i]))}
# predicted chlorine values
for(i in 1:N){w[i]~dnorm(mu[i], tau)}
# prior distributions

theta[1]~ dnorm(1,.001)
theta[2]~ dnorm(1,.001)
tau~dgamma(.0001,.0001)
}

# BOD data
list(y = c(8.3,10.3,19,16,15.6,19.8),
x = c(1,2,3,4,5,7),N = 6)
# initial values BOD

list(theta = c(20,.24))

# chlorine concentration
list(N = 54,
```

```
y = c(17.3,17.6,17.9,18.3,18.5,18.9,19,19.3,19.8,19.9,20.2,
      20.5,20.6,21.1,21.5,21.9,22,22.3,22.6,22.8,23.0,23.2,
      23.4,23.7,24,24.2,24.5,25,25.4,25.5,25.9,25.9,26.3,26.2,
      26.5,26.5,26.6,27,27,27,27,27.3,27.8,28.1,28.1,28.1,28.4,
      28.6,29,29.2,29.3,29.4,29.4, 29.4),
x = c(2.45,2.55,2.65,2.75,2.85,2.95,3.05,3.15,3.25,3.35,3.45,3.55,
3.65,3.75,3.85,3.95,4.05,4.15,4.25,4.35,4.45,4.55,4.65,4.75,
      4,85,4.95,5.05,5.15,5.25,5.35,5.45,5.55,5.65,5.75,5.85,5.95,
      6.05,6.15,6.25,6.35,6.45,6.55,6.65,6.75,6.85,6.95,7.05,
      7.15,7.25,7.35,7.45,7.55,7.65,7.75))
# initial values chlorine
list(theta = c(0,0), tau = 1)
```

The Bayesian analysis is based on Table 8.5, and BUGS CODE 8.3 is executed with 250,000 observations for the simulation, a burn-in of 5000, and a refresh of 100. The posterior analysis is reported in Table 8.6.

The precision about the regression curve is estimated as .7886 (corresponding to a variance of 1.26) with the posterior mean, and the parameter θ_2 of the nonlinear regression (8.12) has a posterior mean of .3322 with a 95% credible interval of (.306, .3613). The other parameter θ_1 has a posterior median of 30.68 and appears to have a symmetric posterior distribution.

In order to determine how well the nonlinear regression Model 8.12 fits the data of Table 8.5, the actual chlorine concentration values are predicted using the statement

```
for (i in 1:N){w[i]~dnorm(mu[i], tau)}
```

in BUGS CODE 8.3. Note that the predicted chlorine concentration values are based on the posterior distribution of

$$Y_n = \theta_1\left(1 - \exp\left(-\theta_2 x_n\right)\right) \tag{8.13}$$

for n = 1, 2, 3, ..., 54. Note that the joint posterior distribution of θ_1 and θ_2 induces the distribution of the predicted values (Equation 8.13).

The goodness-of-fit of Model 8.12 is assessed graphically by plotting the actual concentration values of Table 8.5 versus the corresponding predicted values of Table 8.7.

TABLE 8.6

Posterior Analysis for Chlorine Ion Concentration

Parameter	Mean	SD	Error	2½	Median	97½
Tau	.7886	.1546	.000924	.5153	.7785	1.12
Theta[1]	30.68	.5139	.01836	29.71	30.68	31.71
Theta[2]	.3322	.01418	.000477	.306	.3312	.3613

Based on Figure 8.6 and a Pearson correlation of .985, it appears that the model is a very good fit for the data.

For additional information about nonlinear regression see the study by Bates and Watts,[8] which presents a good general introduction, while Denison et al.'s[9] book is a Bayesian approach.

TABLE 8.7

Predicted Chlorine Ion Concentration

Time (min.)	Concentration (%)	Time (min.)	Concentration (%)	Time (min.)	Concentration (%)
2.45	17.07	4.25	23.19	6.05	26.41
2.55	17.52	4.35	23.42	6.15	26.55
2.65	17.94	4.45	23.67	6.25	26.68
2.75	18.36	4.55	23.89	6.35	26.82
2.85	18.76	4.65	24.11	6.45	26.94
2.95	19.15	4.75	24.33	6.55	27.06
3.05	19.53	4.85	22.54	6.65	27.18
3.15	19.89	4.95	30.68	6.75	27.29
3.25	20.24	5.05	24.74	6.85	27.4
3.35	20.58	5.15	24.93	6.95	27.51
3.45	20.91	5.25	25.12	7.05	27.61
3.55	21.23	5.35	25.29	7.15	27.71
3.65	21.54	5.45	25.47	7.25	27.81
3.75	21.83	5.55	25.64	7.35	27.9
3.85	22.13	5.65	25.81	7.45	27.99
3.95	22.4	5.75	25.97	7.55	28.08
4.05	22.67	5.85	26.12	7.65	28.17
4.15	22.93	5.95	26.27	7.75	28.25

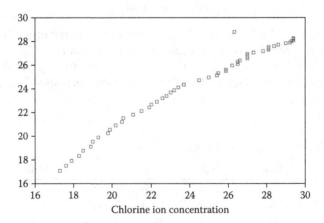

FIGURE 8.6
Predicted chlorine concentration (%) versus actual concentration.

8.5 Repeated Measures Model

Our first encounter with a repeated measures study was an example involving Alzheimer's disease in a study by Hand and Taylor,[10] where two groups of patients were compared. One group received a placebo and the other group received lecithin. Each of the 26 patients in the placebo group and 22 in the treatment group were measured five times, where the measurement was the number of words the subject could recall from a list of words. Note that the same measurement is repeated on the same subject for a fixed number of occasions, and one would expect that the measurements correlate. The unique aspect of a repeated measures study is the presence of correlation between measurements on the same subject. From a statistical point of view, this correlation is taken into account when estimating the other parameters of the model. The data of this study appear in the first list statement of BUGS CODE 8.4.

In order to analyze the Alzheimer's information, the following model is adopted.

Let the observation for the ith subject on occasion j be

$$y_{ij} = \theta + \alpha_i + \beta_j + e_{ij} \tag{8.14}$$

where i = 1, 2, ..., n, j = 1, 2, ..., p, where n is the number of subjects and p the number of time points.

It is assumed that θ is a constant,

$$\alpha_i \sim nid(0, \tau_\alpha), \quad i = 1, 2, ..., n \tag{8.15}$$

$$\beta_j \sim nid(0, \tau_\beta), \quad j = 1, 2, ..., p \tag{8.16}$$

and

$$e_{ij} \sim nid(0, \tau) \tag{8.17}$$

The variance of α_i is $\sigma_\alpha^2 = 1/\tau_\alpha$, of β_j is $\sigma_\beta^2 = 1/\tau_\beta$, and of e_{ij} is $\sigma^2 = 1/\tau$, where all the three taus are positive. The variance component $\sigma_\alpha^2 = 1/\tau_\alpha$ measures the variability of the observations between the various subjects, the component $\sigma_\beta^2 = 1/\tau_\beta$ measures the variability between the several times (occasions), and $\sigma^2 = 1/\tau$ measures the overall variability of the $y(i, j)$ observations. Note that the θ parameter measures the overall mean of the observations.

Note that

$$cov(y_{ij}, y_{ij'}) = \sigma_\alpha^2 \tag{8.18}$$

and

$$cov(y_{ij}, y_{ij}) = \sigma_\alpha^2 + \sigma_\beta^2 + \sigma^2 \tag{8.19}$$

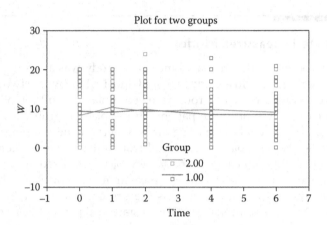

FIGURE 8.7
Alzheimer's study of number of correctly recalled words.

that is, the observations of the same subject are correlated with covariance σ_α^2, and that the common variance is $\sigma_\alpha^2 + \sigma_\beta^2 + \sigma^2$, which implies that the correlation between measurements of the same subject is

$$\rho = \frac{\sigma_\alpha^2}{(\sigma_\alpha^2 + \sigma_\beta^2 + \sigma^2)} \tag{8.20}$$

For the first example, consider a Bayesian analysis for the placebo group of Hand and Taylor's[10] Alzheimer's study. See BUGS CODE 8.4, which closely follows Formulas 8.14 through 8.20. Figure 8.7 portrays the trend of the placebo and treatment groups. The vertical axis is the number of correctly recalled words and the horizontal axis denotes the time periods at times 0, 1, 2, 4, and 6. The two lowess curves corresponding to the two groups can be used to compare the two groups. Group 1 is the placebo group and Group 2 is the treatment group.

BUGS CODE 8.4 is applicable only for the placebo group.

BUGS CODE 8.4

```
model;
{

for(i in 1:n){for(j in 1:p){y[i,j]~dnorm(mu[i,j],tau)}}
for(i in 1:n){for(j in 1:p){mu[i,j]<- theta+alpha[i]+beta[j]}}

for(i in 1:n){alpha[i]~dnorm(0,tau.alpha)}
```

```
for(j in 1:p){beta[j]~dnorm(0,tau.beta)}
for(i in 1:n){for(j in 1:p){z[i,j]~dnorm(mu[i,j],tau)}}
# prior distributions
        theta~dnorm(0,.0001)
tau.alpha~dgamma(.0001,.0001)
tau.beta~dgamma(.0001,.0001)
tau~dgamma(.0001,.0001)

sigma.alpha<-1/tau.alpha
sigma.beta<-1/tau.beta
sigma<-1/tau

}

# example with placebo group only
list(n = 26, p = 5, y = structure(.Data =
c(20,19,20,20,18,
14,15,16,9,6,
7,5,8,8,5,
6,10,9,10,10,
9,7,9,4,6,
9,10,9,11,11,
7,3,7,6,3,
18,20,20,23,21,
6,10,10,13,14,
10,15,15,15,14,
5,9,7,3,12,
11,11,8,10,9,
10,2,9,3,2,
17,12,14,15,13,
16,15,13,7,9,
7,10,4,10,5,
5,0,5,0,0,
16,7,7,6,10,
5,6,9,5,6,
2,1,1,2,2,
7,11,7,5,11,
9,16,17,10,6,
2,5,6,7,6,
7,3,5,5,5,
19,13,19,17,17,
7,5,8,8,6),.Dim = c(26,5)))

# initial values
list(alpha = c(0,0,0,0,0,0,0,0,0,0,0,0,0,0,0,0,0,0,0,0,0,0,0
,0,0),
beta = c(0,0,0,0,0), tau.alpha = 1, tau.beta = 1, tau = 1,
  theta = 0)
```

The analysis is executed with 55,000 observations for the simulation, a burn-in of 5000, and a refresh of 100. Note that vague noninformative prior distributions are placed on the model parameters. A normal distribution is used for the theta parameter and a gamma is used for the three precision parameters.

Bayesian inferences for the Alzheimer's placebo cohort follow:

1. The correlation between observations of the same patient is estimated as .753 via the posterior mean with a 95% credible interval of (.6195, .8629).

2. The variance of the main observations (the number of correctly recalled words) is estimated as 6.978.

3. The variance of the observations between individuals is estimated as 23.2 with a 95% credible interval of (12.56, 41.51).

4. σ_β^2 measures the variability of the observations between time periods, and the posterior mean is given as .1256.

5. Lastly, the average number of correctly recalled words is estimated as 9.315 with a 95% credible interval of (7.41, 11.22).

The Z values of Table 8.8 are the predicted number of correctly recalled words for the placebo group for patients 1, 2, and 26, with five values for each patient corresponding to the five time points 0, 1, 2, 4, and 6.

How well does Model 8.14 fit the data for the placebo group? Figure 8.8 depicts the association between the observed and predicted number of recalled words for the placebo group from Table 8.10. The lowess curve shows a close association implying a very good fit of the model to the data.

In order to compare the placebo with the treatment group, refer to Figure 8.7, which reveals very little difference in the two trend curves of the two groups.

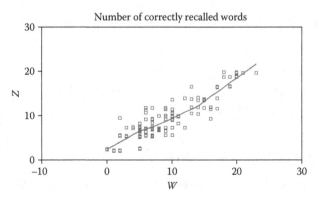

FIGURE 8.8
Goodness-of-fit predicted values versus actual values.

TABLE 8.8

Posterior Analysis of Alzheimer's Study (Placebo Group)

Parameter	Mean	SD	Error	2½	Median	97½
ρ	.7536	.06426	.000403	.6195	.7578	.8629
σ^2	6.978	.9967	.00585	5.296	6.889	9.202
σ_α^2	23.2	7.537	.04688	12.56	21.88	41.51
σ_β^2	.1256	.6324	.01072	.0000977	.005952	.9147
θ	9.315	.9729	.0238	7.41	9.317	11.22
Z[1,1]	18.82	2.884	.01297	13.16	18.84	24.5
Z[1,2]	18.78	2.874	.01169	13.13	18.8	24.4
Z[1,3]	18.9	2.883	.01315	13.22	18.9	24.54
Z[1,4]	18.72	2.907	.0136	13.02	18.72	24.39
Z[1,5]	18.69	2.882	.01401	12.99	18.68	24.36
Z[2,1]	11.91	2.89	.01263	6.272	11.9	17.54
Z[2,2]	11.8	2.887	.01157	6.103	11.82	17.43
Z[2,3]	11.94	2.873	.01424	6.287	11.96	17.57
Z[2,4]	11.77	2.87	.01288	6.177	11.76	17.44
Z[2,5]	11.74	2.878	.01239	6.077	11.75	17.4
.						
.						
.						
Z[26,1]	7.015	2.891	.01264	1.3	7.025	12.66
Z[26,2]	6.938	2.88	.0139	1.315	6.926	12.64
Z[26,3]	7.075	2.892	.01323	1.41	7.063	12.75
Z[26,4]	6.88	2.901	.01274	1.18	6.87	12.55
Z[26,5]	6.902	2.888	.01324	1.247	6.902	12.6

Thus, based on Figure 8.7, it appears that the average value of the number of correctly recalled words is about the same for the placebo and treatment groups. The predicted values of z are computed by executing BUGS CODE 8.4 and reported in Table 8.8.

For the first part of the section of repeated measures, the dependent variable was assumed to be continuous with a normal distribution, but now Poisson regression will be used to analyze a repeated measures study of Thall and Vail,[11] where the dependent variable is the number of seizures of epileptic patients randomly assigned to two groups, a placebo and a treatment group where the patients are given Progabide, an adjuvant to the standard antiepileptic chemotherapy (See Table 8.9). The dependent variable Y has a Poisson distribution with probability mass function

$$P(y \mid \lambda) = \frac{e^{-\lambda}\lambda^y}{y!}, y = 0,1,2,\ldots \tag{8.21}$$

TABLE 8.9

Descriptive Statistics for Epilepsy Study

Group	Variable	Mean	SD	Median	n
1	Seizures	8.6071	10.38	5.00	112
2		7.98	13.91	4.00	124
1	Age	28.96	5.42	29	112
2		28.70	8.82	26	124
1	Baseline	7.69	6.43	4.75	112
2		7.91	6.91	6.00	124

where λ is an unknown positive parameter and y is the number of seizures. Thus, over a given period, the number of seizures can be 0, 1, 2, and so on. It can be shown for the Poisson distribution that the mean and variance are both λ, that is, the average number of seizures over a given period (to be defined by the investigator) is λ, and the variability of the number of seizures over a long sequence of time periods is also λ.

According to Thall and Vail,[11] Progabide is an antiepileptic drug, where the primary mechanism of action is to enhance gamma-aminobutyric (GABA) acid; GABA is the main inhibitory neurotransmitter in the brain. Before receiving treatment, baseline information on the number of epileptic seizures during the preceding 8-week interval are reported. Counts of seizures during 2-week intervals before each of the four following post-randomization clinic visits are known. The information from the epileptic study appears in the first list statement of BUGS CODE 8.6.

Does the intervention have an effect on the number of seizures? If it does, we would expect the average number of seizures to decrease compared with placebo. The number of seizures in a 2-week period is assumed to have a Poisson distribution, and one would expect the distribution to depend on age and the 8-week baseline number of seizures.

Baseline values for the number of seizures in an 8-week period preceding treatment can have an effect on the number of seizures in the subsequent 2-week periods; thus, the baseline measurement (divided by 4) is considered as a covariate in the regression model.

Descriptive statistics for the epileptic study provide insight to the effect of the treatment on the number of seizures.

There is a hint that the number of seizures is reduced by the treatment, but is an important difference only if the other factors (age and baseline) are taken into account, that is adjusted for in the regression analysis.

The Poisson regression model is defined by

$$y[j,k] \sim \text{Poisson}\big(mu[j,k]\big) \tag{8.22}$$

where the log (natural log) of the mean of the number of seizures for the *j*th subject at time *k* is

$$\ln\left(\text{mu}[j,k]\right) = a_0 + alpha..Base * (Base[j] - avg.Base) + alpha..Trt * (Trt[j] - avg.trt)$$
$$+ alpha..Age * (Age[k] - avg.age) + int^* (Trt[j] * age[j] - avg.Trt * avg.Age)$$
$$+ b1[j] + b2[k] \tag{8.23}$$

The subscripts are $j = 1, 2, ..., N$ and $k = 1, 2, ...T$, where N is the number of patients and T the number of time periods.

Terms of the mean are defined as follows: a_0 is a constant that measures the overall log of the average number of seizures, alpha.base is a constant measuring the effect of the baseline measurement Base on the log of the average of the number of seizures, alpha.Trt measures the effect of the treatment variable Trt on the log of the average number of seizures, and alpha. Age measures the effect of age on the log of the average number of seizures. There are two random effects: b1[j], which measures the effect of subject *j*, and b2[k], which measures the effect of the *k*th time point. Note that each variable, Base, Trt, and Age, are centered at the average. Lastly, the variable int is the age–treatment interaction.

The Bayesian analysis is executed with BUGS CODE 8.6, and the code is a revision of Poisson regression example 11 found in volume 1 of WinBUGS®. The program statements closely follow Formulas 8.22 and 8.23 and the first list statement contains the information for the Thall and Vail[11] epilepsy study: a matrix y for the number of seizures, a column for the baseline seizures information, a column that identifies the two groups, and a column for the age of the patient.

BUGS CODE 8.6

```
model;
    {
       for(j in 1 : N) {
        for(k in 1 : T) {
         log(mu[j, k]) <- a0 + alpha.Base * (Base4[j] -
         Base4.bar)
         + alpha.Trt * (Trt[j] - Trt.bar)
         + alpha.Age * (Age[j] - Age.bar) +b1[j]+b2[k]
      # the following is the Poisson regression model
             y[j,k] ~ dpois(mu[j,k])
      # z are the predicted values of y
          z[j,k]~dpois(mu[j,k])

         }
```

```
     * the baseline measurement is divided by 4
     Base4[j] <-(Base[j]/4)
    }
# covariate means:
  Age.bar <- mean(Age[])
  Trt.bar <- mean(Trt[])

  Base4.bar <- mean(Base4[])

# priors:
for(j in 1:N){b1[j]~dnorm(0,taub1)}
for(k in 1:T){b2[k]~dnorm(0,taub2)}
  a0 ~ dnorm(0.0,1.0E-4)
  alpha.Base ~ dnorm(0.0,1.0E-4)
  alpha.Trt ~ dnorm(0.0,1.0E-4);
  taub1~dgamma(.0001,.0001)
  taub2~dgamma(.0001,.0001)
  alpha.Age ~ dnorm(0.0,1.0E-4)
  sigmab1<-1/taub1
  sigmab2<-1/taub2


}
# epileptic information
list(N = 59, T = 4,
y = structure(.Data = c(5, 3, 3, 3,
3, 5, 3, 3,
2, 4, 0, 5,
4, 4, 1, 4,
7, 18, 9, 21,
5, 2, 8, 7,
6, 4, 0, 2,
40, 20, 21, 12,
5, 6, 6, 5,
14, 13, 6, 0,
26, 12, 6, 22,
12, 6, 8, 4,
4, 4, 6, 2,
7, 9, 12, 14,
16, 24, 10, 9,
11, 0, 0, 5,
0, 0, 3, 3,
37, 29, 28, 29,
3, 5, 2, 5,
3, 0, 6, 7,
3, 4, 3, 4,
3, 4, 3, 4,
2, 3, 3, 5,
8, 12, 2, 8,
18, 24, 76, 25,
```

```
          2,  1,  2,  1,
          3,  1,  4,  2,
          13,  15,  13,  12,
          11,  14,  9,  8,
          8,  7,  9,  4,
          0,  4,  3,  0,
          3,  6,  1,  3,
          2,  6,  7,  4,
          4,  3,  1,  3,
          22,  17,  19,  16,
          5,  4,  7,  4,
          2,  4,  0,  4,
          3,  7,  7,  7,
          4,  18,  2,  5,
          2,  1,  1,  0,
          0,  2,  4,  0,
          5,  4,  0,  3,
          11,  14,  25,  15,
          10,  5,  3,  8,
          19,  7,  6,  7,
          1,  1,  2,  3,
          6,  10,  8,  8,
          2,  1,  0,  0,
          102,  65,  72,  63,
          4,  3,  2,  4,
          8,  6,  5,  7,
          1,  3,  1,  5,
          18,  11,  28,  13,
          6,  3,  4,  0,
          3,  5,  4,  3,
          1,  23,  19,  8,
          2,  3,  0,  1,
          0,  0,  0,  0,
          1,  4,  3,  2),.Dim = c(59, 4)),
Trt = c(0,  0,  0,  0,  0,  0,  0,  0,  0,  0,
          0,  0,  0,  0,  0,  0,  0,  0,  0,  0,
          0,  0,  0,  0,  0,  0,  0,  0,  1,  1,
          1,  1,  1,  1,  1,  1,  1,  1,  1,  1,
          1,  1,  1,  1,  1,  1,  1,  1,  1,  1,
          1,  1,  1,  1,  1,  1,  1,  1,  1),
Base = c(11,  11,  6,  8,  66,  27,  12,  52,  23,  10,
          52,  33,  18,  42,  87,  50,  18,111,  18,  20,
          12,  9,  17,  28,  55,  9,  10,  47,  76,  38,
          19,  10,  19,  24,  31,  14,  11,  67,  41,  7,
          22,  13,  46,  36,  38,  7,  36,  11,151,  22,
          41,  32,  56,  24,  16,  22,  25,  13,  12),
    Age = c(31,30,25,36,22,29,31,42,37,28,
          36,24,23,36,26,26,28,31,32,21,
          29,21,32,25,30,40,19,22,18,32,
```

```
     20,30,18,24,30,35,27,20,22,28,
     23,40,33,21,35,25,26,25,22,32,
     25,35,21,41,32,26,21,36,37))

list(a0 = 1, alpha.Base = 0, alpha.Trt = 0,
alpha.Age = 0, taub1 = 1,taub2 = 1)
```

A Bayesian analysis is executed with 55,000 observations for the simulation, 5000 for the burn-in, and a refresh of 100. When the interaction term is included in Model 8.23, the analysis is reported in Table 8.10. Note the uninformative prior distributions given to the model parameters: normal for the coefficient of base, treatment, and age, and gammas for the precision parameters. The two random effects are given normal distributions.

It appears that interaction, treatment, and age have very little effect on the log of the average number of seizures; however, the effect of the baseline values does seem to have some effect. In view of the results of Table 8.12, a model with no interaction term is included for the Bayesian analysis reported in Table 8.11. BUGS CODE 8.6 needs to be revised accordingly.

As one would expect, when interaction is not included, there is very little change in the estimates (via the posterior mean) of the remaining coefficients of the model. Inspecting Table 8.11 leads one to a model that only has the Base and Treatment variables in the model; thus, BUGS CODE 8.6 needs to be revised.

Based on Table 8.12, one would conclude that there is very little difference in placebo compared to treatment, but that the baseline values do indeed affect the number of seizures. By keeping the Base variable in the model, the treatment effect is adjusted for the baseline measurement of the number of seizures before treatment. Note that orig.level is the average value of the observations adjusted for base and treatment and is estimated as 5.148, the average number of seizures per 2-week period.

An introduction to the topic of repeated regression was presented in this section, but the topic is vast and the reader is invited to read some of the following references. A good introduction is the study by Crowder and Hand,[12] which

TABLE 8.10

Bayesian Analysis for Epileptic Study with Interaction in the Model

Parameter	Mean	SD	Error	2½	Median	97½
a_0	1.625	.1073	.00304	1.416	1.627	1.828
alpha.Age	.01528	.01994	.000516	−.0241	.01527	.05453
alpha.Base	.1092	.01243	.000484	.08456	.1092	.1336
alpha.Trt	−.2145	.7994	.01918	−1.811	−.2196	1.364
iat	−.001964	.02759	.000653	−.05669	−.00119	.05283
sigmab1	.3361	.08221	.000817	−.05669	.00193	.05283
sigmab2	.01467	.06997	.0013	.000160	.0046	.0867

TABLE 8.11

Bayesian Analysis for Epileptic Study without Interaction in the Model

Parameter	Mean	SD	Error	2½	Median	97½
a_0	1.63	.09597	.00238	1.438	1.631	1.817
alpha.Age	.01548	.01322	.000285	−.01169	.01463	.04068
alpha.Base	.1111	.01212	.000487	.08734	.111	.1352
alpha.Trt	−.2747	.71621	.00376	−.5976	−.2732	.03985
sigmab1	.3283	.0792	.000843	.2039	.3178	.5111
sigmab2	.0114	.03461	.000353	.000155	.00448	.06297

TABLE 8.12

Bayesian Analysis for Epileptic Study with Treatment and Base

Parameter	Mean	SD	Error	2½	Median	97½
a_0	1.632	.1058	.00392	1.436	1.632	1.831
alpha.Base	.1055	.01165	.0048	.0830	.1055	.1284
alpha.Trt	−.2947	.1606	.00345	−.6166	−.2945	.01833
orig.level	5.148	.5446	.02189	4.227	5.121	6.232
sigmab1	.3304	.0787	.00067	.2055	.3204	.5129
sigmab2	.01605	.1349	.00254	.0001508	.00452	.08206

contains the theory and has many good examples. A similar treatment, but from a more practical viewpoint, is the study by Hand and Crowder,[13] which again has many instructive examples from a variety of scientific endeavors.

I recommend Jones,[14] whose presentation is somewhat different from the previous two references, in that the correlation structure between observations from the same individual is based on time-series techniques. The most comprehensive book among repeated measures books is by Lindsey[15] and the book contains a list of references that account for most of the relevant publications in 1993. Of course, since that time the topic has moved forward and has become more specialized as seen in Davidian and Giltinana,[16] whose main focus is on the nonlinear repeated measures model. The student is introduced to nonlinear regression in repeated measures by one of the examples taken from the book and analyzed via Bayesian methods.

8.6 Spatial Models for Epidemiology

Before introducing spatial models used in epidemiology, it is interesting to note the definition of epidemiology given by Mausner and Kramer (p. 1)[17]: "Epidemiology may be defined as the study of the distribution and the determinants of disease and injuries in the human population. That is, epidemiology is concerned with the frequencies and types of illnesses and injuries in groups of people and with the factors that influence their distribution." As will be seen in this section, spatial models allow us to investigate the distribution of disease in the spatial domain and its association with various exposure to risk factors.

The subject is introduced with an example of the number of cases of lip cancer diagnosed in 56 counties in Scotland, where the relative risk of the disease is estimated for each county, and the association between the number employed in outdoor jobs and the incidence of lip cancer is explored. Essentially, the model is a Poisson regression model where the spatial correlation between the lip cancer rates of one county and its neighbors is taken into account by a conditional autoregressive (CAR) model. One would expect the incidence of lip cancer of a county to be related to the incidence of its neighboring counties. One would also expect the same sort of association between an exposure in one county and its neighbors. That is, one would expect the incidence of lip cancer in a county to be more related to the incidence with neighboring counties than to the incidence with counties that are not neighbors. Thus, one sees the similarity between spatial models in epidemiology and the repeated measures models of the previous section, in which adjacent observations tend to be correlated.

In epidemiology, maps of disease rates and disease risks state that a spatial population is small and the rate and risk estimates may be unstable; thus, our Bayesian analysis will be based on regression models where the spatial correlation is modeled by a CAR process. The effect will be to spatially smooth disease rates and risk estimates by allowing each site to borrow strength from its neighbors. Covariates that measure exposure to risk can also be included in the model in such a way that a possible association between risk factors (exposures) and disease may be established. Thus, the lip cancer incidences will be smoothed over counties via a Bayesian analysis that employs MCMC techniques to estimate the model parameters.

Geographical epidemiology and medical geography are terms used for mapping the distribution of disease with respect to place and time. Maps take into account the spatial relationship that may be missed in descriptive tables. Good examples of this are the Palm[18] study of the spatial distribution of rickets, which established the association of the disease with the lack of sunlight, and in a similar fashion the study of Lancaster[19] for the association between exposure to sunlight and melanoma. The maps revealed the spatial distribution of the diseases, which established the association between the disease and the relevant exposure.

A major concern is that the data values being mapped including estimates of relative risk can be very unstable when dealing with disease clusters, rare diseases, and small populations. The numbers of observed cases of disease are usually assumed to have a Poisson distribution, but extra-Poisson variability usually occurs (recall that the mean and variance of the Poisson are the same) and the extra variability is accounted for by including variables that follow a CAR distribution.

The topic of spatial models is vast and there have been many approaches to analyze such data. For example, empirical Bayesian approaches with regression and CAR processes for estimating the association between disease and risk factors have been pursued by Clayton and Kaldor,[20] Cressie,[21] and Mollie and Richardson,[22] and this approach with some alteration will be implemented for the lip cancer example.

The rates of lip cancer in 56 counties in Scotland have been analyzed by Clayton and Kaldor.[20] The form of the data includes the observed and expected cases (expected numbers based on the population and its age and sex distribution in the county), a covariate measuring the percentage of the population engaged in agriculture, fishing, or forestry, and the "position" of each county expressed as a list of adjacent counties.

We may smooth the raw standardized mortality rates (SMRs) by fitting a random-effects Poisson model allowing for spatial correlation, using the intrinsic CAR prior. For the lip cancer example, the model may be written as

$$O_i \sim \text{Poisson(mu}_i) \tag{8.24}$$

$$\log (\text{mu}_i) = \frac{\log E_i + \alpha_0 + \alpha_1 x_i}{10 + b_i} \tag{8.25}$$

where α_0 is an intercept term representing the baseline (log) relative risk of disease across the study region, x_i is the covariate "percentage of the population engaged in agriculture, fishing, or forestry" in district i, with associated regression coefficient α_1, and b_i is an area-specific random effect capturing the residual or unexplained (log) relative risk of disease in area i. We often think of b_i as representing the effect of latent (unobserved) risk factors. Note that the O_i is the observed number of lip cancer cases and E_i the corresponding expected number of cases (expected numbers based on the population and its age and sex distribution in the county).

To allow for spatial dependence between the random effects b_i in nearby areas, we may assume a CAR prior for these terms. We give a brief description of the CAR process as follows:

Let the b_i, i = 1, 2 ,..., n be normal random variables where the index i denotes the ith site (ith county) such that the conditional distribution of b_i given b_j is denoted by

$$b_i \mid b_j \sim \text{norm}(\mu_i + \rho \sum_{j=1}^{j=n} w_{ij}(b_j - \mu_j), \sigma^2) \tag{8.26}$$

where $w_{ii} = 0, w_{ij} = 1$ if i and j are adjacent neighbors, otherwise $w_{ij} = 0$, and ρ is a constant. The information for the lip cancer study is portrayed in Table 8.13, where the second column is the number of observed cases, the third is the

TABLE 8.13

Lip Cancer for Counties of Scotland

County	Observed, O	Expected, E	Percent Working Outdoors, X	Number of Neighbors
1	9	1.4	16	3
2	39	8.7	16	2
3	11	3	10	1
4	9	2.5	24	3
5	15	4.3	10	3
6	8	2.4	24	0
7	26	8.1	10	5
8	7	2.3	7	0
9	6	2	7	5
10	20	6.6	16	4
11	13	4.4	7	0
12	5	1.8	16	2
13	3	1.1	10	3
14	8	3.3	24	3
15	17	7.8	7	2
16	9	4.6	16	6
17	2	1.1	10	6
18	7	4.2	7	6
19	9	5.5	7	5
20	7	4.4	10	3
21	16	10.5	7	3
22	31	22.7	16	2
23	11	8.8	10	4
24	7	5.6	7	8
25	19	15.5	1	3
26	15	12.5	1	3
27	7	6.0	7	4
28	10	9.0	7	4
29	16	14.4	10	11
30	11	10.2	10	6
31	5	4.8	7	7
32	3	2.9	24	3
33	7	7.0	10	4
34	8	8.5	7	9
35	11	12.3	7	4
36	9	10.1	0	2
37	11	12.7	10	4
38	8	9.4	1	6
39	6	7.2	16	3

TABLE 8.13 (*Continued*)

Lip Cancer for Counties of Scotland

County	Observed, O	Expected, E	Percent Working Outdoors, X	Number of Neighbors
40	4	5.3	0	4
41	10	18.8	1	5
42	8	15.8	16	5
43	2	4.3	16	4
44	6	14.6	0	5
45	19	50.7	1	4
46	3	8.2	7	6
47	2	5.6	1	6
48	3	9.3	1	4
49	28	88.7	0	9
50	6	19.6	1	2
51	1	3.4	1	4
52	1	3.6	0	4
53	1	5.7	1	4
54	1	7.0	1	5
55	0	4.2	16	6
56		1.8	10	5

Source: Example 1 of GeoBUGS of WinBUGS®. www.mrc-bsu.cam.ac.uk/bugs.

number of expected cases, the percent of the population with outdoor jobs is the fourth column, the last column is the number of neighbors of each county.

BUGS CODE 8.7 will execute the Bayesian analysis, using 55,000 observations for the simulation, 5000 for the burn-in, and a refresh of 100. Note that the first list statement contains the data of Table 8.13 and a matrix called adj that gives the neighbors of each county

BUGS CODE 8.7

```
model {
                # Likelihood
                for (i in 1 : N) {
                        O[i] ~ dpois(mu[i])
                        log(mu[i]) <- log(E[i]) + alpha0 +
                        alpha1 * X[i]/10 + b[i]
# predicted cases        u[i] ~ dpois(mu[i])
                        # Area-specific relative risk (for
```

```
                                maps)
                                RR[i] <- exp(alpha0 + alpha1
*X[i]/10 + b[i])


                       }

                # CAR prior distribution for random effects:
                       b[1:N] ~ car.normal(adj[], weights[],
                       num[], tau)
                       for(k in 1:sumNumNeigh) {
                              weights[k] <- 1


                       }

                # Other priors:
                       alpha0 ~ dflat()
                       alpha1 ~ dnorm(0.0, 1.0E-5)
                       tau ~ dgamma(0.5, 0.0005)              #
                       prior on precision
                       sigma <- sqrt(1/tau)              # standard
                       deviation
                  b.mean <- sum(b[])


                  }


# lip cancer
                  list(N = 56,
O = c( 9, 39, 11, 9, 15, 8, 26, 7, 6, 20,
            13, 5, 3, 8, 17, 9, 2, 7, 9, 7,
            16, 31, 11, 7, 19, 15, 7, 10, 16, 11,
            5, 3, 7, 8, 11, 9, 11, 8, 6, 4,
            10, 8, 2, 6, 19, 3, 2, 3, 28, 6,
            1, 1, 1, 1, 0, 0),
E = c(1.4, 8.7, 3.0, 2.5, 4.3, 2.4, 8.1, 2.3, 2.0, 6.6,
            4.4, 1.8, 1.1, 3.3, 7.8, 4.6, 1.1, 4.2, 5.5, 4.4,
            10.5,22.7, 8.8, 5.6,15.5,12.5, 6.0, 9.0,14.4,10.2,
            4.8, 2.9, 7.0, 8.5,12.3,10.1,12.7, 9.4, 7.2, 5.3,
            18.8,15.8, 4.3,14.6,50.7, 8.2, 5.6, 9.3,88.7,19.6,
            3.4, 3.6, 5.7, 7.0, 4.2, 1.8),
X = c(16,16,10,24,10,24,10, 7, 7,16,
            7,16,10,24, 7,16,10, 7, 7,10,
            7,16,10, 7, 1, 1, 7, 7,10,10,
            7,24,10, 7, 7, 0,10, 1,16, 0,
            1,16,16, 0, 1, 7, 1, 1, 0, 1,
            1, 0, 1, 1,16,10),
num = c(3, 2, 1, 3, 3, 0, 5, 0, 5, 4,
0, 2, 3, 3, 2, 6, 6, 6, 5, 3,
3, 2, 4, 8, 3, 3, 4, 4, 11, 6,
7, 3, 4, 9, 4, 2, 4, 6, 3, 4,
```

```
5,  5,  4,  5,  4,  6,  6,  4,  9,  2,
4,  4,  4,  5,  6,  5

),

adj = c(
19,  9,  5,
10,  7,
12,
28,  20,  18,
19,  12,  1,

17,  16,  13,  10,  2,

29,  23,  19,  17,  1,
22,  16,  7,  2,

5,  3,
19,  17,  7,
35,  32,  31,
29,  25,
29,  22,  21,  17,  10,  7,
29,  19,  16,  13,  9,  7,
56,  55,  33,  28,  20,  4,
17,  13,  9,  5,  1,
56,  18,  4,
50,  29,  16,
16,  10,
39,  34,  29,  9,
56,  55,  48,  47,  44,  31,  30,  27,
29,  26,  15,
43,  29,  25,
56,  32,  31,  24,
45,  33,  18,  4,
50,  43,  34,  26,  25,  23,  21,  17,  16,  15,  9,
55,  45,  44,  42,  38,  24,
47,  46,  35,  32,  27,  24,  14,
31,  27,  14,
55,  45,  28,  18,
54,  52,  51,  43,  42,  40,  39,  29,  23,
46,  37,  31,  14,
41,  37,
46,  41,  36,  35,
54,  51,  49,  44,  42,  30,
40,  34,  23,
52,  49,  39,  34,
53,  49,  46,  37,  36,
51,  43,  38,  34,  30,
42,  34,  29,  26,
49,  48,  38,  30,  24,
```

```
55, 33, 30, 28,
53, 47, 41, 37, 35, 31,
53, 49, 48, 46, 31, 24,
49, 47, 44, 24,
54, 53, 52, 48, 47, 44, 41, 40, 38,
29, 21,
54, 42, 38, 34,
54, 49, 40, 34,
49, 47, 46, 41,
52, 51, 49, 38, 34,
56, 45, 33, 30, 24, 18,
55, 27, 24, 20, 18
),
sumNumNeigh = 234)

# initial values
list(tau = 1, alpha0 = 0, alpha1 = 0,
b = c(0,0,0,0,0,NA,0,NA,0,0,
NA,0,0,0,0,0,0,0,0,0,
0,0,0,0,0,0,0,0,0,0,
0,0,0,0,0,0,0,0,0,0,
0,0,0,0,0,0,0,0,0,0,
0,0,0,0,0,0),
u = c(1,1,1,1,1,1,1,1,1,1,1,1,1,1,1,1,1,1,1,1,1,1,1,1,1,
   1,1,1,1,
1,1,1,1,1,1,1,1,1,1,1,1,1,1,1,1,1,1,1,1,1,1,1,1,1,1,1))
```

The objective is to estimate the relative risk RR[i] of each county, and the parameters α_0 and α_1 of the Poisson regression (Equation 8.24.)

The posterior mean of the intercept α_0, the baseline log relative risk, is −.3025, and the effect of the covariate α_1 of the log of the average number of cases is .4595 with a credible interval of (.227, .6781). Both appear to have an effect on the average number of lip cancer cases (See Table 8.14). The estimated relative risks by county are reported in Table 8.15.

The smoothed estimates of the relative risk for lung cancer are reported in Table 8.15, and for the first county, the posterior mean is 4.898 with a simulation error of .014 and 95% credible interval of (2.557, 8.199). Only six of the counties are listed in the table, but the student will be asked to compute the others in an exercise. These estimates are used to produce the map of Figure 8.9 for the smoothed values of relative risk. Counties

TABLE 8.14

Bayesian Analysis for Parameters of the Poisson Regression

Parameter	Mean	SD	Error	2½	Median	97½
α_0	−.3025	.1127	.001598	−.5202	−.3033	−.0783
α_1	.4595	.1158	.00182	.227	.4624	.6781

bordering a given county are neighbors of that county; thus, the map reveals the geographic distribution of the relative risk of lip cancer. Do the percentage of outdoor jobs increase as one moves from the south to the north of Scotland? If so, this would be the evidence that exposure to sunlight influences lip cancer incidence.

TABLE 8.15

Posterior Distribution of the Relative Risk by County

Parameter	Mean	SD	Error	2½	Median	97½
RR[1]	4.898	1.469	.014	2.557	4.714	8.199
RR[2]	4.356	.6735	.00682	3.135	4.322	5.767
RR[3]	3.537	1.019	.01144	1.86	3.433	5.831
.						
.						
.						
RR[54]	.4245	.1157	.00108	.2314	.4134	.681
RR[55]	.9604	.2599	.00294	.528	.9348	1.532
RR[56]	.9517	.2744	.00252	.5111	.9184	1.578

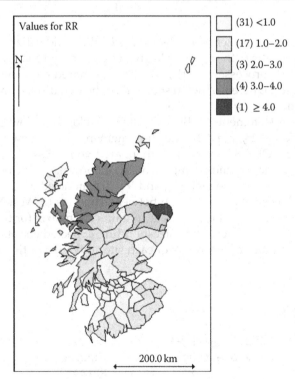

FIGURE 8.9

Estimated relative risk of counties in Scotland. (Example 1 from GeoBUGS of WinBUGS®. www.mrc-bsu.cam.ac.uk/bugs.)

The final example is taken from Example 8 of GeoBUGS of WinBUGS® and is an extension of the lip cancer example, in that the spatial distribution of oral cavity and lung cancer are modeled with two Poisson regressions, with one for oral cancer and the other for lung. There are 126 regions in West Yorkshire, UK. As before, the log of the mean of each Poisson process is a linear function of (1) the log of the age- and sex-standardized expected values for the counts, (2) the baseline (log) relative risk of the disease across the study regions, (3) a component that shares the effect of both diseases, and (4) a random component that accounts for the extra-Poisson variability and is modeled by a CAR process. The model is defined as

$$O_{ik} \sim Poisson(\mu_{ik}) \tag{8.27}$$

$$\log(\mu_{i1}) = \log(E_{i1}) + \alpha_1 + \phi_i * \delta + \Psi_{i1} \tag{8.28}$$

$$\log(\mu_{i2}) = \frac{\log(E_{i2}) + \alpha_2 + \varphi_i}{\delta + \Psi_{i2}} \tag{8.29}$$

where $k = 1, 2$, and the means are the μ_{ik}, the E_{ik} the standardized age and sex expected counts, the α_i the baseline log relative risks, and the Ψ_{ik} the random component CAR processes. Lastly, the ϕ_i is the shared component between the two diseases, where i is the ith site, that is, the two diseases are observed at the same sites, for $i = 1, 2, ..., 126$.

The program statements of BUGS CODE 8.8 closely follow the notation of Formulas 8.27, 8.28, and 8.29, and the statements are somewhat similar to those of BUGS CODE 8.7, except that there are two regressions that are connected through the common components ϕ_i of Formulas 8.28 and 8.29.

The two random components Ψ_{i1} and Ψ_{i2} are modeled as CAR processes and these induce spatial correlation between the sites of West Yorkshire in the UK. The list statement for the data contains the matrix for the counts for oral cavity and lung cancer, the corresponding expected counts for the two diseases, the number of neighbors of each site, and a list of the neighbors of each site.

BUGS CODE 8.8

```
model {
# Likelihood
# See formulas (8.28) and (8.29)
        for (i in 1 : Nareas) {
                for (k in 1 : Ndiseases) {
                        Y[i, k] ~ dpois(mu[i, k])
                        log(mu[i, k]) <- log(E[i,
```

```
                              k]) + alpha[k] + eta[i, k]
                  }
           }


           for(i in 1:Nareas) {
     # Define log relative risk in terms of
       disease-specific (psi) and shared (phi)
     # random effects
     # changed order of k and i index for psi
       (needed because car.normal assumes
     # right hand index is areas)
                  eta[i, 1] <- phi[i] * delta +
                     psi[1, i]
                  eta[i, 2] <- phi[i]/delta +
                     psi[2, i]

           }


     # Spatial priors (BYM) for the disease-
       specific random effects
           for (k in 1 : Ndiseases) {
                  for (i in 1 : Nareas) {
     # convolution prior = sum of unstructured and
       spatial effects
                     psi[k, i] <- U.sp[k, i] +
                        S.sp[k, i]
     # unstructured disease-specific random
       effects
                        U.sp[k, i] ~ dnorm(0, tau.
                           unstr[k])

                  }


     # spatial disease-specific effects
                  S.sp[k,1 : Nareas] ~ car.
                     normal(adj[], weights[],
                     num[], tau.spatial[k])

           }


     # Spatial priors (BYM) for the shared random
       effects
           for (i in 1:Nareas) {
     # convolution prior = sum of unstructured and
       spatial effects
                  phi[i] <- U.sh[i] + S.sh[i]
     # unstructured shared random effects
                  U.sh[i] ~ dnorm(0, omega.
```

```
                          unstr)
     }
# spatial shared random effects
     S.sh[1:Nareas] ~ car.normal(adj[],
        weights[], num[], omega.spatial)
     for (k in 1:sumNumNeigh) {
              weights[k] <- 1


     }


# Other priors
     for (k in 1:Ndiseases) {
                   alpha[k] ~ dflat()
                   tau.unstr[k] ~ dgamma(0.5,
                   0.0005)
                      tau.spatial[k] ~
                   dgamma(0.5, 0.0005)
          }
               omega.unstr ~ dgamma(0.5, 0.0005)
               omega.spatial ~ dgamma(0.5,
               0.0005)
# scaling factor for relative strength of
  shared component for each disease
               logdelta ~ dnorm(0, 5.9)
# prior assumes 95% probability that delta^2
  is between 1/5 and 5;
               delta <- exp(logdelta)
# lognormal assumption is invariant to which
  disease is labelled 1
# and which is labelled 2)
# ratio (relative risk of disease 1 associated
  with shared component) to
# (relative risk of disease 2 associated with
  shared component)
# see Knorr-Held and Best (2001) for further
  details
       RR.ratio <- pow(delta, 2)
# Relative risks and other summary quantities
# The GeoBUGS map tool can only map vectors, so
  need to create separate vector
# of quantities to be mapped, rather than an
  array (i.e. totalRR[i,k] won't work!)
       for (i in 1 : Nareas) {
               SMR1[i] <- Y[i,1]/E[i,1]
                # SMR for disease 1 (oral)
               SMR2[i] <- Y[i,2]/E[i,2]
                # SMR for disease 2 (lung)
                  totalRR1[i] <-
                  exp(eta[i,1])
```

```
# overall RR of disease 1 (oral) in area i
        totalRR2[i] <-
        exp(eta[i,2])
# overall RR of disease 2 (lung) in area i
        # residual RR specific to
        disease 1 (oral cancer)
            specificRR1[i]<-
            exp(psi[1,i])
        # residual RR specific to disease
        2 (lung cancer)
            specificRR2[i]<-
            exp(psi[2,i])
        # shared component of risk common
        to both diseases
            sharedRR[i] <- exp(phi[i])
        # Note that this needs to be
        scaled by delta or 1/delta if the
        # absolute magnitude of shared RR
        for each disease is of interest
            logsharedRR1[i] <- phi[i] *
            delta
            logsharedRR2[i] <- phi[i] /
            delta

    }

            # empirical variance of shared
            effects (scaled for disease 1)
                var.shared[1] <- sd(logshare
                dRR1[])*sd(logsharedRR1[])
            # empirical variance of shared
            effects (scaled for disease 2)
                var.shared[2] <-
                sd(logsharedRR2[])*sd(logsha
                redRR2[])
            # empirical variance of disease 1
            specific effects
                var.specific[1] <-
                sd(psi[1,])*sd(psi[1,])
            # empirical variance of disease 2
            specific effects
                var.specific[2] <-
                sd(psi[2,])*sd(psi[2,])
            # fraction of total variation in
            relative risks for each disease
            that is explained
            # by the shared component
                frac.shared[1] <- var.
                shared[1]/(var.shared[1] +
                var.specific[1])
```

388 Bayesian Methods in Epidemiology

```
                              frac.shared[2] <- var.
                              shared[2]/(var.shared[2] +
                              var.specific[2])

        }

list(Nareas = 126, Ndiseases = 2,
Y = structure(.Data = c(7, 103, 3, 160, 6, 97, 5, 156, 3, 88,
8, 168, 1, 88, 12, 157, 8, 110,
        5, 134, 4, 74, 11, 162, 8, 136, 7, 81, 4, 108, 4,
100, 7, 137,
        7, 130, 5, 176, 8, 182, 7, 161, 4, 86, 6, 169, 10,
154, 7, 121,
        4, 247, 10, 179, 10, 219, 10, 88, 4, 108, 6, 211,
3, 107, 3, 120,
        4, 70, 6, 216, 6, 221, 4, 142, 8, 136, 11, 246,11,
209, 4, 130,
        8, 90, 9, 117, 3, 121, 6, 157, 14, 300, 5, 160, 6,
107, 10, 233,
        10, 270, 7, 141, 7, 110, 14, 158, 12, 203, 5, 74,
12, 149, 3, 98,
        3, 110, 9, 151, 7, 157, 8, 111, 2, 84, 8, 149, 9,
166, 3, 205,
        9, 114, 3, 68, 7, 115, 5, 179, 2, 101, 1, 73, 6,
109, 4, 118,
        9, 73, 7, 108, 7, 98, 9, 137, 3, 79, 4, 126, 8,
134, 4, 108, 5,
        119, 3, 144, 2, 89, 2, 103, 6, 99, 4, 74, 3, 75,
3, 90, 3, 69,
        6, 137, 3, 129, 5, 122, 8, 114, 7, 115, 4, 73, 5,
94, 2, 110,
        3, 69, 7, 94, 6, 133, 4, 120, 7, 112, 5, 118, 9,
129, 9, 128,
        7, 91, 8, 141, 4, 93, 9, 107, 13, 85, 5, 81, 7,
98, 7, 85, 3,
        89, 4, 82, 5, 98, 4, 127, 4, 105, 5, 119, 6, 133,
7, 82, 5, 73,
        5, 77, 3, 109, 4, 80.),.Dim = c(126,2)),
E = structure(.Data = c(5.7993, 124.2824, 9.8345, 199.3086,
6.5001, 138.3337, 9.0741, 191.0307, 6.4746, 136.3728,
        9.9171, 208.4544, 5.515, 113.2827, 9.8557,
205.0383, 5.7956, 118.5219, 8.299,
        168.9939, 4.8596, 99.1749, 8.8737, 182.1041,
5.3387, 115.9417, 6.2182, 128.2199,
        5.368, 111.3884, 5.4669, 115.3283, 8.3298,
172.3197, 8.5829, 176.3519, 8.0016,
        172.5571, 9.8149, 217.5105, 6.8331, 137.3765,
5.4115, 111.0865, 5.6492, 119.7825,
        8.7562, 178.6785, 6.3436, 132.6835, 7.8424,
165.7296, 7.9467, 163.6806, 8.2651,
```

```
           179.5329, 4.9209, 99.8676, 4.8173, 97.8898,
8.2088, 172.8071, 5.1494, 111.4239,
           6.1063, 125.3408, 4.2812, 89.5027, 7.439,
154.4872, 6.4687, 138.0246, 8.648,
           173.9356, 5.4981, 112.8979, 8.1535, 172.1137,
8.1074, 167.9885, 5.0473, 103.0942,
           5.337, 110.1, 4.9104, 95.0673, 4.899, 100.6339,
8.2719, 166.6911, 9.0572, 192.7174,
           8.3271, 170.1211, 5.2504, 109.0334, 9.2167,
189.3181, 8.2078, 173.5003, 6.3403,
           130.4997, 5.2363, 106.8044, 5.7656, 119.1885,
6.349, 134.6238, 3.1955, 63.4971,
           4.8059, 96.6801, 4.3527, 92.6587, 5.6589,
113.8283, 7.3528, 158.2496, 6.4591,
           134.9901, 5.0808, 103.3924, 3.6248, 73.217,
8.1897, 167.9081, 8.29, 167.5654,
           7.3986, 153.8624, 6.3236, 127.4593, 4.261,
86.9421, 5.5404, 111.7691, 9.0518,
           186.5931, 5.6972, 117.9196, 4.1577, 90.4632,
3.7059, 75.4029, 4.5956, 99.7396,
           4.0921, 85.8003, 5.0072, 99.5433, 5.7554,
118.2132, 5.4192, 113.6951, 4.4614,
           95.4255, 5.3266, 113.212, 5.9185, 120.4505,
5.0662, 99.3974, 5.5931, 113.8424,
           4.7091, 100.6834, 5.7394, 118.5351, 5.4592,
110.8379, 5.6211, 113.0633, 4.2743,
           91.6488, 4.2199, 87.8078, 5.0116, 106.9507,
3.9801, 84.0584, 5.2632, 107.4919,
           5.1299, 103.2763, 5.5062, 114.7298, 5.684, 119.65,
5.033, 102.6666, 3.6625,
           74.6299, 4.2702, 92.1799, 5.9149, 121.8377, 4.048,
84.8317, 5.8719, 121.1919,
           6.6767, 141.6445, 5.3587, 110.0901, 5.4904,
111.204, 4.7363, 96.3014, 5.3231,
           108.2957, 6.5753, 137.6387, 5.457, 111.0015,
5.8437, 120.9779, 5.2844, 106.7001,
           5.0919, 101.9814, 6.5409, 135.9033, 5.3966,
106.1728, 5.1725, 102.7724, 6.066,
           129.0506, 5.9403, 123.6225, 5.4443, 117.149,
5.5746, 115.9296, 5.4766, 113.8003,
           5.3397, 108.0343, 5.8259, 124.6071, 5.7783,
125.2003, 6.0794, 127.5812, 5.6695,
           116.8155, 5.5123, 114.1313, 5.1238, 104.6006,
5.46, 112.2663),.Dim = c(126,2)),
num = c(3, 2, 3, 6, 6, 5, 7, 5, 4, 9,
6, 5, 4, 6, 7, 6, 7, 8, 6, 5,
4, 8, 6, 8, 6, 5, 6, 7, 4, 4,
5, 4, 6, 4, 5, 7, 7, 10, 6, 5,
6, 4, 6, 3, 6, 8, 6, 5, 6, 6,
7, 5, 6, 5, 9, 8, 7, 6, 5, 5,
```

```
6,  5,  6,  6,  7,  8,  7,  6,  6,  6,
1,  3,  6,  5,  4,  4,  6,  6,  6,  5,
3,  5,  8,  8,  5,  7,  4,  6,  5,  9,
6,  5,  5,  6,  7,  4,  6,  5,  6,  5,
6,  6,  5,  7,  7,  6,  5,  5,  5,  7,
6,  6,  6,  6,  9,  7,  4,  4,  4,  4,
6,  5,  3,  5,  2,  4
),
adj = c(
9,  7,  3,
10,  4,
7,  5,  1,
20, 17, 12, 10,  6,  2,
16, 14,  8,  7,  6,  3,
18, 12,  8,  5,  4,
22, 16, 13,  9,  5,  3,  1,
18, 15, 14,  6,  5,
13, 11,  7,  1,
79, 77, 75, 45, 37, 21, 17,  4,  2,
57, 55, 34, 22, 13,  9,
20, 19, 18,  6,  4,
22, 11,  9,  7,
25, 23, 16, 15,  8,  5,
30, 29, 24, 23, 18, 14,  8,
25, 23, 22, 14,  7,  5,
35, 27, 26, 21, 20, 10,  4,
31, 28, 24, 19, 15, 12,  8,  6,
32, 28, 27, 20, 18, 12,
27, 19, 17, 12,  4,
37, 26, 17, 10,
55, 41, 33, 25, 16, 13, 11,  7,
38, 29, 25, 16, 15, 14,
51, 47, 44, 38, 31, 30, 18, 15,
38, 33, 23, 22, 16, 14,
39, 37, 35, 21, 17,
36, 35, 32, 20, 19, 17,
46, 40, 36, 32, 31, 19, 18,
38, 30, 23, 15,
38, 29, 24, 15,
47, 40, 28, 24, 18,
36, 28, 27, 19,
43, 42, 41, 38, 25, 22,
88, 71, 57, 11,
39, 36, 27, 26, 17,
50, 46, 39, 35, 32, 28, 27,
63, 50, 45, 39, 26, 21, 10,
53, 51, 44, 43, 33, 30, 29, 25, 24, 23,
50, 46, 37, 36, 35, 26,
49, 47, 46, 31, 28,
58, 55, 48, 42, 33, 22,
```

```
48, 43, 41, 33,
53, 52, 48, 42, 38, 33,
51, 38, 24,
84, 79, 77, 63, 37, 10,
59, 54, 50, 49, 40, 39, 36, 28,
56, 51, 49, 40, 31, 24,
58, 52, 43, 42, 41,
64, 59, 56, 47, 46, 40,
63, 54, 46, 39, 37, 36,
60, 56, 53, 47, 44, 38, 24,
61, 58, 53, 48, 43,
61, 60, 52, 51, 43, 38,
65, 63, 59, 50, 46,
73, 72, 67, 62, 58, 57, 41, 22, 11,
70, 68, 66, 64, 60, 51, 49, 47,
88, 87, 78, 62, 55, 34, 11,
67, 61, 55, 52, 48, 41,
65, 64, 54, 49, 46,
68, 61, 56, 53, 51,
68, 67, 60, 58, 53, 52,
78, 73, 72, 57, 55,
84, 65, 54, 50, 45, 37,
69, 66, 65, 59, 56, 49,
84, 82, 69, 64, 63, 59, 54,
86, 85, 80, 76, 70, 69, 64, 56,
83, 74, 73, 68, 61, 58, 55,
74, 70, 67, 61, 60, 56,
95, 85, 82, 66, 65, 64,
90, 76, 74, 68, 66, 56,
34,
73, 62, 55,
83, 78, 72, 67, 62, 55,
90, 83, 70, 68, 67,
92, 81, 77, 10,
90, 86, 70, 66,
92, 91, 79, 75, 45, 10,
89, 87, 83, 73, 62, 57,
94, 91, 84, 77, 45, 10,
98, 93, 86, 85, 66,
103, 92, 75,
102, 95, 84, 69, 65,
99, 97, 90, 89, 78, 74, 73, 67,
102, 101, 94, 82, 79, 65, 63, 45,
95, 93, 80, 69, 66,
106, 100, 98, 90, 80, 76, 66,
89, 88, 78, 57,
116, 99, 89, 87, 57, 34,
99, 88, 87, 83, 78,
106, 104, 97, 96, 86, 83, 76, 74, 70,
110, 103, 94, 92, 79, 77,
```

```
103, 91, 81, 77, 75,
105, 98, 95, 85, 80,
112, 110, 101, 91, 84, 79,
109, 105, 102, 93, 85, 82, 69,
107, 104, 97, 90,
111, 107, 99, 96, 90, 83,
105, 100, 93, 86, 80,
116, 111, 97, 89, 88, 83,
115, 106, 105, 98, 86,
112, 110, 108, 102, 94, 84,
109, 108, 101, 95, 84, 82,
118, 110, 92, 91, 81,
121, 120, 114, 107, 106, 96, 90,
115, 113, 109, 100, 98, 95, 93,
120, 115, 104, 100, 90, 86,
114, 111, 104, 97, 96,
113, 112, 109, 102, 101,
113, 108, 105, 102, 95,
119, 118, 112, 103, 101, 94, 91,
117, 116, 114, 107, 99, 97,
119, 113, 110, 108, 101, 94,
123, 115, 112, 109, 108, 105,
122, 121, 117, 111, 107, 104,
126, 124, 123, 121, 120, 113, 106, 105, 100,
126, 124, 122, 117, 111, 99, 88,
122, 116, 114, 111,
125, 119, 110, 103,
125, 118, 112, 110,
121, 115, 106, 104,
124, 122, 120, 115, 114, 104,
124, 121, 117, 116, 114,
126, 115, 113,
126, 122, 121, 116, 115,
119, 118,
124, 123, 116, 115
),
sumNumNeigh = 710
)


list(U.sp = structure(.Data = c(-0.45, 0.14, -1.2, 2.5, -0.24,
    0.31, 0.46, -1.7, 0.99, 0.92, 1.3, -2., -0.23, 1.6, -0.41,
    0.49, -0.35, -0.4,
-0.94, -0.034, -0.13, 0.028, -0.24, 0.97, -1.3, 5.4, -0.83,
    0.69, -0.0039, 0.5, -1.2, 0.65, 4.1, -0.07, -0.45, 0.86,
-0.96, -0.46, -2.5, 0.2,
-1.0, 1.3, 1.6, 0.16, 0.22, -2., -0.051, 0.12, 1.4, -1.8, 2.2,
    -0.16, -0.038, -0.47, 2.2, 1.4,-1.1, -0.16, 0.8, 0.42,
-0.62, 0.11, -0.57, -0.86,
0.26, 0.83, -0.52, -0.76, -1.3, 1.7, -0.33, -0.81, 4.3, 1.7,
```

```
    0.32, 0.36, -0.57, -2.6, 1.5, -0.13, 0.89, -1.1,
-0.044,-1.5, 2.5, -0.94, -0.24,
2.9, 2.4, 0.33, 0.56, 0.83, -1., -0.89, 3., -2., -1.1, -4.4,
    0.021, -0.5, -1.7, 1.7, -0.71, 0.51, -1.4, 1.3, 0.47, -0.14,
-0.3, 1.7, 2., 0.58, 0.6,
-0.95, 0.76, -0.043, -0.8, 0.62, -0.0084, -0.0028, 0.21, 1.8,
    2.3, 2.9, -1.5, -0.47, -1.4, 0.91, 0.27, -1.1, 0.27, 0.33,
-0.2, -2.2, -0.39,
0.037, -1.2, 1., 0.41, 0.22, -0.14, 0.99, -0.21, -0.2, 0.63,
    0.1, 0.97, 0.034, -1.5, -1.2, -0.077, -0.78, 1.2, 6., 0.29,
-0.77, 0.21, 0.76, 1.6,
-0.42, 0.35, -0.29, -0.23, 2.1, -3.8, -1.5, 1.1, 2.9, 0.0046,
    -0.47, -0.71, 4.2, 1.5, -1., -0.0046, -0.55, -1.1, -0.073,
0.64, -0.17, -1.8, 0.45,
0.69, -0.49, -0.18, 0.09, -0.92, -0.84, -0.1, 0.23, -2., -1.7,
    -0.43, -1.6, -1.2, -0.019, 1.1, -0.084, 0.29, -4.9, -1.1,
2.8, -1.2, 1.2, -0.95,
-0.87, 2.1, -0.8, 1.5, -1.4, 0.32, 1.2, -0.061, -0.15, 1.2,
    -0.27, 1.2, -0.066, -0.86, 0.42,-0.22, 0.6, -0.6, 2.2,
-0.085, -4.3, -0.13, -2.4, -1.2,
1.8, -1.8, -0.8, -0.46, 4., -0.35, 0.26, 0.85, 1.5, 0.79,
    -1.9, 1.9, -0.53, 1.9, -1.1, 0.66, 0.71, -0.94, -0.23,
    0.048, -0.7, 4.4, 0.63),.Dim = c(2, 126)),
S.sp = structure(.Data = c(-1.3, 0.53, -0.59, -0.19, 0.47,
    -3.1, 1.1, -0.24, -0.62, -0.49, 1.3, -4.1, -0.11, 0.47,
    -0.54, -2.1, -0.46, 0.63,
-1.5, -1.1, -2.0, -0.016, 0.17, 1.3, -1.1, 0.6, -1.3,
-1.2, -1.4, -0.67, -0.69, 6.1, 0.77, -0.43, -0.42, 0.99,
    0.28, 1.1, 0.82, -0.89, 0.75, 1.2, 0.21, 0.56, 0.87, 0.33,
    0.039, -0.15, -1.1, 1.9, 1.3, 0.36, 2.1, -1.1, 5.3, 0.46,
    -1.8, 0.14, 0.83, -0.1, -0.089, -0.58, -0.77, -0.52, 0.93,
    0.43, 1.3, 1.5, -0.17, -0.027, 0.79, -0.11, 0.28, 1., -0.9,
    1.1, 0.42, 0.78, -0.37, -0.39, 0.74, -0.019, -0.17, 0.14,
    2., -0.15, 2.9, -0.85, -0.33, 0.35, 0.12, -2.1, -0.027,
    0.64, 0.17, -0.97, -1.4, 1.3, 0.56, -2.8, 0.71, 3.2, -0.57,
    0.11, -1.5, 0.62, 2.9, -0.19, -0.36, -0.052, -0.54, -1.2,
    -1.4, 0.89, -2.6, -2.1, -0.34, -0.8, -0.49, -2., 0.067, 2.5,
    -1.5, 0.53, 1.9, -0.2, 5.2, 0.71, 0.33, 0.49, 1.2, 2.2,
    0.25, -0.64, -0.046, 0.48, 1.6, -0.51, -3.9, 0.39, 0.79,
    -0.52, -0.61, 1.5, 0.95, -1.5, 0.97, 1.6, 0.46, -0.97, 0.69,
    -5.1, -0.16, -0.66, -0.19, -1.2, 0.22, 1., -0.39, -1.2,
    0.16, 1.3, -0.99, 1.4, 0.83, 2.1, 1.2, -2.3, 0.02, 1.2, 1.4,
    -0.77, 1.1, -0.18, -1.1, 1.3, 0.94, -0.26, -1., 1.4, -1.9,
    -1.3, 0.21, 0.3, -0.27, -0.87, -0.25, -2.5, 2.6, 0.12,
    -0.21, -0.2, 1.5, 1.7, 1.5, -0.22, 1.3, 0.42, 0.91, -1.7,
    1.9, -1.2, -1.6, 0.83, -0.43, -1.2, 0.96, 0.9, -3.2, 4.3,
    -0.029, 1.9, -0.54, 0.83, 1.1, -1.3, -0.73, -0.97, -0.22,
    -1.2, 1.1, 3.6, -2.0, 0.23, 0.88, -2.6, -0.73, -0.24, -2.,
    0.00026, -1.2, -0.46, 0.25, 0.075, -0.5, -0.38, -0.42,
    -0.047, -2.4, -0.27, -1.1, 0.18, 1., -2.4, 1., -2., 1.4,
```

```
     2.1,  1.1,  -1.,  -1.2,  -0.77),.Dim = c(2, 126)),
U.sh = c(1.4,  -0.042,  -0.19,  0.5,  2.5,  0.43,  -4.3,  -0.84,  1.,
     -1.1,  1.,  -1.7,  0.44,  -4.7,  0.58,  0.11,  0.15,  0.96,  -0.28,
     0.8,  0.083,  -1.5,  -0.78,
0.33,  0.37,  -0.6,  0.66,  -0.52,  -0.81,  -1.6,  2.9,  1.8,  -0.28,
     0.67,  -0.35,  0.17,  1.8,  -0.039,  2.,  -0.62,  0.17,  -1.6,  -3.3,
     0.52,  -0.13,  1.8,  0.55,
-0.99,  1.,  -0.58,  -0.4,  3.2,  -0.14,  2.3,  -0.017,  -0.79,  0.54,
     0.45,  1.5,  -0.33,  0.69,  1.2,  1.7,  0.48,  1.3,  1.4,  -1.7,  1.3,
     0.4,  -0.66,  -0.69,
-0.06,  -0.19,  -0.13,  2.2,  0.84,  -0.46,  -0.63,  0.35,  -0.45,
     1.2,  0.82,  -0.74,  -0.063,  -0.37,  -0.71,  -0.057,  -0.64,  0.11,
     -0.39,  0.11,  0.53,  1.3,
-0.17,  0.21,  0.89,  0.98,  -2.,  1.2,  -0.59,  1.6,  0.036,  -0.54,
     -0.68,  -1.7,  -1.,  1.7,  0.12,  -0.81,  -0.17,  -0.18,  1.1,  0.38,
     -0.16,  -0.46,  -0.26,
-1.4,  -0.49,  -0.35,  0.52,  0.59,  2.8,  -0.61,  2.,  0.94,  0.98),
S.sh = c(-0.99,  -0.21,  0.7,  -0.15,  -0.52,  0.62,  0.13,  0.077,
     0.63,  0.086,  -1.6,  0.24,  0.55,  -0.98,  0.37,  0.95,  -0.33,
     1.9,  0.44,  1.,  -0.15,
-0.49,  -2.1,  -1.2,  -1.3,  0.22,  1.6,  -1.1,  -1.5,  -1.2,  0.51,
     -0.062,  -0.32,  0.068,  0.24,  -0.18,  -0.24,  -0.12,  1.3,  0.28,
     -0.099,  0.65,  1.2,  -0.64,
-4.9,  1.3,  -0.096,  -2.7,  -0.75,  -0.4,  -1.4,  -0.098,  -1.,  1.6,
     -1.7,  -0.046,  -0.45,  -0.82,  -2.5,  -1.2,  0.31,  -0.33,  2.2,
     0.44,  0.47,  1.2,  2.2,
-0.63,  0.058,  0.027,  0.084,  0.22,  1.7,  0.94,  -1.3,  -0.24,
     -1.6,  0.012,  -0.94,  -0.059,  -0.33,  -1.2,  -0.33,  0.42,  -0.16,
     -0.42,  -3.2,  1.9,  1.9,
0.47,  0.32,  -0.16,  -1.9,  0.99,  -0.57,  -0.38,  -1.1,  0.11,  1.,
     -0.42,  0.32,  -0.22,  0.41,  -0.4,  -0.68,  1.4,  1.5,  1.4,  -1.2,
     -0.55,  -1.5,  -1.5,  -2.0,
0.83,  1.5,  0.47,  -0.24,  0.16,  0.12,  -2.,  -0.8,  0.087,  2.6,
     3.5,  -1.6,  0.31),
alpha = c(0.14,  -0.78),
tau.unstr = c(0.25,  2.7),
tau.spatial = c(7.5,  9.8),
omega.unstr = 9.4,
omega.spatial = 4.0,
logdelta = 1.34

)
```

A Bayesian analysis is executed with 55,000 observations for the simulation, a burn-in of 5000, and a refresh of 100, and the posterior analysis is reported in Table 8.16 for the two intercept parameters. Note that the model parameters are given noninformative prior distributions.

A 95% credible interval for α_1 is (−.09333, .05687), which implies that the first intercept for oral cavity cancer baseline relative risk is negligible. A map of the 126 smoothed relative risks appears in Figure 8.10.

There are 126 regions or sites in West Yorkshire and thus there are 126 posterior means for relative risks of oral cavity cancer and 126 for lung cancer, but they are not reported here; they are left as an exercise (see Exercise 12). The literature for spatial models in epidemiology is voluminous and the following books should be of interest to the student. The book by Getis and Boots[23] is one of the earlier books in spatial geography, while the one by Cliff and Ord[24] is one of the earliest for statisticians as is the Ripley[25] volume, which deals with spatial point processes, a topic not introduced here.

Another book dealing specifically with point processes is by Diggle,[26] but the most comprehensive and interesting is Cressie's[27] book, which made a

TABLE 8.16

Posterior Analysis for Baseline (log) Relative Risk, Oral Cavity and Lung Cancer, West Yorkshire, UK

Parameter	Mean	SD	Error	2½	Median	97½
α_1	−.017230	.03848	.0004454	−.09333	−.016690	.05687
α_2	−.023350	.01153	.0000914	−.04599	−.02328	−.0003964

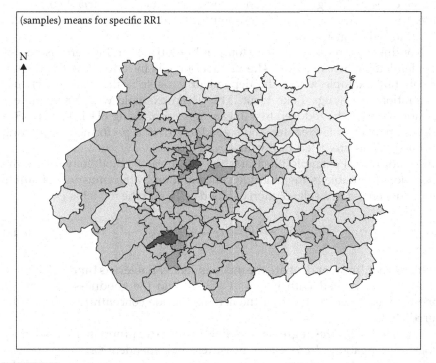

(samples) means for specific RR1

N

FIGURE 8.10

Estimated relative risks for West Yorkshire UK. Oral cavity cancer for 126 sites. (Example 5 of GeoBUGS from WinBUGS®. www.mrc-bsu.cam.ac.uk/bugs.)

major impact in the field. Lastly, for the Bayesian approach, I recommend the study by Lawson,[28] whose focus is on spatial mapping of the disease process. This book is of interest to the epidemiologist who wants to learn more about the Bayesian approach, and the present book is a good companion to the Lawson work.

8.7 Comments and Conclusions

The final chapter of the book has introduced the epidemiologist to a variety of regression models useful in studying the association between disease and exposure to various risk factors. Chapter 8 is an extension of some of the regression models used in earlier chapters and introduces the reader to some specialized but useful regression models.

Categorical models are appropriate when the dependent variable (disease status) assumes a small number of discrete values and where the independent variable measures the exposure of various risk factors. An example involving melanoma metastasis introduces the reader to the methodology of the categorical regression model where the dependent variable measures the extent of metastasis on a five-point scale and the independent variables are four radiologists.

Nonlinear regression is a vast topic in biostatistics and is very useful for epidemiologic investigation. The methodology is introduced with an environmental example where the dependent variable is the PCB (toxic) concentration in Cayuga Lake trout. This is an example where a nonlinear relationship can be reduced to a linear regression. Using BUGS CODE 8.2, a Bayesian analysis is executed and a goodness-of-fit is performed by plotting the actual and predicted PCB values.

An additional example illustrates the nonlinear methodology where the dependent variable is the chloride concentration in the transport of sulfite ions suspended in a salt solution of blood cells, and the model is

$$Y_n = \theta_1 \left(1 - \exp\left(-\theta_2 x_n\right)\right) + e_n \tag{8.12}$$

which is nonlinear and where the independent variable x_n is time. A Bayesian analysis is executed with BUGS CODE 8.3 and the goodness-of-fit of the model is assessed by plotting the actual chloride concentration versus the predicted.

Repeated measures regression models are quite appropriate for epidemiologic investigations. In repeated measures, the dependent variable is repeatedly measured on the same subject, and the repeated measurements on the same subject are correlated; thus, the model takes the correlation into account when assessing the association between disease status and exposure to various risk factors. An example from the study of Hand and Taylor[10] illustrates the repeated measures regression. It is a randomized study of two groups of Alzheimer's patients, where one group receives a placebo and the other receives a treatment for the disease. The dependent variable is the number of correctly recalled words y_{ij} and the model is

$$y_{ij} = \theta + \alpha_i + \beta_j + e_{ij} \tag{8.14}$$

where the α_i and β_j are random effects that induce the same correlation between observations of the same subject. A Bayesian analysis estimates the correlation and generates future observations for a goodness-of-fit test.

One of the most useful models in epidemiology is the spatial regression model where the spatial association between a disease and exposure can be assessed. The lip cancer study of Clayton and Kaldor[20] illustrates the methodology with the spatial model is

$$O_i \sim \text{Poisson}(mu_i) \tag{8.24}$$

$$\log(mu_i) = \frac{\log E_i + \alpha_0 + \alpha_1 x_i}{10 + b_i} \tag{8.25}$$

where the dependent value O_i for each county is the dependent variable and the independent variable is x, the percent of the population with outdoor jobs.

The mean of the regression model is a function of the expected number E_i of lip cancer cases (standardized by gender and age) and the baseline relative risk α_0. The effect of the exposure variable x is measured by α_1, while the remaining term is a random effect b_i, which is modeled by a CAR model. The latter takes into account the neighbors of each county and induces a spatial correlation between each county and its neighbors.

BUGS CODE 8.8 will execute the Bayesian analysis from which a map (Figure 8.9) of the smoothed relative risks can be portrayed. By doing Exercise 12, the student can investigate the goodness-of-fit of the Models 8.24 and 8.25. An additional example involving the joint occurrence of two diseases (oral cavity cancer and lung cancer) in West Yorkshire of the UK presents the student with another tool to study the spatial association between disease and various exposures.

Exercises

1. Based on Table 8.1 and BUGS CODE 8.1, execute a Bayesian analysis with 55,000 observations for the simulation, a burn-in of 5000, and a refresh of 100.

   a. Validate Table 8.2.

   b. What is the MCMC simulation error for the ROC area?

   c. Plot the posterior density of the ROC area.

   d. What does component a1 of the ROC area measure?

2. Based on Table 8.3 and BUGS CODE 8.2, generate 55,000 observations for the simulation, with a burn-in of 5000 and a refresh of 100.

   a. Verify Table 8.4.

   b. What does the parameter θ_2 measure?

   c. Validate Table 8.5.

   d. What does the parameter β_2 measure?

   e. Explain the differences and similarities between Models 8.9 and (8.10).

   f. Which model gives the "best fit"? Explain your answer.

3. Based on Table 8.5 and BUGS CODE 8.3, perform a Bayesian analysis with 250,000 observations for the simulation, a burn-in of 5000, and a refresh of 100.

   a. Validate Table 8.6.

   b. What is the posterior mean and 95% credible interval for θ_2?

   c. Refer to Equation 8.12. What is the interpretation of the parameter θ_1?

   d. Validate Table 8.8.

   e. Reproduce Figure 8.6, the predicted chloride concentration values.

4. Consider a nonlinear regression example that considers the biochemical oxygen demand of stream water. A sample of water is taken, injected with soluble organic matter, inorganic nutrients, and dissolved oxygen, and subdivided into BOD (biochemical oxygen demand) bottles.

   Each bottle is inoculated with a mixed culture of microorganisms, sealed, and incubated at a constant temperature, then the bottles are unsealed periodically and the dissolved oxygen concentration is measured. The BOD is then calculated in milligrams per liter (mg/L). For additional information about this study, see Bates and Watts.[8]

The data are analyzed with the nonlinear model

$$Y_n = \theta_1 \left(1 - \exp(-\theta_2 x_n)\right) + e_n$$

where Y_n is the biochemical oxygen demand at time x_n with n = 1, 2, 3, 4, 5, 6, thus, the average value of the oxygen demand increases exponentially.

The two unknown parameters θ_1 and θ_2 will be estimated with a nonlinear Bayesian regression. Using BUGS CODE 8.3, the two parameters θ_1 and θ_2 are estimated with a Bayesian analysis with 150,000 observations for the simulation, a burn-in of 5000, and a refresh of 100. When using BUGS CODE 8.3, remove the # sign appearing before the code of the BOD nonlinear regression, and place a # before each statement of the chloride concentration regression. Remember that a # appearing before an executable statement deactivates that statement. The # in front of comments (nonexecutable statements) should remain.

5. Based on BUGS CODE 8.4, generate 55,000 observations for the simulation, a burn-in of 5000, and a refresh of 100.

   a. Verify Table 8.8.

   b. σ_β^2 is much smaller than σ_α^2. Explain why.

   c. Plot the posterior density of ρ.

   d. Estimate the correlation between observations of the same individual.

   e. Complete Table 8.8 for the predicted values for patients 3–25. The following statement from BUGS CODE 8.4 predicts the number of correctly recalled words for the 26 patients of the placebo group:

   `for(i in 1:n){for(j in 1:p){z[i,j]~dnorm(mu[i,j],tau)}}`

6. Using the predicted values from problem 5 part e, verify Figure 8.8 for the goodness-of-fit for the placebo group of the Alzheimer's study.

   a. Does Model 8.14 give a good fit to the data? Why?

   b. What is the third observation for the second patient? What is the corresponding predicted value Z[2,3]?

   c. What is the 95% prediction interval for Z[2,3]?

   d. What is the residual corresponding to the third observation of the second patient?

7. The first list statement of BUGS CODE 8.5 below records the Alzheimer's data for the two groups, where the first 26 patients correspond to the placebo group and the remaining 22 correspond to the treatment group. The five observations per patient correspond to the five time points.

BUGS CODE 8.5

```
model;

{

for(i in 1:n){for(j in 1:p){y[i,j]~dnorm(mu[i,j],tau)}}
for(i in 1:n){for(j in 1:p){mu[i,j]<- theta+alpha[i]+beta[j]}}
for(i in 1:n){for(j in 1:p){z[i,j]~dnorm(mu[i,j],tau)}}
theta~dnorm(0,.001)
for(i in 1:n){alpha[i]~dnorm(0,tau.alpha)}
for(j in 1:p){beta[j]~dnorm(0,tau.beta)}

tau.alpha~dgamma(.001,.001)
tau.beta~dgamma(.001,.001)
tau~dgamma(.001,.001)

sigma.alpha<-1/tau.alpha
sigma.beta<-1/tau.beta
sigma<-1/tau

rho<-sigma.alpha/(sigma.alpha+sigma.beta+sigma)
}
# data for both groups of Alzheimer's study
list(n = 48, p = 5, y = structure(.Data =
c(20,19,20,20,18,
14,15,16,9,6,
7,5,8,8,5,
6,10,9,10,10,
9,7,9,4,6,
9,10,9,11,11,
7,3,7,6,3,
18,20,20,23,21,
6,10,10,13,14,
10,15,15,15,14,
5,9,7,3,12,
11,11,8,10,9,
10,2,9,3,2,
17,12,14,15,13,
16,15,13,7,9,
7,10,4,10,5,
5,0,5,0,0,
16,7,7,6,10,
5,6,9,5,6,
2,1,1,2,2,
7,11,7,5,11,
9,16,17,10,6,
2,5,6,7,6,
7,3,5,5,5,
19,13,19,17,17,
```

```
7,5,8,8,6,
9,11,14,11,14,
6,7,9,12,16,
13,18,24,20,14,
9,10,9,8,7,
6,7,4,5,4,
11,11,5,10,2,
7,10,11,8,5,
8,18,19,15,14,
3,3,3,1,3,
4,10,9,17,10,
11,10,5,15,16,
1,3,2,2,5,
6,7,7,6,7,
0,3,2,0,0,
18,19,15,17,20,
15,15,15,14,12,
14,11,8,10,8,
6,6,5,5,8,
10,10,6,10,9,
4,6,6,4,2,
4,13,9,8,7,
14,17,18,10,6),.Dim = c(48,5)))

# initial values
list(alpha = c(0,0,0,0,0,0,0,0,0,0,0,0,0,0,0,0,0,0,0,0,0,0,0,0,0
    ,0,0,0,
0,0,0,0,0,0,0,0,0,0,0,0,0,0,0,0,0,0,0,0,0,0), theta = 0,
beta = c(0,0,0,0,0), tau.alpha = 1, tau.beta = 1, tau = 1))
```

The Bayesian analysis is executed with 55,000 observations, a burn-in of 5000, and a refresh of 100, and the results are portrayed in Table 8.17. Verify the posterior analysis given in Table 8.17

a. How does the posterior mean for θ compare with the posterior mean of θ reported in Table 8.10? The results of Table 8.8 are based on the placebo group, whereas the reports of Table 8.17 are based on both groups of the Alzheimer's study.

TABLE 8.17

Posterior Analysis for Alzheimer's Study

Parameter	Mean	SD	Error	2½	Median	97½
ρ	.7231	.0502	.000326	.6192	.7252	.8149
σ^2	7.637	.79	.004457	6.254	7.579	9.38
σ_α^2	21.01	4.848	.02741	13.43	20.36	32.39
σ_β^2	.1487	.622	.00901	.0008	.02951	.9294
θ	9.238	.7043	.01519	7.846	9.242	10.6

   b. Generate the predicted values for all 48 patients of the study. Use BUGS CODE 8.5.

   c. How well does Model 8.14 fit the data? Plot the predicted values versus the actual values.

   d. Why is σ_β^2 much smaller than σ_α^2?

8. Using BUGS CODE 8.6, generate 55,000 observations for the simulation, with a burn-in of 5000, and a refresh of 100 and verify Tables 8.10, 8.11, and 8.12.

   a. Table 8.10 reveals the results using age, treatment, base, and the interaction between age and treatment. Should the interaction term be dropped from Model 8.23? Why or why not?

   b. Table 8.11 reports the Bayesian analysis without the interaction term. Is age an important predictor of the number of seizures? Why? Is base an important independent variable? Why? Is treatment an important predictor? Why?

   c. Table 8.12 shows the analysis using base and treatment as independent variables. Is there a difference in the two treatments? Why?

   d. What does sigmab1 measure?

   e. Refer to BUGS CODE 8.6, where orig.level is defined as exp(a0). What does orig.level measure?

   f. What does sigmab2 measure?

   g. As estimated by the posterior mean, would you expect sigmab1 to be larger than sigmab2?

9. Using BUGS CODE 8.6 and Model 8.23, generate future observations for the number of epileptic seizures for each patient over the four time periods.

   a. Plot the predicted values versus the actual values.

   b. Does the model provide a good fit? Why?

10. Consider the Poisson regression Model 8.23. What is the correlation between observations of the same individual (See Table 8.17)?

11. Based on Table 8.13 and BUGS CODE 8.7, perform a Bayesian analysis for the Scottish lip cancer data. With 55,000 observations for the simulation, a burn-in of 5000, and a refresh of 100, verify Tables 8.14 and 8.15 (See Table 8.18).

   a. Plot the posterior density of α_0.

   b. Plot the posterior density of α_1.

   c. Compute the relative risk of lip cancer for all 56 counties of Scotland.

   d. Is there an association between the occurrence of lip cancer and the percentage of workers who work in outside occupations? Why?

TABLE 8.18

Posterior Analysis of Future Values of Lip Cancer

Parameter	Mean	SD	Error	2½	Median	97½
u[1]	6.875	3.343	.02381	2	6	14
u[2]	37.94	8.441	.08546	23	38	56
u[3]	10.6	4.476	.03446	3	10	21

   e. What is the posterior distribution of the relative risk of lip cancer RR[33] for county 33?

   f. Plot the posterior density of RR[33].

12. Using BUGS CODE 8.8, execute a Bayesian analysis with 55,000 observations, a burn-in of 5000, and a refresh of 100.

   a. Verify Table 8.16.

   b. Estimate the relative risk for the 126 regions for oral cavity cancer. Note that specificRR1[i] is the node for the relative risk of site $i, i = 1, 2, ..., 126$.

   c. Estimate the relative risk for the 126 regions for lung cancer. Note that specificRR2[i] is the node for the relative risk of site $i, i = 1, 2, ..., 126$.

   d. Plot the posterior density of α_2, the baseline (log) relative risk for lung cancer.

   e. Plot the posterior density of specificRR2[5], the relative risk for lung cancer for site 5. As a check, I found the posterior mean of specificRR2[5] is .9085 with a credible interval of (.6858, 1.084).

   f. Plot the posterior density of specificRR1[120], the relative risk for oral cavity cancer for site 120. As a check, I found that the posterior mean of specificRR1[120] is 1.064 with a 95% credible interval (.7696, 1.48).

   g. From the first list statement of BUGS CODE 8.8, identify the neighbors of site 2.

13. Refer to the lip cancer data in Table 8.13. Using BUGS CODE 8.7, generate future values for the number of cases of lip cancer. There are 56 counties, and thus 56 predicted values. Use 55,000 for the simulation, with a burn-in of 5000, and a refresh of 100.

   a. Provide the posterior analysis for the 56 future values u[i], $i = 1, 2, ..., 56$. The posterior analysis for the first three counties is:

      a. Complete the posterior analysis for the remaining counties.

      b. Perform a goodness-of-fit by plotting the predicted lip cancer cases u[i] versus the actual cases O[i].

c. How well does the model fit the data in Table 8.16?
d. Plot the posterior density of u[1]. See Table 8.18 for the three first future values.

References

1. Freireich, E.J., Gehan, E.A., Frei, E., et al. The effect of 6-mercaptopurince on the duration of remission in acute leukemia: A model for the evaluation of other potentially useful therapies, *Blood*, 21, 699–716, 1963.
2. Congdon, P. *Applied Bayesian Modeling*, John Wiley & Sons Inc., 2003, New York.
3. Broemeling, L.D. *Advanced Bayesian Methods for Medical Test Accuracy*, CRC Taylor & Francis, 2012, Boca Raton, FL.
4. Johnson, V.E. and Albert, J.H. *Ordinal Data Modeling*, Springer-Verlag, 1999, New York.
5. Albert, J.H. and Chib, S. Bayesian analysis of binary and polychotomous response data, *J. Am. Stat. Assoc.* 88, 669–679, 1993.
6. Bache, C.A., Serum, J.W., Youngs, D.W., and Lisk, D.J. Polychlorinated biphenyl residues: Accumulation in Lake Cayuga trout with age, *Science*, 117, 1192–1193, 1972.
7. Sredni, J. Problems of Design, Estimation, and Lack of Fit in Model Building, Ph.D. Thesis, University of Wisconsin Madison, 1970.
8. Bates, D.M. and Watts, D.G. *Nonlinear Regression Analysis and Its Applications*, John Wiley and Sons Inc., 1988, New York.
9. Denison, T., Holmes, C.C., Mallick, B.K., and Smith, A.F.M. *Bayesian Methods for Nonlinear Classification and Regression*, John Wiley & Sons. Inc., 2002, New York.
10. Hand, D.J. and Taylor, C.C. *Multivariate Analysis of Variance and Repeated Measures*, Chapman & Hall, 1987, London, UK.
11. Thall, P.F. and Vail, S.C. Some covariance models for longitudinal count data with overdispersion, *Biometrics*, 46, 657–671, 1990.
12. Crowder, M.J. and Hand, D.J. *Analysis of Repeated Measures*, Chapman and Hall, 1990, London, UK.
13. Hand, D. and Crowder, M. *Practical Longitudinal Data Analysis*, Chapman and Hall/CRC, 1996, Boca Raton, FL.
14. Jones, R.H. *Longitudinal Data with Serial Correlation, A State-Space Approach*, Chapman and Hall, 1993, London, UK.
15. Lindsey, J.K. *Models for Repeated Measurements*, Oxford University Press, 1993, Oxford, UK.
16. Davidian, M. and Giltinan, D.M. *Nonlinear Models for Repeated Measurement Data*, Chapman & Hall, 1995, London, UK.
17. Mausner, J.S. and Kramer, S. *Epidemiology—An Introductory Text*, Second Edition, W.B. Saunders Company, 1985, Philadelphia, PA.
18. Palm, T.A. The geographical distribution and aetiology of rickets, *Practitioner*, 45, 1890, 270–279.
19. Lancaster, H.O. Some geographical aspects of the mortality from melanoma in Europeans, *Med. J. Aust.*, 1, 1082–1087, 1956.

20. Clayton, D. and Kaldor, J. Empirical Bayes estimates of age-standardized relative risks for use in disease mapping, *Biometrics*, 43, 671–681, 1987.
21. Cressie, N.A.C. Regional mapping of incidence rates using spatial Bayesian models, *Med. Care*, 31(supplement), YS60–YS65, 1993.
22. Mollie, A. and Richardson, S. Empirical Bayes estimates of cancer mortality rates using spatial models, *Stat. Med.* 10, 95–112, 1991.
23. Getis, A. and Boots, B. *Models of Spatial Processes*, Cambridge University Press, 1978, Cambridge.
24. Cliff, A.D. and Ord, J.K. *Spatial Autocorrelation*, Pion Limited, 1973, London, UK.
25. Ripley, B.D. *Statistical Inference for Spatial Processes*, Cambridge University Press, 1988, Cambridge.
26. Diggle, P.J. *Statistical Analysis of Spatial Point Processes*, Academic Press, 1983, London, UK.
27. Cressie, N.A. *Statistics for Spatial Data*, John Wiley & Sons Inc., 1993, New York.
28. Lawson, A. *Bayesian Disease Mapping: Hierarchical Modeling in Spatial Epidemiology*, CRC/Chapman & Hall, 2008, New York.

Appendix A: Introduction to Bayesian Statistics

A.1 Introduction

Bayesian methods will be employed to design and analyze studies in epidemiology, and this chapter will introduce the theory that is necessary in order to describe Bayesian inference. Bayes' theorem, the foundation of the subject, is first introduced and followed by an explanation of the various components of Bayes' theorem: prior information, information from the sample given by the likelihood function, the posterior distribution, which is the basis of all inferential techniques, and lastly, the Bayesian predictive distribution. A description of the main three elements of inference, namely, estimation, tests of hypotheses, and forecasting future observations, follows.

The remaining sections refer to the important standard distributions for Bayesian inference, namely, the Bernoulli, beta, multinomial, Dirichlet, normal, gamma, normal–gamma, multivariate normal, Wishart, normal–Wishart, and multivariate t-distributions. As will be seen, the relevance of these standard distributions to inferential techniques is essential for understanding the analysis of methods used in epidemiology.

Of course, inferential procedures can be applied only if there is adequate computing. If the posterior distribution is known, often analytical methods are quite sufficient to implement Bayesian inferences and this will be demonstrated for the binomial, multinomial, and Poisson populations, and several cases of normal populations. For example, when using a beta prior distribution for the parameter of a binomial population, the resulting beta posterior density has well-known characteristics, including its moments. In a similar fashion, when sampling from a normal population with unknown mean and precision and with a vague improper prior, the resulting posterior t-distribution for the mean has known moments and percentiles, which can be used for inferences.

Posterior inferences by direct sampling methods are easily done if the relevant random number generators are available. On the other hand, if the posterior distribution is quite complicated and not recognized as a standard distribution, other techniques are needed. To solve this problem, Monte Carlo Markov chain (MCMC) techniques have been developing for the last 25 years and have been a major success in providing Bayesian inferences for quite complicated problems. This has been a great achievement in this field and will be described in later sections.

Minitab, S-Plus, and WinBUGS are packages that provide random number generators for direct sampling from the posterior distribution for many standard distributions, such as binomial, gamma, beta, and t-distributions. These will be used on occasion; however, my preference is WinBUGS, because it has been well accepted by other Bayesians. This is also true for indirect sampling, where WinBUGS is a good package. WinBUGS is the software of choice for the book, and it is introduced in Appendix B. Many institutions provide special purpose software for specific Bayesian routines. For example, at MD Anderson Cancer Center, where Bayesian applications are routine, several special purpose programs are available for designing (including sample size justification) and analyzing clinical trials, and will be described. The theoretical foundation for MCMC is introduced in the following sections.

Inferences for studies in epidemiology consist of testing hypotheses about unknown population parameters, estimating those parameters, and forecasting future observations. When a sharp null hypothesis is involved, special care is taken in specifying the prior distribution for the parameters. A formula for the posterior probability of the null hypothesis is derived, via Bayes' theorem, and illustrated for Bernoulli, Poisson, and normal populations. If the main focus is estimation of parameters, the posterior distribution is determined, and the mean, median, standard deviation, and credible intervals are found, either analytically or by computation with WinBUGS. For example, when sampling from a normal population with unknown parameters and using a conjugate prior density, the posterior distribution of the mean is a t and will be derived algebraically. On the other hand, in observational studies, the experimental results are usually portrayed in a 2 × 2 table that gives the cell frequencies for the four combinations of exposure and disease status, where the consequent posterior distributions are beta for the cell frequencies, and posterior inferences are provided both analytically and with WinBUGS. Of course, all analyses should be preceded by checking to determine if the model is appropriate, and this is where predictive distribution comes into play. By comparing the observed results of the experiment (e.g., a case–control study) with those predicted, the model assumptions are tested. The most frequent use of Bayesian predictive distribution is for forecasting future observation in time-series studies, and time series in the form of cohort studies (repeated measures) are part of many epidemiologic studies.

A.2 Bayes' Theorem

Bayes' theorem is based on the conditional probability law:

$$P[A\,|\,B] = \frac{P[B\,|\,A]P[A]}{P[B]} \tag{A.1}$$

where P[A] is the probability of A before one knows the outcome of the event B, P[B|A] is the probability of B assuming what one knows about the event A, and P[A|B] is the probability of A knowing that event B has occurred. P[A] is called the prior probability of A, while P[A|B] is called the posterior probability of A.

Another version of Bayes' theorem is to suppose X is a continuous observable random vector and $\theta \in \Omega \subset R^m$ is an unknown parameter vector, and suppose the conditional density of X at given θ is denoted by $f(x\,|\,\theta)$. If $x = (x_1, x_2, \ldots, x_n)$ represents a random sample of size n from a population with density $f(x\,|\,\theta)$, and $\xi(\theta)$ is the prior density of θ, then Bayes' theorem (A.1) expresses the posterior density as

$$\xi(\theta\,|\,x) = cf(x_i\,|\,\theta)\xi(\theta), \ x_i \in R \text{ and } \theta \in \Omega \tag{A.2}$$

where the proportionality constant is c, and the term $\prod_{i=1}^{i=n} f(x_i\,|\,\theta)$ is called the likelihood function. The density $\xi(\theta)$ is the prior density of θ and represents the knowledge one possesses about the parameter before one observes X. Such prior information is most likely available to the experimenter from other previous related experiments. Note that θ is considered a random variable and that Bayes' theorem transforms one's prior knowledge of θ, represented by its prior density, to the posterior density, and that the transformation is the combination of the prior information about θ with the sample information represented by the likelihood function.

"An essay toward solving a problem in the doctrine of chances" by the Reverend Thomas Bayes[1] is the beginning of our subject. He considered a binomial experiment with n trials, assumed that the probability θ of success was uniformly distributed (by constructing a billiard table), and presented a way to calculate $\Pr(a \le \theta \le b\,|\,x = p)$, where x is the number of successes in n independent trials. This was a first in the sense that Bayes was making inferences via $\xi(\theta\,|\,x)$, the conditional density of θ at given x. Also, by assuming the parameter as uniformly distributed, he was assuming vague prior information for θ. The type of prior information, where very little is known about the parameter, is called noninformative or vague information.

It can well be argued that Laplace[2] is the greatest Bayesian, because he made many significant contributions to inverse probability (he did not know of Bayes), beginning in 1774 with "Memorie sur la probabilite des causes par la evenemens," with his own version of Bayes' theorem, and over a period of some 40 years culminating in "Theorie analytique des probabilites." See Stigler[3] and chapters 9 through 20 of Hald[4] for the history of Laplace's contributions to inverse probability.

It was in modern times that Bayesian statistics began its resurgence with Lhoste,[5] Jeffreys,[6] Savage,[7] and Lindley.[8] According to Broemeling and Broemeling,[9] Lhoste was the first to justify noninformative priors by invariance principals, a tradition carried on by Jeffreys. Savage's book was a major contribution in that Bayesian inference and decision theory was put on a sound theoretical footing as a consequence of certain axioms of probability and utility, while Lindley's two volumes showed the relevance of Bayesian inference to everyday statistical problems; they were quite influential and set the tone and style for later books such as Box and Tiao,[10] Zellner,[11] and Broemeling.[12] Box and Tiao and Broemeling were essentially works that presented Bayesian methods for the usual statistical problems of the analysis of variance and regression, while Zellner focused Bayesian methods primarily on certain regression problems in econometrics. During this period, inferential problems were solved analytically or by numerical integration. Models with many parameters (such as hierarchical models with many levels) were difficult to use because at that time numerical integration methods had limited capability in higher dimensions. For a good history of inverse probability see Chapter 3 of Stigler, and Hald,[13] who presented a comprehensive history and are invaluable as a reference. Dale[14] gives a complete and very interesting account of Bayes' life.

The last 20 years is characterized by the rediscovery and development of resampling techniques, where samples are generated from the posterior distribution via MCMC methods, such as Gibbs sampling. Large samples generated from the posterior make it possible to make statistical inferences and to employ multilevel hierarchical models to solve complex, but practical problems. See the studies by Leonard and Hsu,[15] Gelman et. al.,[16] Congdon,[17,18,19] Carlin and Louis,[20] and Gilks, Richardson, and Spiegelhalter,[21] who demonstrated the utility of MCMC techniques in Bayesian statistics.

A.3 Prior Information

Where do we begin with prior information, a crucial component of Bayes' theorem? Bayes assumed that the prior distribution of the parameter is uniform, namely

$$\xi(\theta) = 1, \ 0 \le \theta \le 1$$

where θ is the common probability of success in n independent trials and

$$f(x \mid \theta) = \binom{n}{x} \theta^x (1-\theta)^{n-x} \qquad \text{(A.3)}$$

where x is the number of successes $= 0, 1, 2, \ldots, n$. The distribution of x, the number of successes, is binomial and denoted by $x \sim$ Binomial (θ, n). The uniform prior was used for many years; however, Lhoste[5] proposed a different prior, namely

$$\xi(\theta) = \theta^{-1}(1-\theta)^{-1}, 0 \le \theta \le 1 \qquad \text{(A.4)}$$

to represent information that is noninformative and is an improper density function. Lhoste based the prior on certain invariance principals, quite similar to Jeffreys.[6] Lhoste also derived a noninformative prior for the standard deviation σ of a normal population with density

$$f(x \mid \mu, \sigma) = \left(\frac{1}{\sqrt{2\pi}\sigma} \right) \exp - \left(\frac{1}{2\sigma} \right)(x - \mu)^2, \mu \in R \text{ and } \sigma > 0 \qquad \text{(A.5)}$$

He used invariance as follows: he reasoned that the prior density of σ and the prior density of $1/\sigma$ should be the same, which leads to

$$\xi(\sigma) = \frac{1}{\sigma} \qquad \text{(A.6)}$$

Jeffreys' approach is similar in developing noninformative priors for binomial and normal populations, but he also developed noninformative priors for multiparameter models, including the mean and standard deviation for the normal density as

$$\xi(\mu, \sigma) = \frac{1}{\sigma}, \mu \in R \text{ and } \sigma > 0 \qquad \text{(A.7)}$$

Noninformative priors were ubiquitous from the 1920s to the 1980s and were included in all the textbooks of that period, for example, see the studies by Box and Tiao, Zellner, and Broemeling. Looking back, it is somewhat ironic that noninformative priors were almost always used, even though informative prior information was almost always available. This limited the utility of the Bayesian approach, and people saw very little advantage over the conventional way of doing business. The major strength of the Bayesian way is that it is a convenient, practical, and logical method of utilizing informative prior information. Surely, the investigator knows informative prior information from previous related studies.

How does one express informative information with a prior density? For example, suppose one has informative prior information for the binomial population. Consider

$$\xi(\theta) = \left[\frac{\Gamma(\alpha + \beta)}{\Gamma(\alpha)\Gamma(\beta)} \right] \theta^{\alpha-1}(1-\theta)^{\beta-1}, 0 \le \theta \le 1 \qquad \text{(A.8)}$$

as the prior density for θ. The beta density with parameters α and β has mean $[\alpha/(\alpha+\beta0]$ and variance $[\alpha\beta/(\alpha+\beta)^2(\alpha+\beta+1)]$ and can express informative prior information in many ways. For example, suppose from a previous cohort study with 20 exposed subjects, there were six subjects who developed disease and 14 who did not develop disease. Then the probability mass function for the observed number of successes $x = 6$ is

$$f(6|\theta) = \binom{20}{6}\theta^6(1-\theta)^{14}, 0 \leq \theta \leq 1 \qquad (A.9)$$

As a function of θ (the incidence rate of disease for those exposed), Equation A.8 is a beta distribution with parameter vector (7,15) and expresses informative prior information, which is combined with Equation A.3, via Bayes' theorem, in order to make inferences (estimation, tests of hypotheses, and predictions) about the incidence rate θ for the exposed subjects. The beta distribution is an example of a conjugate density, because the prior and posterior distributions for θ belong to the same parametric family. Thus, the likelihood function based on previous sample information can serve as a source of informative prior information. The binomial and beta distributions occur quite frequently in cohort and case–control studies, where the investigator is examining the association between disease and exposure (say between smoking and lung cancer).

Of course, the normal density (Equation A.5) also plays an important role as a model in epidemiology. For example, as seen in preceding chapters, the normal distribution will model the distribution of observations that occur in screening studies, and others. For example, the measured value of blood glucose can be considered as a continuous measurement for diagnosing diabetes. How is informative prior information expressed for the parameters μ and σ (the mean and standard deviation)? Suppose a previous study has m observations $X = (x_1x_2,...,x_m)$. Then the density of X at given μ and σ is

$$f(x|\mu,\sigma) \propto \left[\frac{\sqrt{m}}{\sqrt{2\pi\sigma^2}}\right]\exp-\left(\frac{m}{2\sigma^2}\right)(\bar{x}-\mu)^2$$
$$\left[(2\pi)^{\frac{-(n-1)}{2}}\sigma^{-(n-1)}\right]\exp-\left(\frac{1}{2\sigma^2}\right)\sum_{i-1}^{i=m}(x_i-\bar{x})^2 \qquad (A.10)$$

This is a conjugate density for the two-parameter normal family and is called the normal-gamma density. Note that it is the product of two functions, where the first, as a function of μ and σ, is the conditional density of μ at a given σ, with mean $\bar{x}$ and variance σ^2/m, while the second is a function of σ only and is an inverse gamma density. Or equivalently, if the normal is parameterized with μ and the precision $\tau = 1/\sigma^2$, the conjugate distribution is as follows: (a) the conditional distribution of μ at given τ is normal with mean $\bar{x}$ and precision $m\tau$, and (b) the marginal distribution of τ is gamma with parameters $(m + 1)/2$ and $\sum_{i=1}^{i=m}(x_i - \bar{x})^2/2 = (m-1)S^2/2$, where S^2 is the sample variance. Thus, if one knows the results of a previous experiment (say from related studies for Type II diabetes), the likelihood function for μ (the population mean blood glucose) and τ provides informative prior information for the normal population.

A.4 Posterior Information

The preceding section explains how prior information is expressed in an informative or in a noninformative way. Several examples are given and will be revisited as illustrations for the determination of the posterior distribution of the parameters. In the Bayes example, where $X \sim$ Binomial (θ,n), a uniform distribution for the incidence rate (of a cohort study) θ is used. What is the posterior distribution? By Bayes' theorem,

$$\xi(\theta \mid x) \propto \binom{n}{x}\theta^x(1-\theta)^{n-x} \tag{A.11}$$

where x is the observed number of subjects with disease among n exposed subjects. Of course, this is recognized as a beta $(x + 1, n - x + 1)$ distribution, and the posterior mean is $(x + 1)/(n + 2)$. On the other hand, if the Lhoste prior density (2.3) is used, the posterior distribution of θ is beta $(x, n-x)$ with mean x/n, the usual estimator of θ. The conjugate prior results in a beta $(x+\alpha, n-x+\beta)$ with mean $(x+\alpha)/(n+\alpha+\beta)$. Suppose the prior is informative in a previous study with 10 subjects with disease among 30 subjects. Then $\alpha = 11$ and $\beta = 21$, and the posterior distribution is beta $(x+11, n-x+21)$. If the current cohort study has 40 exposed subjects and 15 have disease, the posterior distribution is beta $(26, 46)$ with mean $26/72 = .361$, which is the estimated incidence rate compared with a prior estimated incidence rate of .343.

Consider a random sample $X = (x_1, x_2, ..., x_n)$ of size n from a normal $(\mu, 1/\tau)$ population, where $\tau = 1/\sigma^2$ is the inverse of the variance, and suppose the prior information is vague and the Jeffreys–Lhoste prior $\xi(\mu, \tau) \propto 1/\tau$ is appropriate. Then the posterior density of the parameters is

$$\xi(\mu, \tau \mid \text{data}) \propto \tau^{\frac{n}{2}-1} \exp\left(-\frac{\tau}{2}\right)\left[n(\mu - \bar{x})^2 + \sum_{i=1}^{i=n}(x_i - \bar{x})^2\right] \tag{A.12}$$

Using the properties of the gamma density, τ is eliminated by integrating the joint density with respect to τ to give

$$\xi(\mu \,|\, \text{data}) \propto \frac{\left\{ \dfrac{\Gamma\left(\dfrac{n}{2}\right) n^{\frac{1}{2}}}{(n-1)^{\frac{1}{2}} S\pi^{\frac{1}{2}} \Gamma\left(\dfrac{n-10}{2}\right)} \right\}}{\left[1 + \dfrac{n(\mu-\bar{x})^2}{(n-1)S^2} \right]^{\frac{(n-1+1)}{2}}} \tag{A.13}$$

which is recognized as a t-distribution with $n-1$ degrees of freedom, location $\bar{x}$, and precision n/S^2. Transforming to $(\mu-\bar{x})\sqrt{n}/S$, the resulting variable has a Student's t-distribution with $n-1$ degrees of freedom. Note that the mean of μ is the sample mean, while the variance is $[(n-1)/n(n-3)]S^2$, $n > 3$.

Eliminating μ from Equation A.12 results in the marginal distribution of τ as

$$\xi(\tau \,|\, S^2) \propto \tau^{\left[\frac{(n-1)}{2}\right]-1} \exp{-\frac{\tau(n-1)S^2}{2}}, \tau > 0 \tag{A.14}$$

which is a gamma density with parameters $(n-1)/2$ and $(n-1)S^2/2$. This implies that the posterior mean is $1/S^2$ and the posterior variance is $2/(n-1)S^4$.

The Poisson distribution often occurs as a population for a discrete random variable with mass function

$$f(x \,|\, \theta) = \frac{e^{-\theta}\theta^x}{x!} \tag{A.15}$$

where the gamma density

$$\xi(\theta) = \left[\frac{\beta^\alpha}{\Gamma(\alpha)} \right] \theta^{\alpha-1} e^{-\theta\beta} \tag{A.16}$$

is a conjugate distribution that expresses informative prior information. For example, in a previous experiment with m observations, the prior density would be as in Equation A.16 with the appropriate values of alpha and beta. Based on a random sample of size n, the posterior density is

$$\xi(\theta \,|\, \text{data}) \propto \theta^{\sum_{i=1}^{i=n} x_i + \alpha - 1} e^{-\theta(n+\beta)} \tag{A.17}$$

which is identified as a gamma density with parameters $\alpha' = \sum_{i=1}^{i=n} x_i + \alpha$ and $\beta' = n + \beta$. Remember, the posterior mean is α'/β', median $(\alpha'-1)/\beta'$, and variance $\alpha'/(\beta')^2$.

A.5 Inference

A.5.1 Introduction

In a statistical context, by inference one usually means estimation of parameters, tests of hypotheses, and prediction of future observations. With the Bayesian approach, all inferences are based on the posterior distribution of the parameters, which in turn is based on the sample, via the likelihood function and the prior distribution. We have seen the role of the prior density and likelihood function in determining the posterior distribution, and presently will focus on the determination of point and interval estimation of the model parameters. Later we will emphasize how the posterior distribution determines a test of hypothesis. Lastly, the role of the predictive distribution in testing hypotheses and in goodness of fit will be explained.

When the model has only one parameter, one would estimate that parameter by listing its characteristics, such as the posterior mean, median, and standard deviation, and plotting the posterior density. On the other hand, if there are several parameters, one would determine the marginal posterior distribution of the relevant parameters and as above, calculate its characteristics (e.g., mean, median, mode, standard deviation, etc.) and plot the densities. Interval estimates of the parameters are also usually reported and are called credible intervals.

A.5.2 Estimation

Suppose we want to estimate θ of the binomial example of the previous section, where the number of people with disease is X, which has a binomial distribution with θ, the incidence rate of those subjects exposed to the risk factor, and the posterior distribution is beta (21,46) with the following characteristics: mean = .361, median = .362, standard deviation = .055, lower 2½ percent point = .254, and upper 2½ percent point = .473. The mean and median are the same, while the lower and upper 2½ percent points determine a 95% credible interval of (.254,.473) for θ.

Inferences for the normal (μ, τ) population are somewhat more demanding, because both parameters are unknown. Assuming the vague prior density $\xi(\mu, \tau) \propto 1/\tau$, the marginal posterior distribution of the population mean μ is a t-distribution with $n - 1$ degrees of freedom, mean $\bar{x}$, and precision n/S^2; thus, the mean and the median are the same and provide a natural estimator of μ, and because of the symmetry of the t-density, a $(1 - \alpha)$ credible interval for μ is $\bar{x} \pm t_{\alpha/2, n-1} S/\sqrt{n}$, where $t_{\alpha/2, n-1}$ is the upper $100\alpha/2$ percent point of the t-distribution with $n - 1$ degrees of freedom. To generate values from the $t\left(n-1, \bar{x}, n/S^2\right)$ distribution, generate values from Student's t-distribution with $n - 1$ degrees of freedom, multiply each by $S/\sqrt{n}$, and then add $\bar{x}$ to each. Suppose $n = 30$, and $x = (7.8902, 4.8343, 11.0677, 8.7969, 4.0391, 4.0024, 6.6494,$

8.4788,0.7939,5.0689,6.9175,6.1092,8.2463,10.3179,1.8429,3.0789,2.8470,5.1471,6.3
730,5.2907,1.5024,3.8193,9.9831,6.2756,5.3620,5.3297,9.3105,6.5555,0.8189,0.4713);
then $\bar{x} = 5.57$ and $S = 2.92$.

Using the same dataset, the following WinBUGS code is used to analyze
the problem.

BUGS CODE A.1

```
Model;
{ for( i in 1:30) { x[i]~dnorm(mu,tau) }
mu~dnorm (0.0,.0001)
tau ~dgamma( .0001,.0001) (4.17)
sigma <- 1/tau }
list( x = c(7.8902,4.8343,11.0677,8.7969,4.0391,4.0024,6.6494,8
.4788,0.7939,5.0689,6.9175,6.1092,8.2463,10.3179,1.8429,3.0789,
2.8470,5.1471,6.3730,5.2907,1.5024,3.8193,9.9831,6.2756,5.3620,
5.3297,9.3105,6.5555,0.8189,0.4713))
list( mu = 0, tau = 1)
```

Note that a somewhat different prior was employed here, compared with
previously, in that μ and τ are independent and assigned proper, but noninformative distributions. The corresponding analysis is given in Table A.1.

Upper and lower refer to the lower and upper 2½ percent points of the posterior distribution. Note that a 95% credible interval for mu is (4.47, 6.65) and
the estimation error is .003566. See Appendix B for the details on executing
the WinBUGS statements above.

The program generated 30,000 samples from the joint posterior distribution of μ and σ using a Gibbs sampling algorithm, and used 29,000 for the
posterior moments and graphs, with a refresh of 100.

A.5.3 Testing Hypotheses

An important feature of inference is testing hypotheses. Often in epidemiologic
studies, the scientific hypothesis of that study can be expressed in statistical
terms and a formal test implemented. Suppose $\Omega = \Omega_0 \cup \Omega_1$ is a partition of the
parameter space; then the null hypothesis is designated as H: $\theta \in \Omega_0$ and the
alternative by A: $\theta \in \Omega_1$, and a test of H versus A consists of rejecting H in favor
of A if the observations $x = (x_1, x_2, ..., x_n)$ belong to a critical region C. In the
usual approach, the critical region is based on the probabilities of type I errors,
namely $\Pr(C \mid \theta)$, where $\theta \in \Omega_0$, and of type II errors $1 - \Pr(C \mid \theta)$, where $\theta \in \Omega_1$. This

TABLE A.1

Posterior Distribution of μ and $\sigma = 1/\sqrt{\tau}$

Parameter	Mean	SD	Error	Median	Lower	Upper
μ	5.572	.5547	.003566	5.571	4.4790	6.656
σ	9.15	2.570	.01589	8.733	5.359	15.37

approach to testing hypothesis was developed formally by Neyman and Pearson and can be found in many of the standard references, such as Lehmann.[22] Lee[23] presents a good elementary introduction to testing and estimation in a Bayesian context. Our approach is to reject the null hypothesis if the 95% credible region for θ does not contain the set of all θ such that $\theta \in \Omega_0$. In the special case that H: $\theta = \theta_0$ versus the alternative A: $\theta \neq \theta_0$, where θ is a scalar, H is rejected when the 95% confidence interval for θ does not include θ_0.

A.6 Predictive Inference

A.6.1 Introduction

Our primary interest in the predictive distribution is to check for model assumptions. Is the adopted model for an analysis the most appropriate?

What is the predictive distribution of a future set of observations Z? It is the conditional distribution of Z given $X = x$, where x represents the past observations, which when expressed as a density is

$$g(z\,|\,x) = \int_{\Omega} f(z\,|\,\theta)\xi(\theta\,|\,x)d\theta, \; z \in R^m \qquad \text{(A.18)}$$

where the integral is with respect to θ, and $f(x\,|\,\theta)$ is the density of $X = (x_1, x_2, ..., x_n)$, given θ. This assumes that given θ, Z and X are independent. Thus, the predictive density is the posterior average of $f(z\,|\,f(z\,|\,\theta))$.

The posterior predictive density will be derived for the binomial and normal populations.

A.6.2 The Binomial

Suppose the binomial case is again considered, where the posterior density of the binomial parameter θ is

$$\xi(\theta\,|\,x) = \left[\frac{\Gamma(\alpha+\beta)\Gamma(n+1)}{\Gamma(\alpha)\Gamma(\beta)\Gamma(x+1)\Gamma(n-x+1)} \right] \theta^{\alpha+x-1}(1-\theta)^{\beta+n-x-1}$$

a beta with parameters $\alpha + x$ and $n - x + \beta$, and x is the sum of the set of n observations. The population mass function of a future observation Z is $f(z/\theta) = \theta^z(1-\theta)^{1-z}$; thus the predictive mass function of Z, called the beta-binomial, is

$$g(z\,|\,x) = \frac{\Gamma(\alpha+\beta)\Gamma(n+1)\Gamma(\alpha+\sum_{i=1}^{i=n}x_i+z)\Gamma(1+n+\beta-x-z)}{\Gamma(\alpha)\Gamma(\beta)\Gamma(n-x+1)\Gamma(x+1)\Gamma(n+1+\alpha+\beta)} \qquad \text{(A.19)}$$

where $z = 0,1$. Note that this function does not depend on the unknown parameter, and that the n past observations are known, and that if $\alpha = \beta = 1$, one is assuming a uniform prior density for θ.

A.6.3 Forecasting from a Normal Population

Moving on to the normal density with both parameters unknown, what is the predictive density of Z, with a noninformative prior density

$$\xi(\mu,\tau) = \frac{1}{\tau}, \mu \in R \text{ and } \tau > 0?$$

The posterior density is

$$\xi(\mu,\tau \,|\, \text{data}) = \left[\frac{\left(\tau^{\frac{n}{2}-1} \right)}{(2\pi)^{\frac{n}{2}}} \right] \exp - \left(\frac{\tau}{2} \right) [n(\mu - \bar{x})^2 + (n-1)S_x^2]$$

where $\bar{x}$ and S_x^2 are the sample mean and variance, based on a random sample of size n, $x = (x_1.x_2,...,x_n)$. Suppose z is a future observation; then the predictive density of Z is

$$g(z \,|\, x) = \int \int [\tau^{\frac{(n)}{2}-1} / (2\pi)^{\frac{(n)}{2}}] \exp - \left(\frac{\tau}{2} \right) [n(\mu - \bar{x})^2 + (n-1)S_x^2] \qquad (A.20)$$

where the integration is with respect to $\mu \in R$ and $\tau \geq 0$. This simplifies to a t density with $d = n - 1$ degrees of freedom, location $\bar{x}$, and precision

$$p = \frac{n}{(n+1)S_x^2} \qquad (A.21)$$

Recall that a t density with d degrees of freedom, location $\bar{x}$, and precision p has density

$$g(t) = \left\{ \frac{\Gamma\left[\frac{(d+1)}{2} \right] p^{\frac{1}{2}}}{\Gamma\left(\frac{d}{2} \right) (d\pi)^{\frac{1}{2}}} \right\} \left\{ 1 + \frac{(t - \bar{x})^2 p}{d} \right\}^{\frac{-(d+1)}{2}}$$

where $t \in R$, the mean is $\bar{x}$, and the variance is $d/(d-2)p$.

The predictive distribution can be used as an inferential tool to test hypotheses about future observations, to estimate the mean of future observations, and to find confidence bands for future observations. In the context of epidemiology, the predictive distribution for future normal observations will be employed to check the distribution of blood glucose values for a screening test for diabetes.

A.7 Checking Model Assumptions

A.7.1 Introduction

It is imperative to check model adequacy in order to choose an appropriate model and to conduct a valid study. The approach taken here is based on many sources, including Gelman et al. (ch. 6),[16] Carlin and Louis (ch. 5),[20] and Congdon (ch. 10).[17] Our main focus will be on the likelihood function of the posterior distribution, and not the prior distribution, and to this end, graphical representations such as histograms, box plots, and various probability plots of the original observations will be compared with those observations generated from the predictive distribution. In addition to graphical methods, Bayesian versions of overall goodness-of-fit type operations are taken to check model validity. Methods presented at this juncture are just a small subset of those presented in more advanced works, including Gelman et al., Carlin and Louis, and Congdon.

Of course, the prior distribution is an important component of the analysis, and if one is not sure of the "true" prior, one should perform a sensitivity analysis to determine the robustness of posterior inferences to various alternative choices of prior information. See Gelman et al. or Carlin and Louis for details of performing a sensitivity study for prior information. Our approach is to either use informative or vague prior distributions, where the former is done when prior relevant experimental evidence determines the prior, or the latter is taken if there are no or very few germane experimental studies. In scientific studies, the most likely scenario is that there are relevant experimental studies providing informative prior information.

A.7.2 Sampling from an Exponential, but Assuming a Normal Population

Consider a random sample of size 30 from an exponential distribution with mean 3. An exponential distribution is often used to model the survival times of a screening test.

$$x = (1.9075, 0.7683, 5.8364, 3.0821, 0.0276, 15.0444, 2.3591, 14.9290,$$

$$6.3841, 7.6572, 5.9606, 1.5316, 3.1619, 1.5236, 2.5458, 1.6693, 4.2076,$$

$$6.7704, 7.0414, 1.0895, 3.7661, 0.0673, 1.3952, 2.8778, 5.8272, 1.5335,$$

$$7.2606, 3.1171, 4.2783, 0.2930)$$

The sample mean and standard deviation are 4.13 and 3.739, respectively.

Assume that the sample is from a normal population with unknown mean and variance, with an improper prior density $\xi(\mu, \tau) = 1/\tau$, $\mu \in R$ and $\tau > 0$; then the posterior predictive density is a univariate t with $n - 1 = 29$ degrees of freedom, mean $\bar{x} = 3.744$, standard deviation 3.872, and precision $p = .0645$. This is verified from the original observations x and the formula for the precision. From the predictive distribution, 30 observations are generated:

$$z = (2.76213, 3.46370, 2.88747, 3.13581, 4.50398, 5.09963, 4.39670, 3.24032,$$
$$3.58791, 5.60893, 3.76411, 3.15034, 4.15961, 2.83306, 3.64620, 3.48478,$$
$$2.24699, 2.44810, 3.39590, 3.56703, 4.04226, 4.00720, 4.33006, 3.44320,$$
$$5.03451, 2.07679, 2.30578, 5.99297, 3.88463, 2.52737)$$

which gives a mean of $\bar{z} = 3.634$ and standard deviation $S = .975$. The histograms for the original and predicted observations are portrayed in Figures A.1 and A.2, respectively.

The histograms obviously are different, as for the original observations, a right skewness is depicted; however, this is lacking for the histogram of the predicted observations, which is for a t distribution. Although the example seems trivial, it would not be for the first time that exponential observations were analyzed as if they were generated from a normal population! When indeed the observations are exponential, there are Bayesian procedures that compare the survival times of a screened population with a control population. This is considered in Chapter 7 with the national screening test of breast cancer conducted at the Health Insurance Plan of New York. The sample statistics do not detect the discrepancy, because they are very similar; the mean and standard deviation for the exponential are 3.744 and 3.872, respectively, but are 3.688 and 3.795, respectively, for the predicted observations. Thus, it is important to use graphical techniques to assess model adequacy.

It would be interesting to generate more replicate samples from the predictive distribution in order to see if these conclusions hold firm.

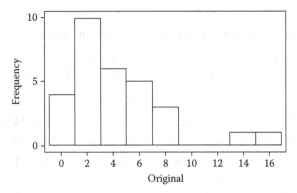

FIGURE A.1
Histogram of original observations.

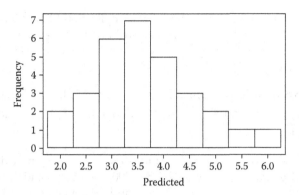

FIGURE A.2
Histogram of predicted observations.

A.7.3 A Poisson Population

It is assumed that the sample is from a Poisson population; however, actually, it is generated from a uniform discrete population over the integers from 0 to 10. The sample of size 25 is x = (8,3,8,2,6,1,0,2,4,10,7,9,5,4,8,4,0,9,0,3,7,10,7, 5,1), with a sample mean of 4.92 and standard deviation of 3.278. When the population is Poisson, $P(\theta)$, and an uninformative prior

$$\xi(\theta) = \frac{1}{\theta}, \theta > 0$$

is appropriate, the posterior density is gamma with parameters alpha = $\sum_{i=1}^{i=25} x_i = 123$ and beta = n = 25. Observations z from the predictive distribution are generated by taking a sample θ from the gamma posterior density, then selecting z from the Poisson distribution $P(\theta)$. This was repeated 25 times to give z = (2,5,6,2,4,3,5,3,2,3,3,6,7,5,5,3,1,5,7,3,5,3,6,4,5), with a sample mean of 4.48 and standard deviation of 1.896.

The most obvious difference shows a symmetric sample from the discrete uniform population, but on the other hand, box plots of the predicted observations reveal a slight skewness to the right. The largest difference is in the interquartile ranges being (2, 8) for the original observations and (3, 5.5) for the predictive sample. Although there are some differences, to declare that the Poisson assumption is not valid might be premature. One should generate more replicate samples from the predictive distribution to reveal additional information.

A.7.4 Measuring Tumor Size

A study of agreement involving the lesion sizes of five radiologists assumed that the observations were normally distributed. Is this assumption valid?

A probability plot of 40 lesion sizes of one replication (there were two) of the reader labeled 1 would show the normal distribution as a reasonable assumption.

Is this the implication from the Bayesian predictive density? The original observations are

$$x = (3.5, 3.8, 2.2, 1.5, 3.8, 3.5, 4.2, 5.4, 7.6, 2.8, 5.0, 2.3, 4.4, 2.5, 5.2, 1.7, 4.5,$$
$$4.5, 6.0, 3.3, 7.0, 4.0, 4.8, 5.02.7, 3.7, 1.8, 6.3, 4.0, 1.5, 2.2, 2.2, 2.6, 3.7, 2.5,$$
$$4.8, 8.0, 4.0, 3.5, 4.8)$$

Descriptive statistics are as follows: mean = 3.920 cm, standard deviation = 1.612 cm, and interquartile range (2.525, 4.800). The basic statistics for the predicted observations are mean = 4.017 and standard deviation = 1.439 cm, with an interquartile range of (4.069, 4.930); these were based on the future observations

$$z = (4.85, 4.24, 3.32, 1.84, 5.56, 3.40, 4.02, 1.38, 4.21, 6.26, 0.55, 4.56, 5.09, 4.51,$$
$$3.28, 3.94, 5.05, 7.23, 4.19, 4.85, 4.24, 2.86, 3.98, 2.00, 2.99, 3.50, 2.53, 1.95,$$
$$6.07, 4.68, 5.39, 1.89, 5.79, 5.86, 2.85, 3.62, 4.95, 4.46, 4.22, 4.33)$$

A comparison of the histograms for the original and predicted observations would show some small difference in the two distributions, but the differences would not be striking; however, the predicted observations appear to be somewhat more symmetric about the mean than those of the original observations. Also, the corresponding sample means, standard deviations, modes, and interquartile ranges are quite alike for the two samples. There is nothing that stands out that implies questioning the validity of the normality assumption. Of course, additional replicates from the posterior predictive density should be computed for additional information about discrepancies between the two.

A.7.5 Testing the Multinomial Assumption

A good example of a multinomial distribution appears as an outcome of the Shields Heart Study carried out in Spokane, Washington, from 1993 to 2001 at the Shields Coronary Artery Center where the disease status of coronary artery disease and various risk factors (coronary artery calcium, diabetes, smoking status, blood pressure, cholesterol, etc.) were measured on each of about 4300 patients (Table A.2). The average age (SD) of 4386 patients was 55.14 (10.757) years, with an average age of 55.42 (10.78) years for 2712 males and 56.31 (10.61) years for 1674 females. The main emphasis of this study was to investigate the ability of coronary artery calcium (as measured by computed tomography) to diagnose coronary artery disease. A typical table from this study investigates the association between having a heart attack and the fraction of subjects with a positive reading for coronary calcium. For additional information about the Shield Heart Study, see the study by Mielke, Shields, and Broemeling.[24] (see Table A.2)

A multinomial model is valid if the 4386 patients are selected at random from some well-defined population, if the responses of an individual are

TABLE A.2

Shields Heart Study

Coronary Calcium	+ Disease (Heart Attack)	Nondisease (No Infarction)
+Positive	$\theta_{++}, n_{++} = 119$	$\theta_{+-}, n_{+-} = 2461$
−Negative	$\theta_{-+}, n_{-+} = 11$	$\theta_{--}, n_{--} = 1798$

independent of the other respondents, and if the probability of an event (the disease status and status of coronary calcium) is the same for all individuals for a particular cell of the table.

In some studies, it is difficult to know if the multinomial population is valid, for example, with so-called chart reviews, where medical records are selected, not necessarily at random, but from some population determined by the eligibility criterion of the study. Thus, in the case of epidemiologic studies, such as above, the best way to determine validity of the multinomial model is to know the details of how the study was designed and conducted. The crucial issue in the multinomial model is independence, that is, given the parameters of the multinomial, the results of one patient are independent of those of another. It is often the case that the details of the study are not available. The other important aspect of a multinomial population is that the probability of a particular outcome is constant over all patients. One statistical way to check is to look for runs in the sequence and so on.

In Table A.2, one could condition on the row totals, and check if the binomial model is valid for each row.

A.8 Computing

A.8.1 Introduction

This section introduces the computing algorithms and software that will be used for the Bayesian analysis of problems encountered in agreement investigations. In the previous sections of the appendix, direct methods (noniterative) of computing the characteristics of the posterior distribution were demonstrated with some standard one-sample and two-sample problems. An example of this is the posterior analysis of a normal population, where the posterior distribution of the mean and variance is generated from its posterior distribution by the *t*-distribution random number generator in Minitab. In addition to some direct methods, iterative algorithms are briefly explained.

MCMC methods (an iterative procedure) of generating samples from the posterior distribution are introduced, where the Metropolis-Hasting algorithm and Gibbs sampling are explained and illustrated with many examples. WinBUGS uses MCMC methods such as the Metropolis-Hasting and Gibbs sampling techniques, and many examples of Bayesian analysis are given. An analysis consists of graphical displays of various plots of the posterior density of the parameters, by portraying the posterior analysis with tables that list the posterior mean, standard deviation, median, and lower and upper 2½ percentiles, and of other graphics that monitor the convergence of the generated observations.

A.8.2 An Example of a Cross-Sectional Study

The general layout of a cross sectional study appears in Table A.3.

Here, a random sample of size $n = n_{++} + n_{+-} + n_{-+} + n_{--}$ subjects is taken from a well-defined population where the disease status and exposure status of each subject are known. Consider the ++ cell; θ_{++} is the probability that a subject will have the disease and will be exposed to the risk factor. It is assumed that n is fixed and that the cell frequencies follow a multinomial distribution with mass function

$$f\left(n_{++}, n_{+-}, n_{-+}, n_{--} \mid \theta_{++}, \theta_{+-}, \theta_{-+}, \theta_{--}\right) \propto \theta_{++}^{n_{++}} \theta_{+-}^{n_{+-}} \theta_{-+}^{n_{-+}} \theta_{--}^{n_{--}} \qquad (A.22)$$

where the thetas are between zero and one and their sum is one, and the values of n are nonnegative integers with sum equal to the sample size n.

As a function of the thetas, Equation A.22 is recognized as a Dirichlet density, Usually, the likelihood for the thetas is combined via Bayes' theorem with a prior distribution for the thetas, and the result is a posterior density for thetas. For example, if a uniform prior is used for the thetas, the posterior distribution of the theta is Dirichlet with parameter vector $\left(n_{++} + 1, n_{+-} + 1, n_{-+} + 1, n_{--} + 1\right)$, but on the other hand, if the improper prior density

$$f\left(\theta_{++}, \theta_{+-}, \theta_{-+}, \theta_{--}\right) \propto \left(\theta_{++}^{n_{++}} \theta_{+-}^{n_{+-}} \theta_{-+}^{n_{-+}} \theta_{--}^{n_{--}}\right)^{-1}$$

is used, the posterior density of the thetas is Dirichlet with parameter vector $\left(n_{++}, n_{+-}, n_{-+}, n_{--}\right)$. When the latter prior is used, the posterior means of the unknown parameters will be the same as the "usual" estimators. For example, the usual estimator of θ_{++} is n_{++}/n, but the posterior distribution of θ_{++} is beta with parameter vector $\left(n_{++}, n - n_{++}\right)$; consequently the posterior mean of θ_{++} is indeed the usual estimator n_{++}/n.

The sampling scheme for a cross-sectional study allows one to estimate the relative risk of disease (the incidence rate of those diseased among those

TABLE A.3

A Cross-Sectional Study

Risk Factor	+ Disease	− Nondisease
+Positive	θ_{++}, n_{++}	θ_{+-}, n_{+-}
−Negative	θ_{-+}, n_{-+}	θ_{--}, n_{--}

exposed divided by the incidence rate of the diseased among those not exposed to the risk) and the odds of exposure among those diseased versus the odds of exposure among the nondiseased.

A good example of a cross-sectional study is the Shields Heart Study carried out in Spokane, Washington, from 1993 to 2001 at the Shields Coronary Artery Center, where the disease status of coronary artery disease and various risk factors (coronary artery calcium, diabetes, smoking status, blood pressure, cholesterol, etc.) were measured on each patient. The average age (SD) of 4386 patients was 55.14 (10.757) years, with an average age of 55.42 (10.78) years for 2712 males and 56.31 (10.61) years for 1674 females. The main emphasis of this study was to investigate the ability of coronary artery calcium (as measured by computed tomography) to diagnose coronary artery disease. A typical table from this study investigates the association between having a heart attack and the fraction of subjects with a positive reading for coronary calcium. For additional information about the Shields Heart Study, see the study by Mielke, Shields, and Broemeling.[24]

In order to investigate the association between infarction and coronary artery calcium, the relative risk and odds ratio are estimated from the information of Table A.4. From a Bayesian viewpoint, assuming an improper prior distribution for the four parameters (Table A.4), the posterior distribution of the four-cell parameters is Dirichlet with hyperparameter vector (119, 2461,11,1798), and the relative risk for heart disease is estimated from the parameter

$$\theta_{RR} = \frac{\theta_{++} / (\theta_{++} + \theta_{+-})}{\theta_{-+} / (\theta_{-+} + \theta_{--})} \tag{A.23}$$

Note that the formula for relative risk has a numerator that is the probability of disease among those exposed (have a positive calcium score) to the risk factor, while the denominator is the probability of disease among those not exposed (negative coronary artery calcium). As for the odds ratio, the posterior distribution of

$$\theta_{OR} = \frac{\theta_{pp}\theta_{mm}}{\theta_{mp}\theta_{pm}} \tag{A.24}$$

will also be determined, and the Bayesian analysis executed with BUGS CODE A.2 using 55,000 observations generated from the posterior distribution with a burn-in of 5000 and a refresh of 100.

TABLE A.4

Shields Heart Study

Coronary Calcium	+ Disease (Heart Attack)	Nondisease (No Infarction)
+Positive	$\theta_{++}, n_{++} = 119$	$\theta_{+-}, n_{+-} = 2461$
−Negative	$\theta_{-+}, n_{-+} = 11$	$\theta_{--}, n_{--} = 1798$

BUGS CODE A.2

```
model;
# the cross-sectional study
{
# below generates observations from the Dirichlet
  distribution
gpp~ dgamma(npp,2)
gpm~dgamma(npm,2)
gmp~dgamma(nmp,2)
gmm~dgamma(nmm,2)
sg<-gpp+gpm+gmp+gmm
thetapp<- gpp/sg
thetapm<-gpm/sg
thetamp<-gmp/sg
thetamm<-gmm/sg
# numerator of RR
nRR<-(thetapp)/(thetapp+thetapm)
# denominator of RR
dRR<-(thetamp)/(thetamp+thetamm)
# the odds ratio
OR<-(thetapp*thetamm)/(thetamp*thetapm)
# the relative rsk
RR<-nRR/dRR
# attributable risk based on the relative risk
ARRR<-(RR-1)/RR
# attributable risk based on the odds ratio
AROR<-(OR-1)/RR
}
# data with improper prior for Shields Heart Study
list(npp=119,npm=2461,nmp=11,nmm=1798)
# initial values generated from the
  specification tool
```

It is interesting that the posterior means of the relative risk and odds ratio are quite similar, as are the posterior means, which implies that the disease (coronary infarction) is quite rare, which of course is obvious from Table 2.8, which indicates a disease rate of 2.96%. If the 4389 patients are actually a random sample, then one is confident that 2.9% is an accurate estimate of the true disease rate.

A.8.3 Monte Carlo Markov Chain

A.8.3.1 Introduction

MCMC techniques are especially useful when analyzing data with complex statistical models. For example, when considering a hierarchical model with many levels of parameters, it is more efficient to use an MCMC technique such as Metropolis-Hasting or Gibbs sampling as an iterative procedure

TABLE A.5

Posterior Distribution for the Shields Heart Study

Parameter	Mean	SD	Error	2½	Median	97½
RR	8.35	2.887	.0133	4.378	7.805	15.49
OR	8.709	3.308	.01397	4.529	8.138	16.21
θ_{pp}	.02711	.00245	<.00001	.0225	.02704	.03215
θ_{mm}	.4096	.007413	<.00001	.395	.4096	.4243
θ_{mp}	.002504	.007553	<.000001	.001246	.002427	.004201
θ_{pm}	.5608	.0075	<.00001	.5461	.5608	.5754

in order to sample from many posterior distributions (Table A.5). It is very difficult, if not impossible, to use noniterative direct methods for complex models.

A way to draw samples from a target posterior density $\xi(\theta|x)$ is to use Markov chain techniques, where each sample only depends on the last sample drawn. Starting with an approximate target density, the approximations are improved with each step of the sequential procedure. Or in other words, the sequence of samples is converging to samples drawn at random from the target distribution. A random walk from a Markov chain is simulated, where the stationary distribution of the chain is the target density, and the simulated values converge to the stationary distribution or the target density. The main concept in a Markov chain simulation is to devise a Markov process whose stationary distribution is the target density. The simulation must be long enough so that the present samples are close enough to the target. It has been shown that this is possible and that convergence can be accomplished. The general scheme for a Markov chain simulation is to create a sequence θ_t, $t = 1, 2$, and so on by beginning at some value of θ_0 and at the tth stage select the present value from a transition function $Q_t(\theta_t|\theta_{t-1})$, where the present value θ_t only depends on the previous one, via the transition function. The value of the starting value θ_0 is usually based on a good approximation to the target density. In order to converge to the target distribution, the transition function must be selected with care. The account given here is a summary of Gelman et al. (ch. 11),[16] who presents a very complete account of MCMC. Metropolis-Hasting is the general name given to methods of choosing appropriate transition functions, and two special cases of this are the Metropolis algorithm and Gibbs sampling.

A.8.3.2 The Metropolis Algorithm

Suppose the target density $\xi(\theta|x)$ can be computed; then the Metropolis technique generates a sequence θ_t, $t = 1, 2$, and so on with a distribution that converges to a stationary distribution of the chain. Briefly, the steps taken to construct the sequence are as follows:

1. Draw the initial value θ_0 from some approximation to the target density.
2. For t = 1, 2, and so on generate a sample θ_* from the jumping distribution

$$G_t\left(\theta_* \,|\, \theta_{t-1}\right)$$

3. Calculate the ratio

$$s = \xi\left(\theta_* \,|\, x\right)\xi\left(\theta_{t-1} \,|\, x\right)$$

4. Let $\theta_t = \theta_*$ with probability min(s,1) or let $\theta_t = \theta_{t-1}$.

To summarize the above, if the jump given by (2) above increases the posterior density, let $\theta_t = \theta_*$; on the other hand, if the jump decreases the posterior density, let $\theta_t = \theta_*$ with probability s, otherwise let $\theta_t = \theta_{t-1}$. One must show that the sequence generated is a Markov chain with a unique stationary density that converges to the target distribution. For more information, see the study by Gelman et al. (p. 325)[16] There is a generalization of the above Metropolis algorithm to the Metropolis-Hasting method.

A.8.3.3 Gibbs Sampling

Another MCMC algorithm is Gibbs sampling, which is quite useful for multidimensional problems and is an alternating conditional sampling way to generate samples from the joint posterior distribution. Gibbs sampling can be thought of as a practical way to implement the fact that the joint distribution of two random variables is determined by the two conditional distributions.

The two-variable case is first considered by starting with a pair (θ_1, θ_2) of random variables. The Gibbs sampler generates a random sample from the joint distribution of θ_1 and θ_2 by sampling from the conditional distributions of θ_1 at given θ_2 and from θ_2 at given θ_1. The Gibbs sequence of size k

$$\theta_2^0, \theta_1^0 ; \theta_2^1, \theta_1^1 ; \theta_2^2, \theta_1^2 ; ... ; \theta_2^k, \theta_1^k \qquad (A.25)$$

is generated by first choosing the initial values θ_2^0, θ_1^0 while the remaining are obtained iteratively by alternating values from the two conditional distributions. Under quite general conditions, for large enough k, the final two values θ_2^k, θ_1^k are samples from their respective marginal distributions. To generate a random sample of size n from the joint posterior distribution, generate the above Gibbs sequence n times. Having generated values from the marginal distributions with large k and n values, the sample mean and variance will converge to the corresponding mean and variance of the posterior distribution of (θ_1, θ_2).

Gibbs sampling is an example of a MCMC, because the generated samples are drawn from the limiting distribution of a 2 by 2 Markov chain. See the study by Casella and George[25] for a proof that the generated values are indeed values from the appropriate marginal distributions. Of course, Gibbs sequences can be generated from the joint distribution of three, four, and more random variables.

The Gibbs sampling scheme is illustrated with two examples: (a) a case of three random variables for the common mean of two normal populations and (b) an example taken from WinBUGS estimating variance components and the intraclass correlation coefficient for agreement between radiologists who are measuring lesion sizes of lung cancer patients aided by MRI images.

A.8.3.4 The Common Mean of Normal Populations

Gregurich and Broemeling[26] described the various steps in Gibbs sampling to determine the posterior distribution of the parameters in independent normal populations with a common mean.

The Gibbs sampling approach can best be explained by illustrating the procedure using two normal populations with a common mean θ. Thus, let $y_{ij}, j = 1, 2, \ldots, n_i$ be a random sample of size n_i from a normal population for $i = 1, 2$.

The likelihood function for θ, τ_1, and τ_2 is

$$L(\theta, \tau_1, \tau_2 \mid \text{data}) \propto$$

$$\tau_1^{\frac{n_1}{2}} \exp{-\frac{\tau_1}{2}\left[(n_1 - 1)s_1^2 + n_1(\theta - \bar{y}_1)^2\right]} * \tau_2^{\frac{n_2}{2}} \exp{-\frac{\tau_2}{2}\left[(n_2 - 1)s_2^2 + n_2(\theta - \bar{y}_2)^2\right]}$$

where $\theta \in \Re, \tau_1 > 0, \tau_2 > 0, s_1^2 = \sum_{j=1}^{n_1}\frac{(y_{1j} - \bar{y}_1)^2}{(n_1 - 1)}$, and $s_2^2 = \sum_{j=1}^{n_2}\frac{(y_{2j} - \bar{y}_2)^2}{(n_2 - 1)}$

The prior distribution for the parameters θ, τ_1, and τ_2 is assumed to be a vague prior defined as

$$g(\theta, \tau_1, \tau_2) \propto \frac{1}{\tau_1}\frac{1}{\tau_2}, \tau_i > 0$$

Then, combining the above gives the posterior density of the parameters as

$$P(\theta, \tau_1, \tau_2 \mid \text{data}) \propto \prod_{i=1}^{2} \tau_i^{\frac{n_i - 1}{2}} \exp{-\frac{\tau_i}{2}\left[(n_i - 1)s_i^2 + n_i(\theta - \bar{y}_i)^2\right]}$$

Therefore, the conditional posterior distribution of τ_1 and τ_2 at given θ is such that

$$\tau_i \mid \theta \sim Gamma\left[\frac{n_i}{2}, \frac{(n_i - 1)s_i^2 + n_i(\theta - \bar{y}_i)^2}{2}\right] \qquad (A.26)$$

for i = 1, 2 and given θ, τ_1 and τ_2 are independent.

The conditional posterior distribution of θ at given τ_1 and τ_2 is normal. It can be shown that

$$\theta \mid \tau_1, \tau_2 \sim N\left[\frac{n_1\tau_1\bar{y}_1 + n_2\tau_2\bar{y}_2}{n_1\tau_1 + n_2\tau_2}, (n_1\tau_1 + n_2\tau_2)^{-1}\right] \qquad (A.27)$$

Given the starting values $\tau_1^{(0)}$, $\tau_2^{(0)}$, *and* $\theta^{(0)}$ where

$$\tau_1^{(0)} = \frac{1}{s_1^2}, \tau_2^{(0)} = \frac{1}{s_2^2}, \text{ and } \theta^{(0)} = \frac{n_1\bar{y}_1 + n_2\bar{y}_2}{n_1 + n_2}$$

draw $\theta^{(1)}$ from the normal conditional distribution (A.27) of θ, given $\tau_1 = \tau_1^{(0)}$ *and* $\tau_2 = \tau_2^{(0)}$. Then, draw $\tau_1^{(1)}$ from the conditional gamma distribution (43), given $\theta = \theta^{(1)}$. And lastly draw $\tau_2^{(1)}$ from the conditional gamma distribution of τ_2 at given $\theta = \theta^{(1)}$. Then generate

$$\theta^{(2)} \sim \theta \mid \tau_1 = \tau_1^{(1)}, \tau_2 = \tau_2^1$$

$$\tau_1^{(2)} \sim \tau_1 \mid \theta = \theta^2$$

$$\tau_2^{(2)} \sim \tau_2 \mid \theta = \theta^2$$

Continue this process until there are *t* iterations $\left(\theta^{(t)}, \tau_1^{(t)}, \tau_2^{(t)}\right)$. For large *t*, $\theta^{(t)}$ would be one sample from the marginal distribution of θ, $\tau_1^{(t)}$ from the marginal distribution of τ_1, and $\tau_2^{(t)}$ from the marginal distribution of τ_2.

Independently repeating the above Gibbs process *m* times produces *m* 3-tuple parameter values $\left(\theta_j^{(t)}, \tau_{1j}^{(t)}, \tau_{2j}^{(t)}\right)$, j = 1, 2, ..., *m*, which represents a random sample of size *m* from the joint posterior distribution of (θ, τ_1, τ_2). The statistical inferences are drawn from the *m* sample values generated by the Gibbs sampler.

The statistical inferences can be drawn from the *m* sample values generated by the Gibbs sampler. The Gibbs sampler will produce 3 columns, where each row is a sample drawn from the posterior distribution of (θ, τ_1, τ_2). The first column is the sequence of the sample *m*, the second column is a random sample of size *m* from the poly-*t*-distribution of θ, and the third and fourth columns are also random samples of size m but from the marginal posterior distributions of τ_1 and τ_2, respectively.

To find the characteristics of the marginal posterior distribution of a parameter such as the mean and variance, it should be noted that the Gibbs sampler generates a sample of values of a marginal distribution from the conditional distributions without the actual marginal distribution. By simulating a large enough sample, the characteristics of the marginal can be calculated. If m is "large," the sample mean of the column is

$$E\left(\frac{\theta}{data}\right) = \sum_{j=1}^{m} \frac{\theta_j^t}{m} = \bar{\theta}$$

and is the mean of the posterior distribution of θ. The sample variance

$$(m-1)^{-1} \sum_{j=1}^{m} \left[\theta_j^t - \bar{\theta}\right]^2$$

is the variance of the posterior distribution of θ.

Additional characteristics such as the median, mode, and the 95% credible region of the posterior distribution of the parameter θ can be calculated from the samples generated by the Gibbs technique. Hypothesis testing can also be performed. Similar characteristics of the parameters $\tau_1, \tau_2, \ldots, \tau_k$ can be calculated from the samples resulting from the Gibbs method.

A.8.3.5 An Example

The example is from Box and Tiao (p. 481).[10] It is referred to as "the weighted mean problem." It has two sets of normally distributed independent samples with common mean and different variances. Samples from the posterior distributions were generated from Gibbs sequences using the statistical software Minitab®. The final value of each sequence was used to approximate the marginal posterior distribution of the parameters $\theta, \tau_1, \ldots, \tau_k$. All Gibbs sequences were generated holding the value of t equal to 50. Each example has the results of the parameters using four different Gibbs sampler sizes, where the sample size m is equal to 250, 500, 750, and 1500 (Table A.6).

The "weighted mean problem" has two sets of normally distributed independent observations with common mean and different variances. The estimated values of θ determined by the Gibbs sampling method are reported in Table 2.10. The mean value of the posterior distribution of θ generated from the 250 Gibbs sequences is 108.42 with 0.07 as the standard error of the mean. The mean value of θ generated from 500 and 750 Gibbs sequences has the same value of 108.31, and the standard errors of the mean equal to 0.04 and 0.03, respectively. The mean value of θ generated from 1500 Gibbs sequences

TABLE A.6

Results from Gibbs Sampler for θ

				95% Credible Region	
M	Mean	SD	SEM	Lower	Upper
250	108.42	1.04	0.07	106.03	110.65
500	108.31	0.94	0.04	106.35	110.21
750	108.31	0.90	0.03	106.64	110.15
1500	108.36	0.94	0.02	106.51	110.26

is 108.36 and a standard error of the mean of 0.02. Box and Tiao determined the posterior distribution of θ using the t-distribution as an approximation to the target density. They estimated the value of θ to be 108.43. This is close to the value generated using the Gibbs sampler method. The exact posterior distribution of θ is the poly-t-distribution. The effect of m appears to be minimal indicating that 500 to 750 iterations of the Gibbs sequence are sufficient.

A.9 Comments and Conclusions

Beginning with Bayes' theorem, introductory material for the understanding of Bayesian inference is presented in this chapter. Many examples from epidemiology illustrate the Bayesian methodology. For example, Bayesian methods for cohort studies and cross-sectional studies are analyzed using the theory and methods unique with the Bayesian approach. Inference for the standard populations is introduced. The most useful population for the case-control and cohort studies is the binomial population, which models the distribution of the number of patients among the exposed who will develop disease. It is shown how to analyze a cohort study with the binomial population, where the posterior distribution of the incidence rate for those exposed is a beta distribution. For the analysis of cross-sectional studies, it is shown that the multinomial distribution models four-cell frequencies for the risk and disease status of a subject and that the corresponding posterior distribution of the cell frequencies is a Dirichlet.

There are many books that introduce Bayesian inference and the computational techniques that will execute a Bayesian analysis, and the reader is encouraged to read Ntzoufras.[27] The material found in Ntzoufras is an excellent introduction to Bayesian inference and to WinBUGS, the computing software that is employed in this book. WinBUGS is also introduced in Appendix B of this book; thus, together with Ntzoufras, the reader should be able to execute the Bayesian analyses for the various epidemiologic investigations that are presented in the chapters of this book.

Exercises

1. For the beta density (A.8) with parameters α and β, show that the mean is $[\alpha/(\alpha+\beta)]$ and the variance is $[\alpha\beta/(\alpha+\beta)^2(\alpha+\beta+1)]$.

2. From Equation A.10, show the following. If the normal distribution is parameterized with μ and the precision $\tau = 1/\sigma^2$, the conjugate distribution is as follows: (a) the conditional distribution of μ at given τ is normal with mean $\bar{x}$ and precision $n\tau$, and (b) the marginal distribution of τ is gamma with parameters $(n-1)/2$ and $\sum_{i=1}^{i=n}(x_i - \bar{x})^2/2 = (n-1)S^2/2$, where S^2 is the sample variance.

3. Verify Table A.1, which reports the Bayesian analysis for the parameters of a normal population.

4. Verify the following statement:
   To generate values from the $t(n-1, \bar{x}, n/S^2)$ distribution, generate values from the Student's t-distribution with $n-1$ degrees of freedom, multiply each by $S/\sqrt{n}$, and then add $\bar{x}$ to each.

5. Verify Equation A.21, the predictive density of a future observation Z from a normal population with both parameters unknown.

6. Suppose $x_1, x_2, ..., x_n$ are independent and that $x_i \sim gamma\ (\alpha_i, \beta)$, and show that $y_i = x_i / (x_1 + x_2 + ... + x_n)$ jointly have a Dirichlet distribution with parameter $(\alpha_1, \alpha_2, ..., \alpha_n)$. Describe how this can be used to generate samples from the Dirichlet distribution.

7. Derive the conditional posterior distribution of the two precisions given the common mean, that is verify Equation A.26.

8. Derive the conditional posterior distribution of the common mean given the two precisions, that is verify Equation A.27.

9. Suppose $(X_1, X_2, ..., X_k)$ is multinomial with parameters n and $(\theta_1, \theta_2, ..., \theta_k)$, where $\sum_{i=1}^{i=k} X_i = n, 0 < \theta_i < 1$, and $\sum_{i=1}^{i=k} \theta_i = 1$. Show that

   $$E(X_i) = n\theta_i, Var(X_i) = n\theta_i(1-\theta_i), \text{ and } Cov(X_i, X_j) = -n\theta_i\theta_j$$

   What is the marginal distribution of θ_i?

10. Suppose $(\theta_1, \theta_2, ..., \theta_k)$ is Dirichlet with parameters $(\alpha_1, \alpha_2, ..., \alpha_k)$, where $\alpha_i > 0, \theta_i > 0$, and $\sum_{i=1}^{i=k}\theta_i = 1$. Find the mean and variance of θ_i and covariance between θ_i and θ_j, $i \neq j$.

11. Show that the Dirichlet family is conjugate to the multinomial family.

12. Suppose $(\theta_1, \theta_2, ..., \theta_k)$ is Dirichlet with parameters $(\alpha_1, \alpha_2, ..., \alpha_k)$. Show that the marginal distribution of θ_i is beta and give the parameters of the beta. What is the conditional distribution of θ_i given θ_j?

13. For the exponential density

$$f(x \mid \theta) = \theta \exp{-\theta x}, \; x > 0$$

where x is positive and θ is a positive unknown parameter, suppose the prior density of θ is

$$g(\theta) \propto \theta^{\alpha-1} \exp{-\beta\theta}, \theta > 0$$

What is the posterior density of θ? In epidemiology, the exponential distribution is often used as the distribution of survival times in screening tests.

References

1. Bayes, T. An essay towards solving a problem in the doctrine of chances, *Phil. Trans. Roy. Soc. London*, 53, 370, 1764.
2. Laplace, P.S. Memorie des les probabilities [Memory of the probabilities], *Memories de l'Academie des sciences de Paris*, 227, 1778.
3. Stigler, M. *The History of Statistics. The Measurement of Uncertainty before 1900*, The Belknap Press of Harvard University Press, 1986, Harvard.
4. Hald, A. *A History of Mathematical Statistics from 1750–1930*, Wiley Interscience, 1990, London, U.K.
5. Lhoste, E. Le calcul des probabilités appliqué a l'artillerie, lois de probabilite a prior [The calculation of probabilities applied to artillery], *Revu d'artillirie*, Mai, 405, 1923.
6. Jeffreys, H. *An Introduction to Probability*, Clarendon Press, 1939, Oxford, U.K.
7. Savage, L.J. *The Foundation of Statistics*, John Wiley & Sons Inc., 1954, New York.
8. Lindley, D.V. *Introduction to Probability and Statistics from a Bayesian Viewpoint*, Volumes I and II, Cambridge University Press, 1965, Cambridge.
9. Broemeling, L.D. and Broemeling, A.L. Studies in the history of probability and statistics XLVIII: The Bayesian contributions of Ernest Lhoste, *Biometrika*, 90(3), 728, 2003.
10. Box, G.E.P. and Tiao, G.C. *Bayesian Inference in Statistical Analysis*, Addison Wesley, 1973, Reading, MA.
11. Zellner, A. *An Introduction to Bayesian Inference in Econometrics*, John Wiley & Sons Inc., 1971, New York.
12. Broemeling, L.D. *The Bayesian Analysis of Linear Models*, Marcel Dekker Inc, 1985, New York.
13. Hald, A.A. *History of Mathematical Statistics before 1750*, Wiley Interscience, 1998, London, U.K.

14. Dale, A.I. *A History of Inverse Probability from Thomas Bayes to Karl Pearson*, Springer Verlag, 1991, Berlin, MA.
15. Leonard, T. and Hsu, J.S.J. *Bayesian Methods. An Analysis for Statisticians and Interdisciplinary Researchers*, Cambridge University Press, 1999, Cambridge.
16. Gelman, A., Carlin, J.B., Stern, H.S., and Rubin, D.B. *Bayesian Data Analysis*, Chapman & Hall/CRC, 1997, New York.
17. Congdon, P. *Bayesian Statistical Modeling*, John Wiley & Sons LTD, 2001, London.
18. Congdon, P. *Applied Bayesian Modeling*, John Wiley & Sons Inc., 2003, New York.
19. Congdon, P. *Bayesian Models for Categorical Data*, John Wiley & Sons Inc., 2005, New York.
20. Carlin, B.P. and Louis, T.A. *Bayes and Empirical Bayes for Data Analysis*, Chapman & Hall, 1996, New York.
21. Gilks, W.R., Richardson, S., and Spiegelhalter, D.J. *Markov Chain Monte Carlo in Practice*, Chapman & Hall/CRC, 1996, Boca Raton, FL.
22. Lehmann, E.L. *Testing Statistical Hypotheses*, John Wiley & Sons Inc., 1959, New York.
23. Lee, P.M. *Bayesian Statistics, An Introduction*, Second Edition, Arnold, a member of the Hodder Headline Group, 1997, London, U.K.
24. Mielke, C.H., Shields, J.P., and Broemeling, L.D. Coronary artery calcium scores for men and women of a large asymptomatic population, *CVD Prevention*, 2, 194–198, 1999.
25. Casella, G. and George, E.I. Explaining the Gibbs sampler, *The American Statistician*, 46, 167–174, 2004.
26. Gregurich, M.A. and Broemeling, L.D. A Bayesian analysis for estimating the common mean of independent normal populations using the Gibbs sampler, *Communications in Statistics*, 26(1), 25–31, 1997.
27. Ntzoufras, I. *Bayesian Modeling Using WinBUGS*, John Wiley & Sons Inc., 2009, Hoboken, NJ.

Appendix B: Introduction to WinBUGS

B.1 Introduction

WinBUGS is the statistical package that is used for the book and it is important that the novice be introduced to the fundamentals of working in the language. This is a brief introduction to the package, and for the first-time user it will be necessary to gain more knowledge and experience by practicing with the numerous examples provided in the download. WinBUGS is specifically designed for Bayesian analysis and is based on Monte Carlo Markov chain (MCMC) techniques for simulating samples from the posterior distribution of the parameters of the statistical model. It is quite versatile and once the user has gained some experience, there are many rewards.

Once the package has been downloaded, the essential features of the program are described, first by explaining the layout of the BUGS document. The program itself is made up of two parts, one part for the program statements and the other for the input for the sample data and the initial values for the simulation. Next to be described are the details of executing the program code and what information is needed for the execution. Information needed for the simulation are the sample sizes of the MCMC simulation for the posterior distribution and the number of such observations that will apply to the posterior distribution.

After execution of the program statements, certain characteristics of the posterior distribution of the parameters are computed including the posterior mean, median, credible intervals, and plots of posterior densities of the parameters. In addition, WinBUGS provides information about the posterior distribution of the correlation between any two parameters and information about the simulation. For example, one may view the record of simulated values of each parameter and the estimated error of estimation of the process. These and other activities involving the simulation and interpretation of the output will be explained.

Examples based on accuracy studies show the use of WinBUGS and include estimation of the true- and false-positive fractions (tpf and fpf) for the exercise stress test, and modeling for the receiver operating characteristic (ROC) area. Of course, this is only a brief introduction, but should be sufficient for the beginner to begin the adventure of analyzing data. Because the book's examples provide the necessary code for all examples, the program can easily be executed by the user. After the book is completed by the dedicated student, they will have a good understanding of WinBUGS and the Bayesian approach to measuring test accuracy.

B.2 Download

I downloaded the latest version of WinBUGS at http://www.mrc-bsu.cam. ac.uk/bugs/winbugs/contents.shtml, which you can install in your program files. Then you will be requested to download a decoder that will allow one to activate the full capabilities of the package.

B.3 The Essentials

The essential feature of the package is a WinBUGS file or document that contains the program statements and space for input information.

B.3.1 Main Body

The main body of the software is a WinBUGS document that contains the program statements, the major part of the document, and a list statement or statements, which include the data values and some initial values for the MCMC simulation of the posterior distribution. The document is given a title and saved as a WinBUGS file, which can be accessed as needed.

B.3.2 List Statements

List statements allow the program to incorporate certain necessary information that is required for successful implementation. For example, experimental or study information is usually inputted with a list statement. In a typical agreement study, the cell frequencies of the two raters are contained in the list statement. As another example, consider estimation of the two most basic measures of test accuracy, namely the tpf and fpf, where the medical test scores are binary. The example is taken from Chapter 4 where the exercise stress test is binary and a positive test indicates coronary heart disease. The basic information is given by Table B.1, and the Bayesian analysis is executed with BUGS CODE 4.1.

Thus, for the (0,0) cell, the number of nondiseased subjects that score 0 is 327, and the number of diseased subjects that score 1 is 818.

The execution of the code is explained in terms of the following statements that make up the WinBUGS file or document.

TABLE B.1

Exercise Stress Test and Heart Disease

EST[b]	CAD[a]	
	$D = 0$	$D = 1$
$X = 0$	327	208
$X = 1$	115	818

[a] Coronary artery disease.
[b] Exercise stress test.

WinBUGS Document

```
# Measures of accuracy
# Binary Scores
Model;

{
# Dirichlet distribution for cell probabilities
g00~dgamma(a00,2)
g01~dgamma(a01,2)
g10~dgamma(a10,2)
g11~dgamma(a11,2)

h<-g00+g01+g10+g11
# the theta have a Dirichlet distribution
theta00<-g00/h
theta01<-g01/h
theta10<-g10/h
theta11<-g11/h
# the basic test accuracies are below
tpf<-theta11/(theta11+theta01)
se<-tpf
sp<-1-fpf
fpf<-theta10/(theta10+theta00)
tnf<-theta00/(theta00+theta10)
fnf<-theta01/(theta01+theta11)
ppv<-theta11/(theta10+theta11)
npv<-theta00/(theta00+theta01)
pdlr<-tpf/fpf
ndlr<-fnf/tnf

}

# Exercise Stress Test
# Uniform Prior (add one to each cell of the table frequencies
!)
```

```
list(a00 = 328,a01 = 209,a10 = 116,a11 = 819)

# chest pain history
# Uniform prior
list(a00 = 198,a01 = 55,a10 = 246,a11 = 970)

# initial values
list(g00 = 1,g01 = 1,g10 = 1,g11 = 1)
```

The objective of this example is to determine the posterior distribution of the true- and false-positive rates for the exercise stress test. The program statements are included between the first and last curly brackets {}, and the code for the tpf is

```
tpf<-theta11/(theta11+theta01)
```

where theta11 is the cell frequency for the (1,1) cell of the table, namely, that corresponding to the diseased subjects that score positive. The information from Table B.1 appears in the first list statement and the number 1 is added to each cell frequency from the above table, which places a uniform prior distribution on the four cell parameters. The first nine statements of the code put a posterior Dirichlet distribution on the four cell parameters, theta00, theta01, theta10, and theta11.

The last list statement contains the initial values or the MCMC simulation.

B.4 Executing the Analysis

MCMC can analyze complex statistical models and the following describes the use of drop-down menus from the toolbar for executing the posterior analysis.

B.4.1 The Specification Tool

The toolbar of WinBUGS is labeled as follows, from left to right: file, edit, attributes, tools, info, **model, inference**, doodle, maps, text, windows, examples, manuals, and help. I have highlighted the model and inference labels. When the user clicks on one of the labels, a pop-up menu appears. To execute the program, the user clicks on **model**, then clicks on **specification** (Figure B.1) and the specification tool appears.

The specification tool is used together with the BUGS document as follows: (1) click on the word "model" of the document, (2) click on the check model box of the specification tool, (3) click on the compile box of the specification tool, (4) click on the word "list" of the list statement of the document, and finally, (5) click on the load inits box of the tool. Now close the specification tool and go to sample monitor tool.

B.4.2 The Sample Monitor Tool

The sample tool is activated first by clicking on the inference menu of the toolbar (Figure B.2). Then click on sample, and the sample monitor tool appears as in Figure B.2. Type fpf, then click on "set," then type tpf in the node box and click on "set," and then type an * in the node box. Type 5000 in the beg box, which means the first 5001 observations generated for the posterior distribution of tpf and fpf will not be used when reporting the results of the simulation. The 5000 observations typed in beg are referred to as the "burn-in."

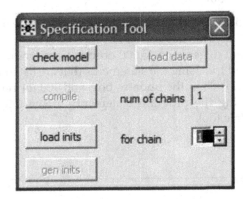

FIGURE B.1
The specification tool.

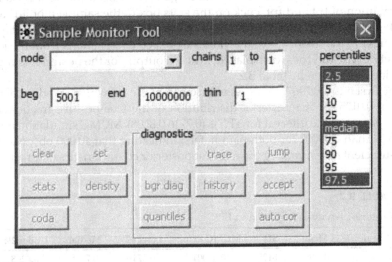

FIGURE B.2
The sample monitor tool.

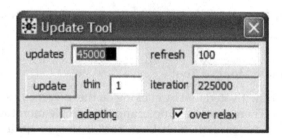

FIGURE B.3
The update tool.

B.4.3 The Update Tool

To activate the update tool, click on the **model** menu of the toolbar, then click on updates, which appears in Figure B.3.

Suppose you want to generate 45,000 observations from the posterior distributions of tpf and fpf, using the statements that are listed in the document above. Then type 45,000 in the updates box, and 100 for refresh. To execute the simulation using the program statements in the document, click on update in the update tool.

B.5 Output

After the 45,000 observations have been generated from the joint posterior distribution of tpf and fpf, click on the stats box of the sample monitor tool, see Figure B.2. Certain characteristics of the joint distribution are displayed. On clicking the history box, the values of 40,000 observations from the joint density of tpf and fpf are displayed, and the output for the posterior analysis will look as shown in Table B.2.

The first entry 0.7967 is the posterior mean of the tpf with a standard deviation of 0.0125 for the posterior distribution of the true-positive fraction TPF and a 95% credible interval for TPF is (0.7716,0.8208). MCMC simulation errors are less than 0.00001 for both parameters, indicating that 40,000 observations are sufficient for estimating the relevant posterior characteristics of tpf and fpf.

TABLE B.2

Posterior Analysis for TPF and FPF

Parameter	Mean	SD	MCMC Error	Lower 2½	Median	Upper 2½
TPF	0.7967	0.0125	<0.00001	0.7716	0.7968	0.8208
FPF	0.2612	0.0208	<0.00001	0.2215	0.2608	0.3033

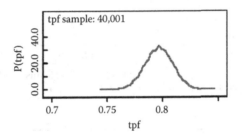

FIGURE B.4
Posterior density of the TPF.

Another feature of the output is that the posterior densities of the nodes can be displayed by clicking on the "density" box of the sample monitor tool. For example, Figure B.4 presents the posterior density of the TPF.

The posterior density of the tpf appears to be symmetric about the mean of 0.79, which is confirmed by the posterior median of 0.79.

B.6 Another Example

To illustrate the use of WinBUGS and to gain additional insight, an example of test accuracy will be presented, which has to do with estimating the area under the ROC curve for the accuracy of mammography for detecting breast cancer.

Consider the results of mammography administered to 60 women, of which 30 had the disease. This example appears in the study of Zhou et al.[1]

Mammography uses scores from 1 to 5, where a score of 1 indicates a normal view, 2 indicates a benign tumor, 3 implies a probably benign tumor, 4 indicates a suspicious lesion, and 5 a malignant tumor; and for those with breast cancer, it seems that the frequency of the scores increase with the increasing score for malignancy, while the opposite is true for those without the disease. A Bayesian analysis will put a uniform prior distribution on the above 10 cell frequencies and will be executed with BUGS CODE 4.2. The formula for the ROC area is given by Broemeling[2] and is coded as

```
auc<- A1+A2/2
A1<-theta2*ph1+theta3*(ph1+ph2)+theta4*(ph1+ph2+ph3)+
theta5*(ph1+ph2+ph3+ph4)
A2<- theta1*ph1+theta2*ph2+theta3*ph3+theta4*ph4+theta5*ph5
```

There are three list statements: the first is for the mammography study, the second is from another study not considered here, and the third contains the initial values for the variables that generate the Dirichlet distribution of the cell probabilities for Table B.3.

TABLE B.3

Mammography Results

Status	Test Result					
	Normal 1	Benign 2	Probably Benign 3	Suspicious 4	Malignant 5	Total
Cancer	1	0	6	11	12	30
No Cancer	9	2	11	8	0	30

WinBUGS Document

```
# Area under the curve
# Ordinal values
# Five values

Model;

{

# generate Dirichlet distribution
g11~dgamma(a11,2)
g12~dgamma(a12,2)
g13~dgamma(a13,2)
g14~dgamma(a14,2)
g15~dgamma(a15,2)

g01~dgamma(a01,2)
g02~dgamma(a02,2)
g03~dgamma(a03,2)
g04~dgamma(a04,2)
g05~dgamma(a05,2)

g1<-g11+g12+g13+g14+g15
g0<-g01+g02+g03+g04+g05

# posterior distribution of probabilities for response of
diseased patients
theta1<-g11/g1
theta2<-g12/g1
theta3<-g13/g1
theta4<-g14/g1
theta5<-g15/g1

# posterior distribution for probabilities of response of
non-diseased patients
ph1<-g01/g0
ph2<-g02/g0
ph3<-g03/g0
```

```
ph4<-g04/g0
ph5<-g05/g0

# auc is area under ROC curve
#A1 is the P[Y>X]
#A2 is the P[Y = X]
# from Broemeling2
auc<- A1+A2/2

A1<-theta2*ph1+theta3*(ph1+ph2)+theta4*(ph1+ph2+ph3)+
theta5*(ph1+ph2+ph3+ph4)

A2<- theta1*ph1+theta2*ph2+theta3*ph3+theta4*ph4
+theta5*ph5

}

# Mammography Example Zhou et al.1
# Uniform Prior
# see Table 4.7
list(a11 = 2,a12 = 1,a13 = 7,a14 = 12,a15 = 13,a01 = 10,a02 =
3,a03 = 12,
a04 = 9,a05 = 1)
# Gallium citrate Example Zhou et al.1
# Uniform Prior
list(a11 = 13,a12 = 7,a13 = 4,a14 = 2,a15 = 19,a01 = 12,a02 =
3,a03 = 4,
a04 = 2,a05 = 4)

# initial values
list(g11 = 1,g12 = 1,g13 = 1,g14 = 1,g15 = 1,g01 = 1,g02 =
1,g03 = 1,
g04 = 1,g05 = 1)
```

To begin the analysis, click on the model menu of the toolbar and pull down the specification tool:

1. Click on the word "model" of the BUGS document.
2. Click on the check model box of the specification tool, see Figure B.1.
3. Activate the word "list" of the first list statement.
4. Click on the compile box of the specification tool.
5. Click on the word "list" of the third list statement of the document. If a mistake is made, the user will be notified, but you are now ready to execute the analysis.

To continue the process, pull down from the inference menu, the sample monitor tool (see Figure B.2) and type auc in the node box, followed by clicking the set box, then repeat the operation for nodes A1 and A2. For the final operation, put an * in the node box, and type 5000 in the beg box for the burn-in.

TABLE B.4

Posterior Distribution of Area under the ROC Curve: Mammography Example

Parameter	Mean	SD	MCMC Error	Lower 2½	Median	Upper 2½
Auc	0.7811	0.0514	<0.0001	0.6702	0.7848	0.8709
A1	0.688	0.0635	<0.0001	0.5564	0.6909	0.8036
A2	0.1861	0.0307	<0.0001	0.128	0.1854	0.2484

Pull down the update tool (see Figure B.3) from the model menu and type 45,000 in the updates box, put 100 in the refresh box, and click on the updates box. The simulation now begins with 45,000 observations generated from the posterior distribution of auc, A1, and A2 of the sample monitor tool, and the output will appear as shown in Table B.4.

The ROC area is estimated with a posterior mean of 0.7811 and a 95% credible interval of (0.6702,0.8709), while the other parameters have posterior means of 0.688 and 0.1861 for A1 and A2, respectively. The latter is the posterior probability of a tie between the score of a diseased subject and the score of a nondiseased subject. Note the small MCMC errors of less than 0.001 for the three parameters, which imply that the value 0.7811 is within two decimal places of the "true" posterior mean for the auc parameter, the area under the ROC curve.

B.7 Summary

This appendix introduces the reader to WinBUGS and the novice should be able to begin Chapter 1 and learn the main topic, namely, how a Bayesian analyzes accuracy studies. To gain additional experience, refer to the manual and to the numerous examples that come with the downloaded version of the package. Practice, practice, and more practice is the key to understanding the importance of analyzing actual data with a Bayesian approach. There are many references about WinBUGS, including Broemeling,[2] Woodworth,[3] and the WinBUGS link,[4] which in turn refer to many books and other resources about the package.

References

1. Zhou, X.H., Obuchowsli, N.A., and McClish, D.K. *Statistical Methods in Diagnostic Medicine*, John Wiley & Sons, 2002, New York.
2. Broemeling, L.D. *Bayesian Biostatistics and Diagnostic Medicine*, Chapman & Hall/CRC, Taylor & Francis Group, 1996, Boca Raton, FL.
3. Woodworth, G.G. *Biostatistics: A Bayesian Introduction*, John Wiley & Sons, 2004, Hoboken, NJ.
4. The BUGS Project Resource, available at http://www.mrc-bsu.cam.ac.uk/bugs/winbugs/contents.shtml.

Index

A

Actual chlorine concentration, predicted
    chlorine ion concentration *vs.*,
    363, 364
Adjustment of data
    Bayesian methods, 9
    by regression, 16–20
Age
    mortality in U.S., California, and
        Florida by, 10–14, 84
    and SBP, MI, 96, 97
Alzheimer's study, repeated measures
    model, 365–369
American Journal of Epidemiology, 19
Attributable risk (AR), 8–9, 69–73

B

Baseline hazard function, 31, 248
Bayesian analysis, 55–57, 60, 124, 146
    blood pressure and age, 140
    cigarette consumption data, 148
    codes, 222–223, 235–236
    directly standardized mortality
        rates, 87, 89
    epileptic study, 374, 375
    estimating time to infection, 240
    exercise stress test, 285
    hazard ratio, 251
    for Heart Study, 130
    for HIP study, 331
    indirect standardization, 95
    interaction, 98, 103
    Israeli Heart Disease Study, 343
    logistic regression, 129
    Matched Case–Control Study,
        107–108
    melanoma study, 353
    multiple linear regression, 144, 154
    net probability of death, lung
        cancer, 180
    observed minus expected, 230
    placebo group, 224

    Poisson regression parameters, 382
    treatment *vs.* placebo group, 252, 256
    for UK Early Detection Trial, 293, 296
Bayesian approach, 53, 55, 73, 411
    direct adjustment of mortality, 109
    life tables, *See* Life tables
    posterior distribution, 92, 306
    regression analysis, 122
    survival analysis, *See* Survival
        analysis
Bayesian inferences
    implementation, 407
    indirect adjustment, 92–93
Bayesian Kaplan–Meier method,
    221–225
Bayesian life table analysis
    coronary artery disease, 173
    lung cancer patients, 176, 178
Bayesian posterior analysis, 125
Bayesian predictive distribution, 133
Bayesian statistics, 45–47
Bayes' theorem, 45, 234, 249, 407, 409–410
    posterior information, 413–414
    prior information, 410–413
Beta prior distribution, 46, 407
Biochemical oxygen demand (BOD)
    nonlinear regression, 362
Blood pressure, 14
Breast cancer
    detection, 296–297
    incidence rates, 55, 298
    prevalence rate, 56–57
    screening, 34, 332
    survival experience, diagnosis,
        183, 184
    UK Early Detection Trial,
        292–296, 330

C

California
    for directly standardized mortality
        rates, 87, 89
    mortality by age, 84

California (*Continued*)
    population and mortality rate by age,
        10–14
    posterior density of standardized
        rates, 90
California Tumor registry (1942–1963),
    183–187
CAR model, *See* Conditional
    autoregressive model
Case–control study, 8, 9, 15, 71, 73
    cataracts and diabetes, 72, 78
    design, 7
    matched, 15, 107, 108, 120
    posterior analysis, 63
    risk factor, 5
    sampling procedure, 64
    smoking and respiratory problems, 79
    tonsillectomy and Hodgkin's
        Disease, 62, 63
Categorical regression models, 17, 36,
    122, 346, 396
Cell entries, 107
Cell frequencies, 100
Cell probabilities, 100
Censored observations, 216, 218
    binomial distribution, 221
    number, 220–222, 227
    probability, 222
    survival and, 309–312
Checking model assumptions
    measuring tumor size, 421–422
    sampling from exponential, 419–421
    testing multinomial assumption,
        422–423
Chi-square distribution, 231
Chlorine concentration
    nonlinear regression models, 362–364
    vs. time, 361
Chronic disease, screening, 33, 296, 332
Cigarette consumption, 146–149
Cohort study, 5–6, 9
    risk and disease in, 57–61
Computed tomography (CT), 7, 280
Computing algorithms, 423–426
Conditional autoregressive (CAR)
    model, 42–44, 376, 377, 384
Conditional probabilities, 219, 221–223
Confounder, 97, 99–102
Confounding, 218

factor, 14, 15, 16, 17
Control cohorts, 315, 316–317, 321
Coronary artery calcium, 66
Coronary artery disease, 132
    Bayesian life table analysis, 173
    data adjustment, 82
    predicted number with, 133
Coronary heart disease
    of blacks, 164
    Caucasians with, 134
    life table calculations, 23–24, 171
    probability, 133
    summarization of survival for
        subjects with, 22–23, 171
Cox model, 253–257, 345
Cox proportional hazards model, 213, 218
    analysis of survival data, 346
    with covariates, 253–257
    definition, 31
    estimating survival, 2, 27, 28
    HIP study, 322
    regression methods using, 17
    survival analysis, 232–233, 248–253
    testing, 257–261
Cox regression model, 248, 322–323
Cross-sectional studies, 7, 8, 65–69, 73
CT, *See* Computed tomography
Cube root of age, log PCB *vs.*, 356, 359
Cumulative odds model, 36, 347

D

Data adjustment
    Bayesian methods, 9
    in case–control study, 107–109
    description, 81–82, 109–110
    direct data adjustment, *See* Direct
        data adjustment
    indirect standardization adjustment,
        See Indirect standardization
        adjustment
    Mantel–Haenszel estimator,
        See Mantel–Haenszel estimator
    by regression, 16–20, 121
Dependent variables, 2, 16, 17, 19
Descriptive statistics, 299–304
    epilepsy study, 370
    estimating lead time, 305
Diagnostic likelihood ratios, 287–288

Direct Bayesian approach, survival
    comparison, 188–190
Direct data adjustment, 82
    California, Florida, and U.S.
        populations, 83–90
    "true" mortality, 83
Directly standardized mortality rates of
    California and Florida, 87–90
Dirichlet posterior distributions,
    182–183
Disease
    probability, 5–6
    risk exposure and, 4–9
    and risk exposure, stratification and
        association between, 96–103
    screening, *See* Screening
Disease-specific life tables, 178–180

E

Epidemiology, 42
    spatial models, 375
    statistical methods in, 1–4, 36
    statistical models, *See* Statistical
        models for epidemiology
Epilepsy study, repeated measures
    model, 370–375
Exercise stress test (EST)
    Bayesian analysis, 285
    and heart disease, 439
    true- and false-positive rates, 47, 440

F

False positive fraction (FPF), 283,
    289, 294
Florida
    for directly standardized mortality
        rates, 87, 89
    population and mortality by age,
        10–14, 84
    posterior density of standardized
        rates, 90

G

Gamma-aminobutyric (GABA) acid, 370
Generalization, 21
Geographical epidemiology, 43

Gibbs sampling, 428–429
Gold standard, disease, 281–283
Goodness-of-fit, 344, 363, 396
    predicted *vs.* actual values, 368
    regression model, 133–134, 157

H

Hazard function, 215, 247, 322
    baseline, 248
    Cox model, 257, 262
    parametric models, 233
Hazard ratio (HR), 31–32, 247, 249,
    255–256
    Bayesian analysis, 251
    parameter in survival studies,
        249, 323
Health Insurance Plan of Greater
    New York (HIP) study, 32, 33,
    34, 35, 296–299
    analysis of, 322
    descriptive statistics, 299–304
    lead time, estimating, 304–308
Health insurance plan (HIP) study
    survival, *See* Survival, estimating
        and comparing
Heart disease, probability, 8
Hodgkin's disease, 7, 61–63
HR, *See* Hazard ratio

I

Incidence rates, 54, 55
    of breast cancer, 298
Independent variables, 16, 17
Indirect Bayesian approach, survival
    comparison, 190–194
Indirectly standardized rate (ISR),
    91–92
Indirect standardization adjustment,
    90–91
    Bayesian inferences, 92–93
    ISR, 91–92
    mortality rates for California and
        Florida, 93–96
Inference
    estimation, 415–416
    predictive, 417–419
    testing hypotheses, 416–417

Interaction
in model, epileptic study with/
without, 374, 375
and stratification, 97–101
ISR, *See* Indirectly standardized rate
Israeli Heart Disease Study
Bayesian analysis, 343
weighted estimator of odds ratio, 107
Israeli Ischemic Heart Disease Study, 14,
15, 18, 101–103, 124–130

J

Jeffreys' approach, 411

K

Kaplan–Meier method, 26–27, 29
Kaplan–Meier plots, leukemia patients
recurrence, 225–226
Kaplan–Meier survival curves, 213,
217–221, 261
Kaplan–Meier test
MCMC simulation errors, 197
product-limit survival probabilities,
197–200
survival estimation, 194, 195
survival experience, 196
Kidney dialysis patients
gender, age, and disease status,
241, 242
recurrence times, 237, 238
Weibull regression, 246

L

Lead time, 35–36, 304–308
Least square regression line, 136–137
Leukemia patients recurrence time,
215, 230
comparison of two groups, 216–217
Kaplan–Meier plots, 225–226
placebo group, 219, 220
posterior analysis, 236
treatment group, 218
Life tables
basic, 172
coronary artery disease,
170–171, 173
generalization of, 174–178

description, 169–170, 201–202
direct Bayesian approach, 188–190
disease-specific, 178–180
indirect Bayesian approach, 190–194
medical studies, 181–187
Life table techniques, 2, 22, 33, 170, 279
Likelihood ratio parameter, 231
Linear regression, 19, 134–135
multiple, 141–145, 344
simple, 135–141
weighted, 345
Lip cancer
diagnosed in Scotland, 376, 378–379
incidence of, 42, 43
posterior analysis of future values
of, 403
rates of, 377
relative risk of, 383
study of Clayton and Kaldor, 397
Logistic regression models, 16, 18–19,
123–124
dependent variables, 157
independent variables, 131–132
Israeli Heart Disease Study, 124–130
Log PCB, 358, 359
vs. cube root of age, 356
Log-rank test, 261, 314, 316, 328
Bayesian version, 30, 36, 213, 226
Kaplan–Meir curve and, 27
treatment/placebo groups of
leukemia patients, 226–227,
230–232
Lung cancer, 382, 383, 384, 395
patients, 27
Bayesian life table, 176, 178, 180
deaths and survivors, 191
survival comparison, 190
survival probabilities for death, 25
Lymph nodes, 17

M

Mammography, 34, 289, 291–295, 331
Mantel–Haenszel (MH) estimator, 15,
103–107
Mantel–Haenszel (MH) odds ratio, 26,
191–194
Markov chain Monte Carlo (MCMC), 56,
57, 87

MCMC, *See* Markov chain Monte Carlo; Monte Carlo Markov chain
Median survival, study and control cohorts, 314
Medical geography, 43
Melanoma patients metastasis, 348
Metropolis algorithm, 427–428
Metropolis–Hasting algorithm, 4, 423
MH estimator, *See* Mantel–Haenszel estimator
MH odds ratio, *See* Mantel–Haenszel odds ratio
MI, *See* Myocardial infarction
Minitab, 46, 408, 431
Monte Carlo Markov chain (MCMC)
   errors, 129, 138, 148, 176, 197, 284, 291, 293, 354
   techniques, 408, 423, 426–427
      mean of normal populations, 429–431
Morbidity, incidence and prevalence, 54–57
Mortality
   incidence and prevalence, 54–57
   probabilities, 172–177
   rate for U.S., California, and Florida, 10–14, 87
Multinomial model, 422
Multinomial regression models, 122
Multiple linear regression, 16–20, 122, 141–145, 157, 344
   Bayesian analysis, 154
   simple and, 19, 134
Myocardial infarction (MI), 14, 96, 98
   vs. SBP, 97, 101, 125, 129

N

National Institutes of Health (NIH), 22, 170–171
Negative predictive value (NPV), 286
Net probability of death, lung cancer, 180
NIH, *See* National Institutes of Health
Noninformative priors, 410, 411
Nonlinear regression models, 38–39, 355, 396
   chlorine concentration, 361–364
   PCB concentration, 355–360

Nonparametric approach, 233
Normal distribution, 243
Normal regression models, 345
NPV, *See* Negative predictive value

O

Observed consumption number, predicted consumption number *vs.*, 149
Odds ratio (OR), 18, 19, 62–64, 72, 124
   MH approach, defined, 190–191
   and relative risk, 2, 3, 7–9, 14–16
   weighted estimator, 106
OR, *See* Odds ratio
Ordinal regression models, 20, 36, 37, 156, 346, 347, 348

P

Parametric models, 232
   survival analysis, 233
   Weibull distribution, *See* Weibull distribution
Parametric survival model, 345
PCB concentration, *See* Polychlorinated biphenyls concentration
PH assumption, *See* Proportional hazards assumption
Placebo group
   Bayesian analysis, 224, 225, 252, 256
   leukemia patients recurrence time, 219, 220
Poisson population, 421
Poisson regression models, 42, 43, 369, 370, 376
   parameters, 382
Polychlorinated biphenyls (PCB) concentration, 39–40, 355–360
Population by age, U.S., California, and Florida, 10
Positive predictive value (PPV), 286, 294
Posterior analysis
   Alzheimer's study, 41, 369
   baseline (log) relative risk, oral cavity and lung cancer, 395
   for case–control study, 63
   for Cataract–Diabetes study, 72
   chlorine ion concentration, 363

Posterior analysis (*Continued*)
  leukemia patients recurrence, 236
  linear regression, 138
  PCB study, 357, 359
  race and coronary artery disease, 132
  screened and control groups,
      survival, 329–330
  stroke cohort study, 60–61
Posterior density, 57, 232
  Mantel–Haenszel estimator, 104
  odds ratio, 99
  probability of death, 173, 174
  relative risk, 69
  standardized rates for California and
      Florida, 90
  survival difference, 190
  true positive fraction, 286
Posterior distribution, 45
  directly standardized mortality
      rates, 86
  lead time, 308
  Mantel–Haenszel estimator, 104
  MH odds ratio, lung cancer
      patients, 194
  mortality and survival rates, 315
  probability of death, breast
      cancer, 187
  product-limit survival
      probabilities, 198
  relative risk, 5–6
  unknown parameters, 30
Posterior inferences, 46
Posterior standard deviation, 231
Postoperative survival experience with
      lung cancer, 175, 178–180, 188
PPV, *See* Positive predictive value
Predicted chlorine ion concentration *vs.*
      actual chlorine concentration,
      363, 364
Predicted consumption number *vs.*
      observed consumption
      number, 149
Predicted PCB *vs.* PCB nonlinear
      regression, 360
Predictive inference, 417–419
Prevalence rate, 56, 57
Probability of disease, 5–6
Product-limit method, 219, 221
Progabide, 369, 370

Proportional hazards (PH) assumption,
      28, 259–261, 262
Proportional odds assumption, 347

Q

Quantitative variables, 19

R

Radiologists, 348, 353
Receiver operating characteristic
      (ROC)
  area, 36, 347, 353, 354, 437, 446
  curve, 288–292, 330
Recurrence probabilities, 219, 251
  direct estimation, 221
  posterior distribution, 223
Recurrence times
  kidney patients, 238
  leukemia patients, 215
  log-rank test for difference, 226
Regression analysis, 2, 121, 134
  adjustment, 16–20
Regression coefficients, 17, 18, 19,
      243, 251
Regression function, 243, 248
Regression models, 343
  categorical, 346
  goodness of fit, 133–134
  nonlinear, *See* Nonlinear regression
      models
  normal, 345
  ordinal, 346–348
  Poisson, *See* Poisson regression
      models
  types, 121–122
Relative risk (RR), 5, 6, 9, 44–45
  parameter, 59, 64, 68–69, 72
Repeated measures model, 365, 396
  Alzheimer's study, 365–369
  epilepsy study, 370–375
Residuals *vs.* age, 140, 141, 145
Risk exposure, 4–9
Risk factor, stratification and association
      between, 96–103
ROC, *See* Receiver operating
      characteristic
RR, *See* Relative risk

S

Sample Monitor Tool, WinBUGS, 441
SBP, *See* Systolic blood pressure
Screening
    chronic disease, 296, 332
    description, 279–280
    disease, 32–36
    principals, 280–281
Screening programs
    binary test, 282–283
    classification probabilities, 283–286
    conditions, 280–281
    diagnostic likelihood ratios, 287–288
    predictive values, 286–287
    ROC curve, 288–292
    test accuracy measures, 281–282
    UK Early Detection Trial, 292–296
Screening trials
    breast cancer, 332
    components, 309
Sentinel lymph node biopsy, 38
Shields Heart Study, 15, 66–68, 72
    ARs, 73
Simple linear regression models, 16, 20,
    122, 135–141, 157
    assumptions, 344
    definition, 19, 134
    log PCB on cube root of age, 358, 359
Skewness, 104
Smoking
    AR, 71
    deleterious effects, 146
    and respiratory problems, 78
SMR, *See* Standard mortality ratio
SMRs, *See* Standardized mortality rates
Spatial epidemiology, 3–4
Spatial models for epidemiology, 42,
    43, 375
S-Plus, 46, 408
Standardized mortality rates (SMRs),
    12–14, 43, 377
Standard mortality ratio (SMR), 92,
    93, 95
Statistical methods in epidemiology,
    1–4, 36
Statistical models for epidemiology,
    343–346
    categorical regression models, 346

nonlinear regression models, 355
    repeated measures model, 365
Stratification
    interaction and, 97–101
    Israeli Heart Disease Study, 101–103
Stratified Cox procedure, 28, 214
Stroke cohort study, posterior analysis,
    60, 61
Survival analysis, 232
    Bayesian approach to, 27–32
    Cox proportional hazards model,
        See Cox proportional hazards
        model
    notation and basic table, 214–217
    parametric models, 233
Survival distributions, 232, 234
Survival, estimating and comparing
    life tables, 309
    survival models, 322
Survival experience
    diagnosis of breast cancer, 183, 184
    Kaplan–Meier test, 196
Survival function, 215, 322
    Cox model, 248, 258, 262
    parametric models, 233
Survival model, parametric, 345
Survival probabilities, 173, 174, 177–179
    posterior distribution of product-
        limit, 197–200
Survival time of patient, 213, 214
Systolic blood pressure (SBP), 18, 19, 121,
    124, 138
    vs. actual values predicted, 141
    vs. age, 140
    data adjustment, 82
    myocardial infarction *vs.*, 97, 101, 125
    vs. predicted SBP, 145

T

Time-series techniques, 375
Tonsillectomy, 7, 61–63, 62–63
    posterior RR, 64
TPF, *See* True positive fraction
Treatment group
    Bayesian analysis, 252, 256
    leukemia patients recurrence time,
        218
    posterior distribution, 223

"True" mortality, 83
True positive fraction (TPF), 283, 285
   vs. FPF for mammography, 289

U

UK Early Detection Trial for breast
   cancer, 292–296, 330
United States, population and mortality
   rate by age, 10–14, 84

W

Weibull density, 234, 235
Weibull distribution, 233, 262
   mean and median, 234
   parameters, 236
Weibull model, 345
Weibull regression model, 27, 31, 241, 246

Weibull survival distribution, 234
Weighted liner regression, 345
Weighted regression, 149, 158
WinBUGS, 22, 46, 47, 416, 423, 437
   Bayesian analyses, 28
   direct sampling, 408
   download, 438
   essential feature, 438–440
   example, 443–446
   executing, analysis, 440–442
   output, 442–443
   package, 214
   posterior analysis, 38, 355
   use, 45
W-values, 358

X

X-ray, modality, 280

Printed in the United States
by Baker & Taylor Publisher Services